Using SAP S/4HANA®

SAP PRESS

SAP PRESS is a joint initiative of SAP and Rheinwerk Publishing. The know-how offered by SAP specialists combined with the expertise of Rheinwerk Publishing offers the reader expert books in the field. SAP PRESS features first-hand information and expert advice, and provides useful skills for professional decision-making.

SAP PRESS offers a variety of books on technical and business-related topics for the SAP user. For further information, please visit our website: *www.sap-press.com*.

Bardhan, Baumgartl, Choi, Dudgeon, Górecki, Lahiri, Meijerink, Worsley-Tonks
SAP S/4HANA: An Introduction (4th Edition)
2021, 648 pages, hardcover and e-book
www.sap-press.com/5232

Tritschler, Walz, Rupp, Mucka
Financial Accounting with SAP S/4HANA: Business User Guide
2020, 604 pages, hardcover and e-book
www.sap-press.com/4938

Janet Salmon, Stefan Walz
Controlling with SAP S/4HANA: Business User Guide
2021, 593 pages, hardcover and e-book
www.sap-press.com/5282

James Olcott, Jon Simmonds
Sales and Distribution with SAP S/4HANA: Business User Guide
2021, 434 pages, hardcover and e-book
www.sap-press.com/5263

Karl Liebstückel
Plant Maintenance with SAP S/4HANA: Business User Guide
2021, 665 pages, hardcover and e-book
www.sap-press.com/5180

Wolfgang Fitznar, Dennis Fitznar

Using SAP S/4HANA®

An Introduction for Business Users

Rheinwerk
Publishing

Editor Megan Fuerst
Acquisitions Editor Emily Nicholls
German Edition Editor Maike Lübbers
Translation Sara Miller
Copyeditor Julie McNamee
Cover Design Graham Geary
Photo Credit iStockphoto.com: 1131359487/© izusek, 638377554/© da-kuk
Layout Design Vera Brauner
Production Hannah Lane
Typesetting III-satz, Husby (Germany)
Printed and bound in Canada

ISBN 978-1-4932-1956-8
© 2022 by Rheinwerk Publishing, Inc., Boston (MA)
1st edition 2022
1st German edition published 2021 by Galileo Press, Bonn, Germany

Library of Congress Cataloging-in-Publication Data
Names: Fitznar, Wolfgang, author. | Fitznar, Dennis, author.
Title: Using SAP S/4HANA : an introduction for business users / Wolfgang
   Fitznar and Dennis Fitznar.
Description: 1st Edition. | Boston : Rheinwerk Publishing, 2021. | Includes
   index.
Identifiers: LCCN 2021033876 | ISBN 9781493219568 (hardcover) | ISBN
   9781493219575 (ebook)
Subjects: LCSH: SAP HANA (Electronic resource) | Industrial management. |
   Business enterprises--Finance. | Database management.
Classification: LCC HD31.2 .F58 2021 | DDC 658--dc23
LC record available at https://lccn.loc.gov/2021033876

Contents at a Glance

Dear Reader,

When you move to a new city, the impact is felt in your daily routine. When I arrived in Boston four years ago, I found myself glued to my phone's GPS app to do pretty much anything, from traversing my new commute (difficult) to finding the nearest Dunkin' for coffee (easy—this is Boston!). Between bus stops, train stations, highways, and bike lanes, simply navigating from A to B every day was overwhelming.

Getting started in a new ERP system is similar. When your company moves its processes to SAP S/4HANA, your daily transactions are rerouted, and you have new navigation options for accessing them. However, completing your tasks smoothly and efficiently requires more than a GPS. Once you access your tasks, you need to know how to perform them—to both find the coffee shop and place your order.

SAP professionals Wolfgang Fitznar and Dennis Fitznar have combined their knowledge to do just that. With this book, they guide you step by step (ahem, click by click) through the new suite. You can expect your journey to include key points of interest, anecdotes from the experts, and plenty of good humor!

What did you think about *Using SAP S/4HANA: An Introduction for Business Users*? Your comments and suggestions are the most useful tools to help us make our books the best they can be. Please feel free to contact me and share any praise or criticism you may have.

Megan Fuerst
Editor, SAP PRESS

meganf@rheinwerk-publishing.com
www.sap-press.com
Rheinwerk Publishing · Boston, MA

Contents

PART I   Using SAP GUI

1   First Steps with SAP GUI   41

4   Changing SAP GUI Reports

7   Displaying and Maintaining Data with SAP Fiori Apps

8 Reporting with SAP Fiori Apps

9 Customizing SAP Fiori Reports

10   Personalizing and Optimizing SAP Fiori

13   Financial Accounting: Posting and Evaluating Business Transactions

Foreword

The development of software to handle all of a company's business processes is SAP's area of expertise. Since its launch in 2015, SAP S/4HANA, along with its predecessor, SAP ERP, has been helping companies prepare for their future.

SAP S/4HANA brings significant advantages for all companies, such as real-time reporting, simplified usability, performance improvements, and shorter innovation cycles—just to name a few. SAP S/4HANA helps simplify these processes and procedures, making working with the SAP system much more efficient and saving you the most valuable resource: time!

The key to this efficiency is the new user interface called SAP Fiori. It breathes new life into an old system design by presenting the user with contemporary tiles, very similar to what can be seen today on tablets and smartphones. The SAP Fiori launchpad can also be personalized, which only elevates the advantages of SAP S/4HANA, especially for users who work with it on a daily basis. However, thanks to the high level of user friendliness, the changeover is easy, whether you're a seasoned veteran or new to the program.

This book offers you a well-structured introduction to the program. Whether you work in purchasing, in a warehouse, in sales, or in accounting and controlling, here you will learn step by step what is important and how you can get the most out of the SAP system. After this, Wolfgang and Dennis Fitznar will introduce you to some more advanced options, but all in the same easy-to-follow format.

Truth be told, I would have liked this book myself, when getting started with SAP S/4HANA, since complicated, technical jargon be overwhelming at first. However, this book shows you how to quickly find your way around the system, which makes doing your daily tasks much easier—and more fun—to do!

Jennifer Hauck
Training System Architect
SAP SE

Preface

Welcome to SAP S/4HANA, the most successful standard software for companies! Are you already working with SAP S/4HANA, or are you about to undergo appropriate training? That's great, because SAP S/4HANA is the latest SAP product, and together with its predecessor SAP ERP, SAP S/4HANA is by far the most widely used software in larger companies.

SAP S/4HANA is one program with two user interfaces: classic SAP GUI and the new, hip SAP Fiori, as shown in the image below. We're currently in a transitional phase, and both user interfaces are being used, so we'll also describe both interfaces for you in this book.

An SAP Fiori Window in the Background, and an SAP GUI Window in the Foreground

But before we accompany you on your journey of discovery through SAP S/4HANA, we have one question: Who are you?

- **Are you an SAP newcomer?**
  Then this is the right book for you! You'll receive systematic and comprehensive basic training for all important overarching tools that SAP S/4HANA offers you as a user and that you need for daily practice.

- **Have you already had a short SAP S/4HANA training, or has someone incidentally incorporated you?**
  In both cases, there's a good chance that you don't yet know many valuable tools, methods, and basic settings of this powerful software. Scroll through the table of contents of this book and check for yourself which topics are relevant to you.
- **Are you already familiar with the older SAP GUI, and are you more interested in the new SAP Fiori user interface?**
  We'll show you where to find the well-known SAP GUI functions in SAP Fiori and which exciting new functions are also available for switchers.

SAP S/4HANA with its two interfaces is like your toolbox with many tools that allow you to enter, maintain, and evaluate data. After reading this book, you'll know the most important tools and methods.

Craftsmen also have tools, for example, classic screwdrivers and a state-of-the-art cordless electric screwdriver. If they only use the classic screwdriver instead of the cordless screwdriver, you'll probably think they could work much more efficiently if they use modern tools.

But what does this have to do with SAP? Many SAP users only use the classic screwdrivers because they don't know the cordless screwdrivers of the SAP system. This may happen because money on training has been saved, or because many users only receive a short briefing from a colleague, who in turn has only received a short briefing from a colleague. In this book, we'll not only introduce you to the basic functions in detail but also give you many tips that will help you with your daily tasks.

Target Groups

Whether you're an accountant, controller, buyer, sales employee, dispatcher, or HR manager, if you regularly work with SAP S/4HANA as a user, this book is just right for you.

If you're an SAP consultant, SAP application manager, or SAP support, you're not part of our target group. The book is made for users, and every click and every input are documented here in detail, which is probably too detailed for you. However, if you don't already know the new SAP Fiori interface or SAP GUI theme Belize, you can benefit from this book. As a consultant or application manager, you should have the same level of knowledge as your power users.

[»]

Note

No topics from customizing, administration, or programming are covered here, as this book is only about the application of SAP S/4HANA.

This book doesn't require any SAP knowledge. The only thing you should bring along is basic knowledge of Windows and an Internet browser.

Objectives and Parts of the Book

In the Introduction, we teach a bit of theory. However, it's only as much as you need for the practical handling of the SAP system because we want to become operational to business with you as quickly as possible, that is, move on to practice.

You bought this book to systematically familiarize yourself with the practical handling of the SAP system? Great! Part I: Using SAP GUI and Part II: Using SAP Fiori are pure practice.

In the first three chapters of both parts, you'll find the basics of both the SAP GUI and SAP Fiori user interfaces. This is about how you log in, surf elegantly through the SAP system, quickly find and maintain your data, and use standard evaluations. With these basics of the user interface, you can prepare to get started in practice when everything is optimally preset in your SAP system.

Unfortunately, there are very, very rarely optimal presets in practice. And for this reality, there are the last two chapters in Part I and Part II. If you bought this book to work with the SAP system in a time-saving and efficient way, you should read these chapters. Here, we show you practical presets that will help you save time. You can personalize your SAP system according to your individual wishes and set it up for yourself, just as you define your home address in a new car in the built-in navigation system or save your favorite stations. There is a lot to discover here!

You bought this book to get an understanding of the processes and their handling on the system? In Part III: Running Core Business Processes, you'll find important functions that are used in practice, starting with purchase orders and sales orders through accounting in finance, even if in practice you don't do all these processes yourself: SAP is teamwork! You should therefore think outside the box and develop a basic understanding of what your colleagues are doing and where the data is going.

Now you know what to expect from this book. But how about what you should *not* expect? Do you expect a complete guide for all topics in your department? We have to put a damper on this expectation here. A book with a complete operating manual for all topics of all departments would have thousands of pages.

Even for key and power users, there are still a variety of tips and tricks that didn't fit into this book.

How to Work with This Book

You've already received some tips on how to proceed in the previous few paragraphs, but here's what we recommend:

- The chapters in Parts I and II largely build on each other. As an SAP beginner, keep to the order here as much as possible. You'll then be ready for general operation with the SAP GUI and SAP Fiori user interfaces.
- In Part III, you'll find optional topics for practice. Simply cherry-pick the sections that interest you here.

Throughout this book, we've incorporated text boxes to provide you with additional guidance. Text boxes with notes will have the [»] icon, text boxes with tips will have the [+] icon, text boxes with examples will have the [Ex] icon, and text boxes with warnings will have the [!] icon like the one below.

[!] **Don't Break Anything!**
Never practice in a productive SAP system. If you're unsure, ask your admin or a supervisor.

Data for Training in the System

Learning by doing is this book's goal! Our screenshots come from a standard system. If you play through the processes of this book in an SAP system in your company, our images of the SAP screens and the functions may not be 100% identical to those of your system.

Companies adapt the SAP software and change SAP screen displays. This process is called *customizing* and is carried out by SAP specialists. However, if your employer's SAP S/4HANA system hasn't been completely "bent," most of the SAP screens will be largely the same as the screenshots in our book.

In addition, SAP S/4HANA and the user interfaces are available in different versions. For the book, we used SAP Fiori 3, SAP GUI 7.60, and SAP S/4HANA 1909. Other versions may also cause you to find different screens and other functions.

Test Access for SAP Fiori

Don't have access to a system? At the time of going to press, SAP has a free offer for a trial access to the SAP Fiori interface at *www.sap.com/cmp/oth/crm-s4hana/s4hana-cloud-erp-trial.html*. However, this is a system with limited possibilities.

Acknowledgments

Colleagues have also supported this project. Here you'll find the lovely persons we would like to thank especially for their support.

Georg G. Hauck has been an SAP trainer and consultant since 1999. He works in the field of logistics as a freelance trainer, master trainer, consultant and tester, and on-site supervisor.

Romuald Peters has been working as an SAP consultant and SAP trainer in international training projects since 1995. Special thanks go to him for his humorous contributions, which also make him one of the best SAP trainers in practice!

SAP SE has provided us with SAP Live Access as a fully configured SAP S/4HANA training system. Many thanks to Tihomir Car and Uwe Hafner from SAP SE for the great support! As a digital learning platform, SAP Learning Hub offers access to a variety of learning resources such as guides to learning journeys, self-determined online courses, SAP Learning Rooms, and the option to put what you've learned into practice and train in real time in the SAP Live Access environment. More information can be found at *https://training.sap.com/learninghub*.

A special thank you for the great cooperation goes to the one and only Megan Fuerst, editor at SAP PRESS, and to Gabriel Tsonyev for his support.

Introduction

A promise is a promise: we'll keep this chapter on theory as short as possible. First, you'll get an overview of the structure and areas of application of SAP S/4HANA. Perhaps you're already curious about what the SAP software looks like. After the initial overview, we'll introduce you to the two associated user interfaces: SAP GUI and SAP Fiori.

After that, we'll show you the structure of a company with its different areas and departments in the SAP system. This knowledge will later help you understand what happens as soon as you use the SAP software on the screen. At the same time, in this part of the book, you'll learn the first basic terms of the "SAPish" language, that is, the special technical terms of the SAP system.

What You'll Learn

- For which tasks SAP S/4HANA is used in companies
- How the SAP GUI and SAP Fiori user interfaces differ
- How the structure of a company is represented in the SAP system
- What data is stored in the SAP system

What Does SAP S/4HANA Have to Offer?

In this section, you'll discover what you can do with SAP software and which companies use this software. You'll first get a rough overview here, and we'll get to the details later.

SAP customers include 80% of the Fortune 500 companies and 87% of the Forbes Global 2000 companies, but SAP software is also increasingly being used in public administration to automate, simplify, and accelerate processes.

The Company SAP

SAP was founded in 1972 and has become by far the world's most successful business software manufacturer and the most successful software company in Europe. At the same time, SAP is the most valuable German company according to stock market valuation. Its headquarters is located in the small town of Walldorf in Germany.

The Three Letters: SAP

You probably know that SAP is the name of the company that developed the SAP S/4HANA software. You may not know that SAP stands for "systems, applications, products in data processing." (Such abbreviations naturally tempt people to find other names. Our favorite is "Super Advanced PlayStation.")

Where does the name "S/4HANA" come from? The initial letter S stands for both "Suite" and "Simple." Is the program really simple to use? You'll be able to form an opinion about that after reading the book. The 4 stands for the fourth generation of the software. Finally, HANA is the name of the database technology of this software. SAP S/4HANA has been on the market since 2015.

Why has SAP software become and remained so successful? One of many reasons is that SAP software is the "all-in-one solution" that covers almost all areas of a company, including logistics with purchasing, production, sales and distribution, and the complete internal and external accounting and human resources management.

> **ERP Software**
>
> SAP S/4HANA belongs to the genre of enterprise resource planning (ERP) programs, just as Microsoft Word belongs to the genre of word processing programs.
>
> The resources of a company are everything that the company needs to make money: personnel, machines, materials, or orders. And with ERP software, these resources can be planned and managed. Therefore, companies use ERP software to manage and optimize all the important basic functions and business processes.

With the help of this software, companies can organize and optimize almost all business processes. Business processes are nothing more than workflows in a company. Here is a definition from the SAP documentation (found in the digital help *https://help.sap.com*):

> *A business process is a description of a cross-functional structure within an organization. It consumes resources and can cross internal organizational boundaries.*

Is this description too abstract for you? For us, too! That's why you should read on right away because the following are two basic business processes that can be implemented with SAP S/4HANA as practical examples.

Business Process: Purchase to Pay

Our first example is the purchase-to-pay business process (sometimes abbreviated as P2P or PTP), as shown in Figure 1.

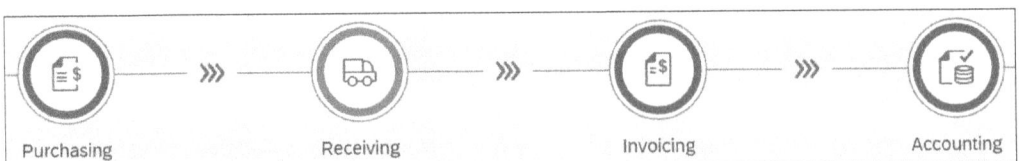

Purchasing — Receiving — Invoicing — Accounting

Figure 1 Flow of the Purchase-to-Pay Business Process

In our example, this business process begins in the purchasing department, where an order is entered into SAP S/4HANA and forwarded to the supplier. Then the company is the recipient of an inbound delivery, and here the goods receipt usually takes place in the warehouse. In the next step, an invoice is created at the supplier, which is entered in the company as an incoming invoice and later paid by the accounting department.

Of course, there are many variants of such a basic process. For example, the purchase order could be preceded by a purchase requisition or an approval procedure, or it could be followed by a confirmation from the supplier.

Business Process: Order to Cash

The order-to-cash business process (sometimes abbreviated as O2C or OTC) begins the example in sales with the creation of a sales order (see Figure 2).

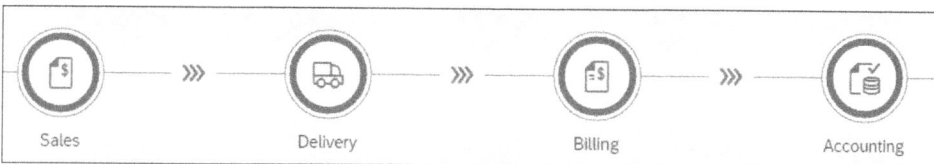

Figure 2 Flow of the Order-to-Cash Process

This is followed by delivery and billing (other terms: invoicing, billing). This process finally ends in accounting with the receipt of payment.

If SAP S/4HANA is used throughout, all data from all departments is stored in a database only once. This provides uniform software for the employees of *all* departments involved and not, for example, one software for sales, another software for shipping, and yet another software for accounting. This ensures that each employee works with the same and always up-to-date data, which is referred to as *real-time processing*. For example, as soon as a colleague from the sales department changes and saves the address data of a customer, this address data is immediately available in the shipping department and in accounting.

Components of SAP S/4HANA

SAP S/4HANA is a huge program package that is divided into the following three major application areas:

- **Accounting**
  Finance (including bookkeeping) and controlling.
- **Logistics**
  Materials management (including purchasing), production, maintenance, quality management, and sales and distribution.

- **Human resources (HR)**
  Personnel administration, payroll, and personnel development.

Each area of the application consists of individual main components, which in practice are also often referred to as *modules*. Each of these modules has an abbreviation consisting of two or three letters. We've listed the most important modules in Table 1.

Area of Application	Module	Abbreviation	Functional Examples
Logistics	Materials management	MM	Inventory management, purchasing, invoice verification
	Production planning and control	PP	Production planning, production control
	Sales and distribution	SD	Sales support, sales, shipping and transportation, billing, credit management
	Quality management	QM	Quality planning, quality notifications
	Enterprise asset management	EAM	Maintenance with repair of faults and servicing, preventive maintenance
Accounting	Financials	FI	General ledger, accounts receivable, accounts payable, fixed asset accounting
	Controlling	CO	Cost center accounting, product cost controlling, profit center accounting
HR	Personnel administration	PA	Master data management
	Personnel time management	PT	Time data recording, time evaluation
	Payroll accounting	PY	Payroll and follow-up activities

Table 1 Areas of Application and Modules of SAP S/4HANA

This classification originates from the predecessor SAP ERP and is used most frequently. You can also find these modules in SAP S/4HANA, for example, in SAP GUI, in the **SAP Menu**, or in the SAP S/4HANA help.

There is a further classification for SAP S/4HANA, as described in Table 2, which takes additional new functionalities into account. Although this classification hasn't yet become generally accepted in practice, we don't want to withhold it from you.

Component	Functional Example
Research and development	Product lifecycle management with product data management, product safety, project management
Sourcing and procurement	Consumption-based planning, purchasing, auditing
Supply chain	Material master data, inventory management and stock-taking, warehouse management, transport management with freight management
Manufacturing	Production planning, production control, quality management
Asset management	Maintenance, deployment planning, workplace safety, environmental protection
Sales	Sales support, sales, invoicing
Service	Service contracts, service orders
Finance	▪ Financial operations: accounts receivable, accounts payable ▪ Accounting and financial close: general ledger, asset accounting ▪ Financial planning: controlling
HR	Personnel administration, time management, payroll accounting

Table 2 Further Components of SAP S/4HANA

SAP GUI and SAP Fiori User Interfaces

There are two different user interfaces for SAP S/4HANA: the older SAP GUI, which is also used with the predecessor SAP ERP, and SAP Fiori. This section provides an overview of the differences between these user interfaces.

Some companies only use SAP GUI, others mainly use SAP Fiori, and others use both user interfaces. Why? They each have advantages and disadvantages.

Old and Trusted: SAP GUI

We'll start with SAP GUI (GUI stands for *graphical user interface*). SAP GUI is also called presentation software because it "presents" data to you, such as a customer list, on your PC. It's therefore a separate SAP software that is normally installed on your PC. Figure 3 shows you the initial screen of SAP GUI in version 7.60.

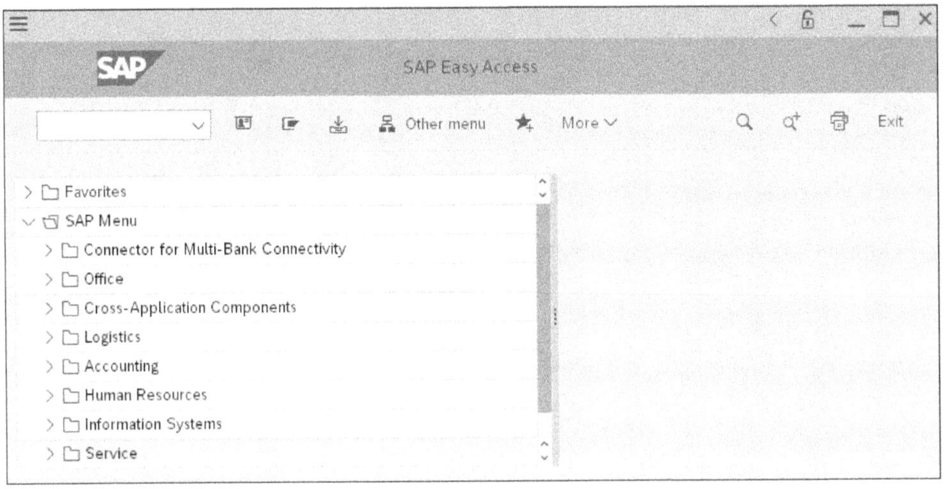

Figure 3 Initial Screen of SAP GUI

SAP GUI offers you menu structures and icons for program operation; Figure 3 uses the Belize theme as an example. At the top left of the initial screen, you'll see the **Favorites** folder where you can add the functions that you need most often.

Under **Favorites**, you'll see the **SAP Menu**, which contains a folder tree, similar to Windows Explorer. In addition to general functions, it includes as folders the three major application areas: **Logistics**, **Accounting**, and **Human Resources**. Via these folders and further subfolders (or from the **Favorites** folder), you can jump to your applications, which are also called *transactions* here. An example of a transaction in logistics is Create Sales Order, and an example from accounting is Post Document.

New and Modern: SAP Fiori

The SAP Fiori user interface design is updated and much clearer with the use of tiles and key figures in the start screen. When you open the user interface, it immediately shows a bouquet of the very latest information: for example, purchasers see the number of open order items, whereas salespeople see the number of open customer inquiries. (The word *Fiori* actually means "flowers" in Italian.)

SAP Fiori is a new world that looks like it was made for a tablet or a smartphone—and it is! Unlike SAP GUI, SAP Fiori can also be used with mobile devices, such as a smartphone. It runs on an internet browser such as Microsoft Edge or Mozilla Firefox. Consequently, unlike SAP GUI, no separate software installation is required on your PC. All you need is a link to access SAP Fiori via the browser.

In SAP Fiori, the SAP Fiori launchpad, as shown in Figure 4, is the counterpart to the **SAP Menu** in SAP GUI. It takes you to your applications, which are called *apps* in SAP Fiori.

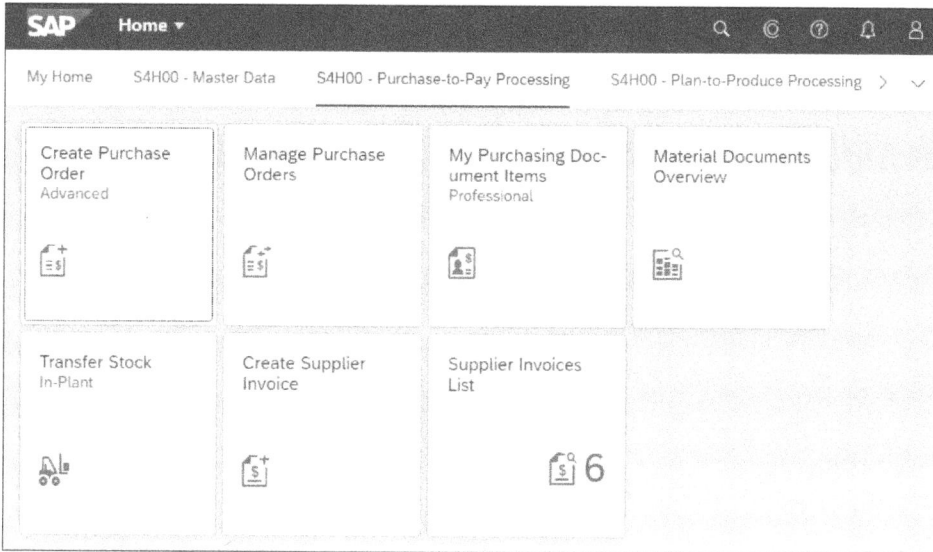

Figure 4 SAP Fiori Launchpad

Instead of the **SAP Menu** with transactions in SAP GUI, you see *tiles* in SAP Fiori, which you can arrange as you wish. You can use the tiles to launch the corresponding apps.

The operation of the two interfaces is very different. For this reason, we've strictly separated the chapters into SAP GUI in Part I of the book and SAP Fiori in Part II of the book.

Although the trend is moving toward SAP Fiori, SAP GUI will continue to be used intensively, at least in the medium term, for several reasons:

- Apps aren't yet available for many SAP GUI functions. As of early 2021, there are more than 100,000 SAP GUI transactions, but only about 12,000 apps. So, for some functions, you definitely need SAP GUI.
- SAP GUI is often the better choice for power users. Experienced users can customize this interface to suit their own needs better.
- There are still functions that can be used more efficiently with transactions in SAP GUI than in SAP Fiori apps. This is especially true for data entry.

On the other hand, SAP Fiori is much easier to use for occasional users and enables significantly better graphical evaluations with charts. You can see an example chart displaying the proportion of purchasing groups in purchasing amounts in Figure 5.

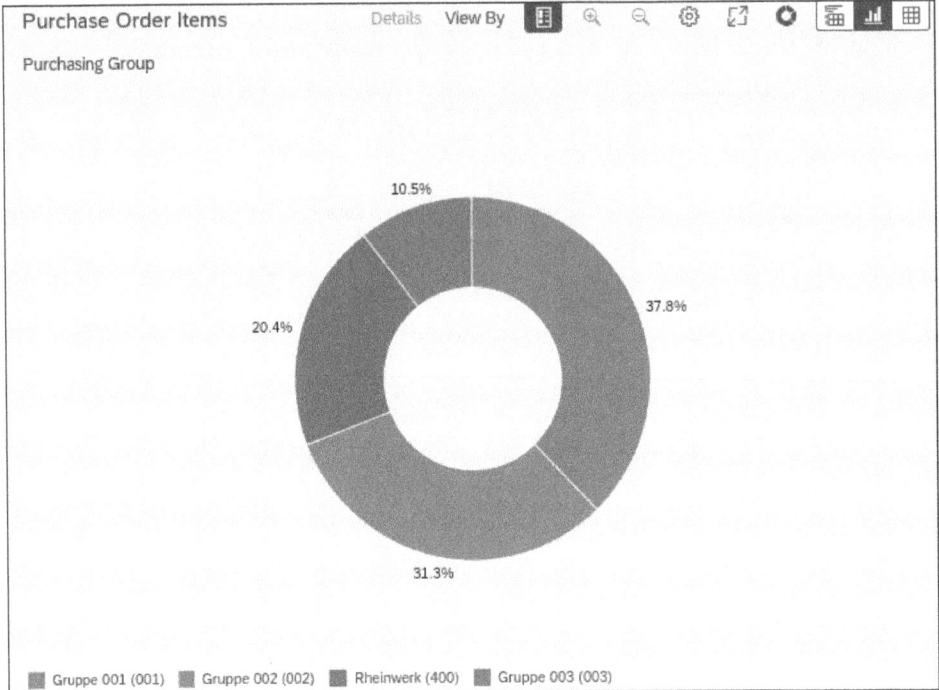

Figure 5 Chart in SAP Fiori

Are both interfaces available to you in your company? Then you can use one or the other or both interfaces, depending on your function. We'll support you in the decision-making process.

Company Structure in the SAP System

To make the explanations in this book as practical as possible, let's imagine that you work in a sample company. Before you start using the data of this company in Part I of the book, you'll get some basic preliminary information on our training data here, which largely corresponds to that of SAP.

Organizational Structure and Organizational Units

This section deals with the structure of a company with its individual components. In the SAP system, this is referred to as the *organizational structure*. The organizational structure is made up of individual subsidiaries (SAPish: company codes) and plants. These are called *organizational units* (other term: organizational elements) in the SAP system. You need organizational units as soon as you enter or display new customers, vendors, materials, sales orders, and purchase orders. And this is exactly what you'll do in the next parts of the book. So now we'll discuss some basic organizational units in preparation. Each organizational unit has a number.

[Ex]

Material Purchase Order

Here's an example from practice. In the *material purchase order*, as shown in Figure 6, you'll find the following organizational units at the bottom right:

- Company code (which company ordered?)
- Purchasing organization (which purchasing department ordered?)

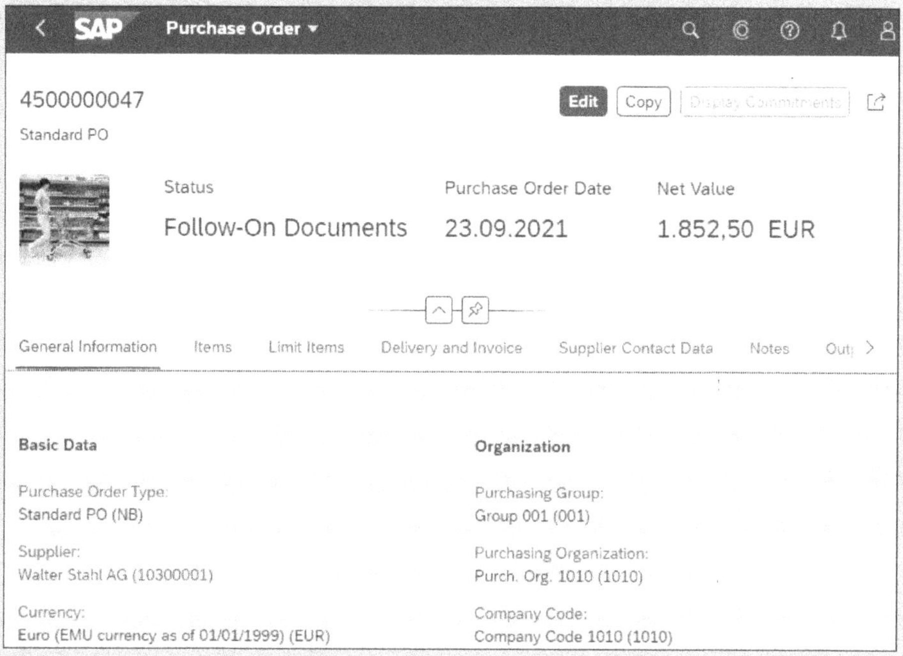

Figure 6 Manage Purchase Order App: Purchase Order with Organizational Unit's Company Code and Purchasing Organization

Client and Company Code

As the top organizational unit, the *client* is the highest hierarchical level in the SAP system and is also referred to as the *highest-level organizational unit*. A client can be a corporate group or a subgroup. The client corresponds to a data pool that contains, for example, all group data. You only enter this once when you log on and don't need it again during the SAP session.

In an SAP system, several clients, that is, subgroups, can be managed in parallel. In our examples, we'll limit ourselves to one client.

[»]
Your Client

Each client has its own three-digit number (e.g., "800"), which you can obtain from your SAP administrators or from your trainers during SAP training.

In Figure 7, you can see that several *company codes* are subordinated to our client. This means that our client has several subsidiaries. A company code is nothing more than a company, for example, a subsidiary with its own balance sheet within a group. Each company that submits a balance sheet to a tax office must be managed as a separate company code in SAP S/4HANA.

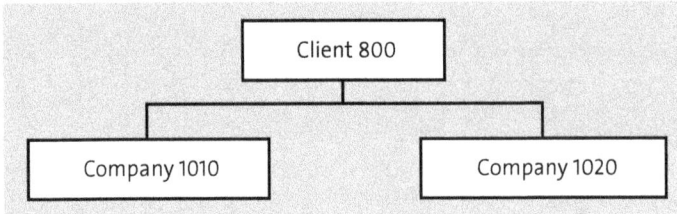

Figure 7 Client with Company Codes 1010 and 1020

[»]
Your Company Code

Our corporation has several subsidiaries or company codes. Your subsidiary has the code 1010.

Plant

A *plant* can be a production facility or a delivery plant, where materials aren't produced but are stored and delivered. A company code can contain several plants. Conversely, a plant always belongs to exactly one company code.

[»]
Your Plant

Your home is the plant 1010, which belongs to company code 1010.

Purchasing Organization

You'll learn in Chapter 11 how to register a purchase order. To do this, you enter the responsible purchasing department into the purchase order. This is referred to as the *purchasing organization* in SAP S/4HANA. There can be different purchasing organizations that are responsible either for only one plant, for only one company code, or as a central purchasing department for all company codes. Figure 8 shows an example in which purchasing organization 1010 handles purchasing for all plants in company code 1010.

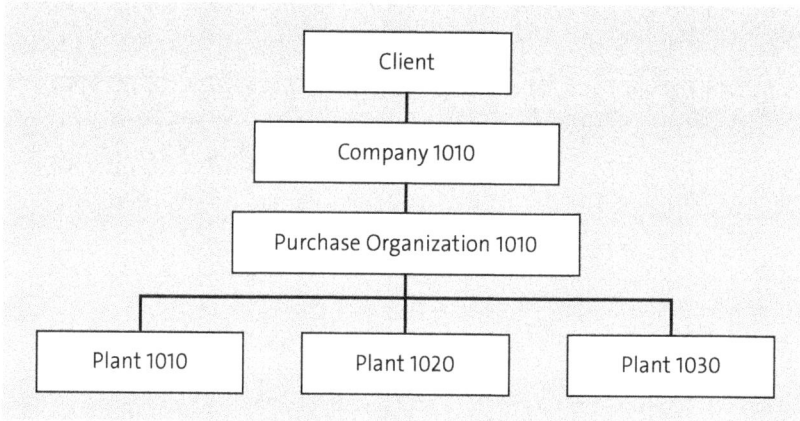

Figure 8 Company Code, Purchasing Organization, and Plants

[«]

Your Purchasing Organization

Exactly one purchasing organization 1010 is assigned to company code 1010. It's responsible for purchasing for all plants of subsidiary 1010.

Sales Organization

Each sales order belongs to a *sales organization*, which is equivalent to a sales department. A sales organization always belongs to exactly one company code. Conversely, a company, that is, a company code, can of course contain several sales organizations.

[«]

Your Sales Organization

Your sales organization has the identifier 1010, which is the same identifier as your purchasing organization and company code.

These are the organizational units you'll need throughout the book. There are many more organizational units, such as *controlling areas* for controlling and *storage locations* for materials management. We'll present these to you in Part III of this book.

Master Data and Documents in the SAP System

You've almost completed the theory! Now let's just go through some basics about the data we'll use in this book, and then you can log in and surf through the SAP system.

The term *master data* may be familiar to you. It's important data that rarely changes and is used over longer periods of time, such as customer, supplier, or material data. Figure 9 shows an example of **Address** master data in the Manage Business Partner Master Data app for a supplier.

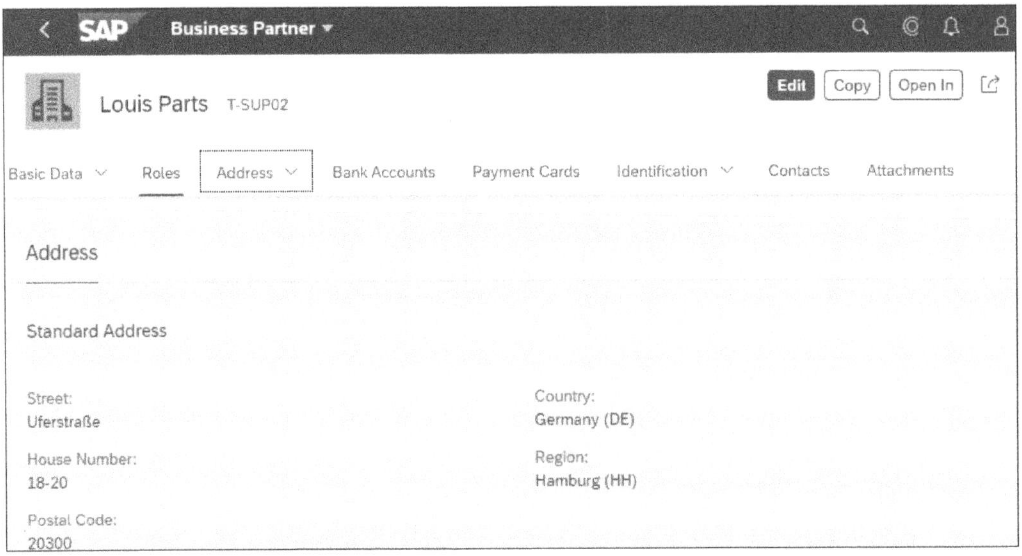

Figure 9 Manage Business Partner Master Data App: Address in Supplier Master Data

Purchase orders, sales orders, and invoices aren't master data because they are transaction-related and only valid for a short period of time. These are called *documents* instead. Figure 10 shows an example of address data in the Manage Purchase Orders app for a purchase order.

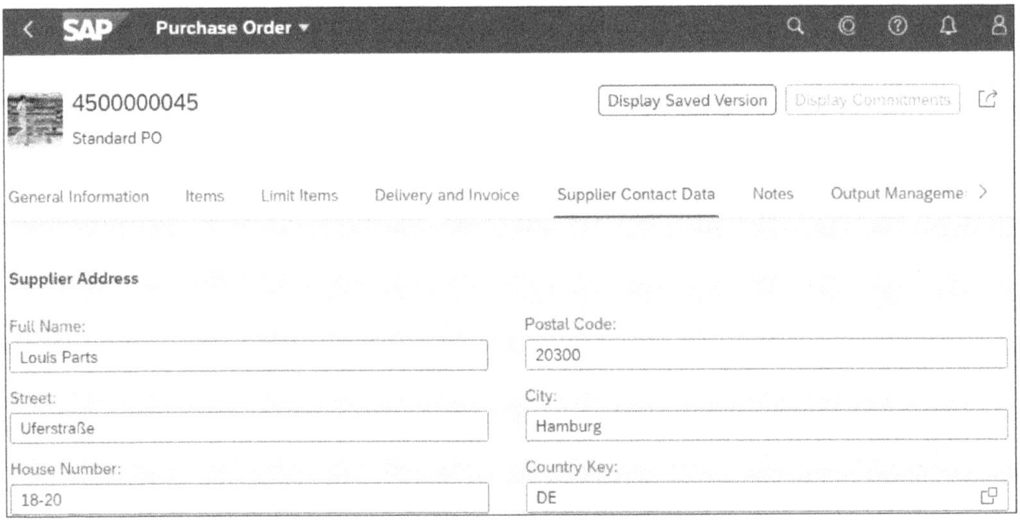

Figure 10 Manage Purchase Orders App: Address in Purchase Order Document

During a business process, different documents are created. At least one document is created for each substep from the purchase-to-pay business process, as you can see in Figure 11.

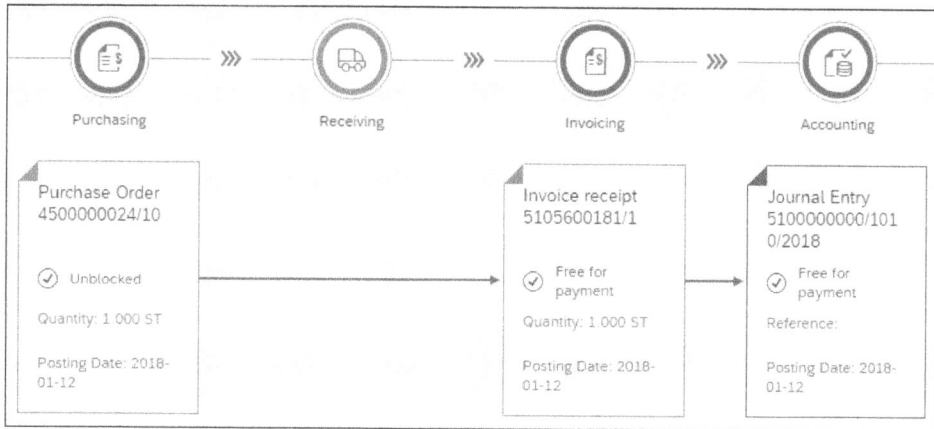

Figure 11 Process Flow: Documents of the Purchase-to-Pay Process

The purchasing department registers a purchase order, the goods receipt department registers a material document, and the invoice verification department creates a document for an incoming invoice. At the end of the process, there is a clearing document in accounting when payment is made, documenting that the invoice has been cleared. Each document has its own document number and document type.

Do you manually enter the vendor address when you create a purchase order? Nope! When you create a purchase order, only the vendor master record number is entered. The SAP system automatically transfers the remaining master data, such as the address or the terms of payment, from the vendor master data to the purchase order. So, if you order from the same vendor 10 times, you save yourself the trouble of entering the vendor address 10 times because SAP S/4HANA copies the up-to-date address from the master data to the purchase order each time. For this reason alone, the splitting of master data and documents is a practical thing to do.

Summary

This brings us to the end of our introduction. Time for a coffee break? You now have enough basic theoretical knowledge of SAP S/4HANA for the following parts of the book, in which you'll learn about the practical handling of the SAP system. This will give you a better understanding of the meaning and effects of all the clicking and typing.

We'll begin with the SAP GUI user interface in the next chapter.

PART I

Using SAP GUI

Chapter 1
First Steps with SAP GUI

Are you motivated to carefully study this part of the book? May I, Wolfgang—the older, grayed-out one of the two authors—briefly tell you about my first steps into the SAP system to begin? I hope you agree because this will increase the probability of you reading this book with maximum motivation!

Unfortunately, I had to learn the SAP software mostly by the trial and error method when I started in 1992. That means by testing. Grrrmpff! I can't number all the lightbulb moments I had afterwards or remember how often I shouted, "Why didn't I know that before!" I spent so much valuable time doing processes with SAP in an awkward and annoying way with too much typing instead of just a few clicks.

This is exactly what we want to spare you from with this book. We want to convey the lightbulb moments of this incredibly powerful software right at the beginning of your SAP career; for this purpose, you should systematically go through the book and enjoy it. The SAP GUI chapters and sections build on each other.

What You'll Learn

- How to log in and off
- What the password rules are and how to change your password
- How to find your way around the SAP GUI screen layout
- How to display transaction codes, which are technical short names for transactions
- How to create your own **Favorites** menu
- How to set user data, such as the login language, the decimal, and the date, to display appropriately

---

Are you only interested in SAP Fiori and wondering whether you can skip this chapter as a beginner? Better not! Many rules from SAP GUI also apply in SAP Fiori, and many terms have been adopted from SAP GUI. Therefore, our recommendation is to read through this part of the book chapter by chapter.

1.1 Log On and Off of SAP GUI

Are you ready? Great! In this section, you'll learn all the essentials about the first and the last step of an SAP session: logging on and logging off.

To log in, you'll need the following four pieces of information, which you'll receive in practice from your SAP contact persons or in training from your trainers:

- **System**
  Each SAP S/4HANA system has its own three-digit identifier.

- **Client**
  Within a system, there can be several clients. One of these clients is your data pool. It also has an identifier consisting of three characters.

- **Username**
  An admin has created a user with a username for you.

- **Initial password**
  This is the password your admin came up with when creating your new user specifically for your first login.

[»]

Logging On and Off of SAP GUI with PKI Cards

In practice, public key infrastructure (PKI) cards or smart cards are often used for logging on. A card reader is connected to the PC for this purpose. A chip on the card contains the access data for the SAP system, so it's no longer necessary to enter the username and password when logging on. The client can also be stored, so that the logon screen in the third step of the following instructions is skipped completely when using PKI cards.

1.1.1   Log On

And here we go with one index finger in the book, and the other hand on the mouse! One last note before you start: the design of your windows may look slightly different from the one shown here, but the operation of calling the functions and commands is almost the same. You'll learn how to change your design to the one in the book immediately after logging in and logging out in Section 1.2.

Follow these steps to log on to the system:

1. In Windows, there are two ways to start with SAP GUI in SAP S/4HANA:
   - Double-click ⬛ (**SAP Logon**) on the Windows desktop.
   - Choose the **SAP Front End** • **SAP Logon** path or the **SAP Login** tile in the Windows **Start** menu.

   In both cases, the **SAP Logon** screen appears, as shown in Figure 1.1.

2. In the **SAP Logon** window, you can see the release status of SAP GUI at the top, in our case, 7.60. Double-click the SAP system that you want to log on to, and this will take you to the logon screen—the "Open Sesame" of SAP GUI—as shown in Figure 1.2.

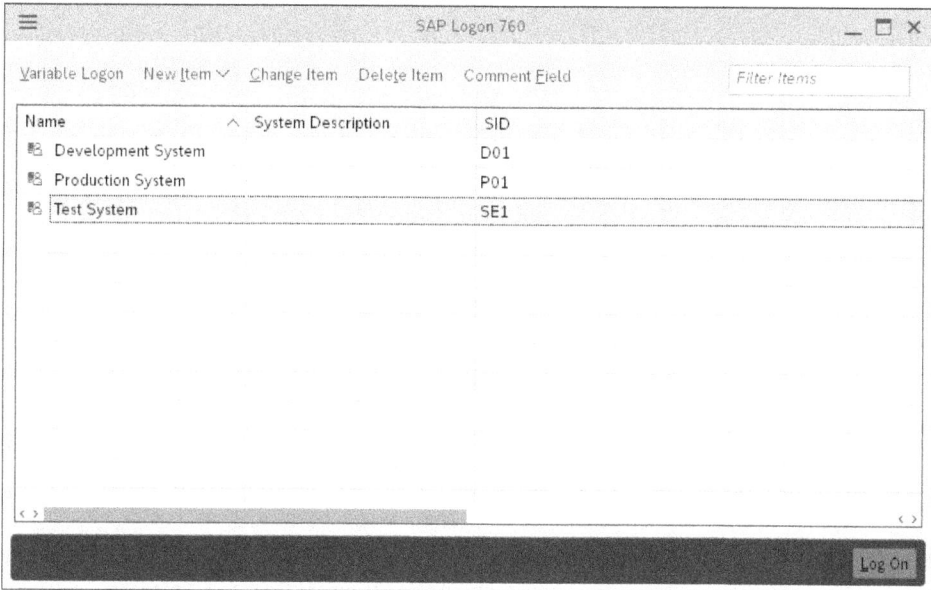

Figure 1.1 SAP Logon Window

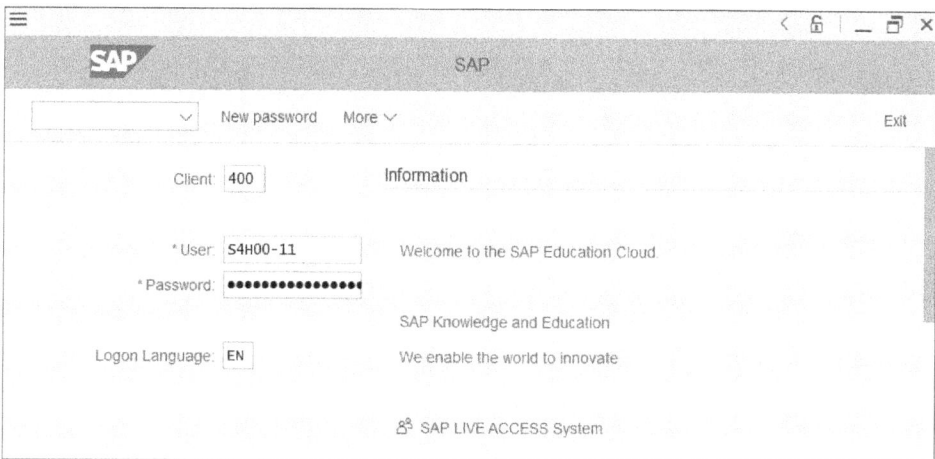

Figure 1.2 SAP Login Parameters on the Login Screen

3. On the login screen shown in Figure 1.2, enter the following information:

    – **Client**: Enter the client number, if not already present.

    – **User**: Enter your username.

    – **Password**: Enter your password in correct uppercase and lowercase letters. This input is masked by asterisks in case someone is looking over your shoulder.

    – **Logon Language**: Enter "EN" for English, if not already set by default. Feel free to try "ZH" for Chinese, "AR" for Arabic, "HE" for Hebrew, or "HI" for Hindi.

[»]

Initial Login

First time? During your first logon with a new user to the SAP S/4HANA system, there are two differences compared to a normal logon: After confirming by pressing the Enter key, the system prompts you to create a new password of your own. Furthermore, there is an additional welcome window here.

4. Press the Enter key. Now SAP S/4HANA will check your data.

   If you've passed the test, the password change window will appear, as shown in Figure 1.3, but only when you log in for the first time. (Failed? It can happen! Read the note box next.)

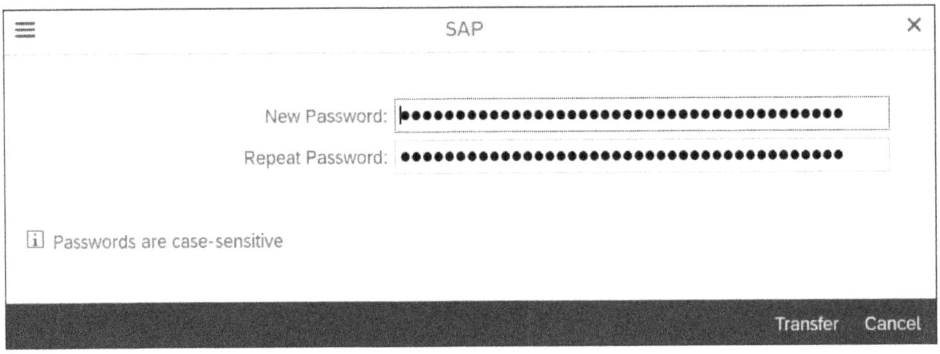

≡	SAP	×
	New Password: ●●	
	Repeat Password: ●●	
	ⓘ Passwords are case-sensitive	
		Transfer Cancel

Figure 1.3 Setting Your Password

[»]

If the Login Fails

In the last window line of the logon screen at the bottom left, the SAP system tells you the reason, for example, an incorrect client number or an incorrect combination of username and password.

5. In contrast to a subsequent "normal" logon, you must enter a personal password in the **New Password** field when logging on for the first time. For security reasons, you repeat this again below.

6. Press Enter or click ✔ (**Transfer**) at the bottom right, and the next window will appear.

   Can't go any further? Again, SAP S/4HANA will tell you why it won't accept a new password in the lower-left corner of the window. You can also find information about password rules after these instructions.

7. As shown in Figure 1.4, the **Copyright** window tells you that you're logging in for the first time (just in case you had already forgotten). Press the Enter key to skip this window.

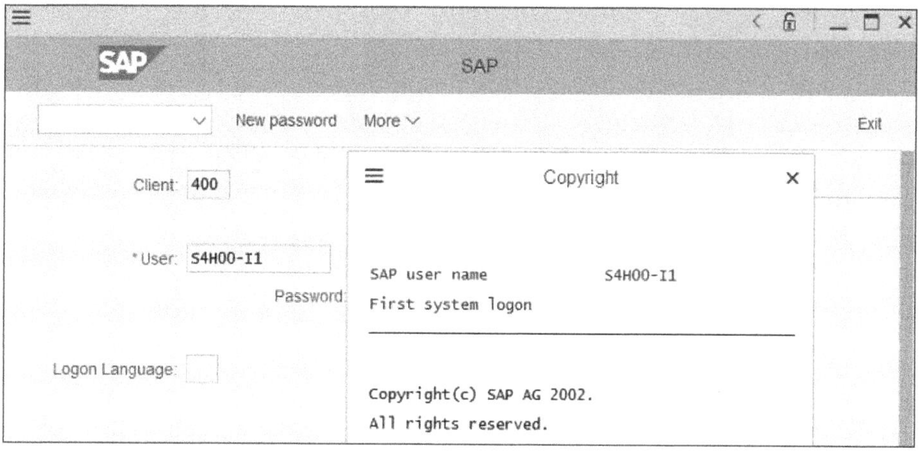

Figure 1.4 Copyright Window

And now for the big moment ... the SAP GUI welcome screen appears in the form of the **SAP Easy Access** window, as shown in Figure 1.5. Congratulations, the login was successful!

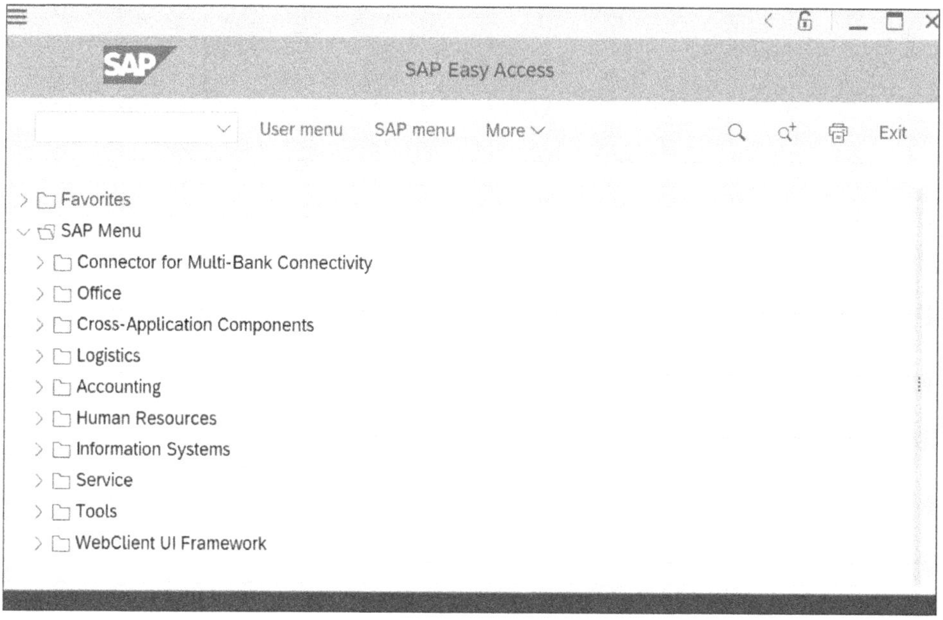

Figure 1.5 SAP Easy Access Screen

Seven steps for one registration? Well, for the next login, there are only four steps because three of the seven steps are only required for the first login.

[»]

> **What Password Rules Do I Need to Know?**
>
> Technically, you don't need to know any password rules because if a password doesn't comply with the rules of your SAP system, the SAP system will notify you with a message in the bottom-left corner of the last line. If this happens, just try it again, taking this message into account. Nevertheless, we've compiled the most important rules here, which always apply in every SAP S/4HANA system:
>
> - A distinction is made between uppercase and lowercase letters.
> - The first character can't be a space, question mark, or exclamation mark.
> - The first three characters can't be identical, such as "uuups123".
>
> In addition, there are many other rules that an admin can set individually, and which can therefore vary from company to company. These include, for example:
>
> - Minimum length of the password.
> - Minimum number of characters that must be different when changing a password from an old one to a new one.
> - Validity, that is, the period until the next forced password change.
> - Number of unsuccessful login attempts before the user is automatically locked. Often the following rule applies: if you mistype three times, the logon screen closes. You may start again at the SAP logon. After another three unsuccessful logon attempts, SAP S/4HANA gets uncomfortable and locks the user. Administrators must then unlock this user for the next logon.
> - Blacklist of disallowed passwords such as company product names.

1.1.2   Change Your Password

Usually, as an additional security measure, the SAP system might have defined how long your password can be valid, for example, three months. So, after three months, you're forced to change your password to be able to log in again. Follow these steps:

1. Do this voluntarily in the logon screen in step 3 of the last section. On the logon screen, first enter the client, username, and your still valid password.

2. Click the **New Password** button, and SAP S/4HANA will allow you to enter the new password (including security retry), as shown in Figure 1.6.

3. Click ✅ (**Transfer**) at the bottom left.

You want to try it out right now? Well, it won't work for you right now because you logged on to the SAP system for the first time today and changed the password in the process. The SAP system only allows one password change per day, so you have to wait at least until midnight to change the password again. Until then, you can continue reading.

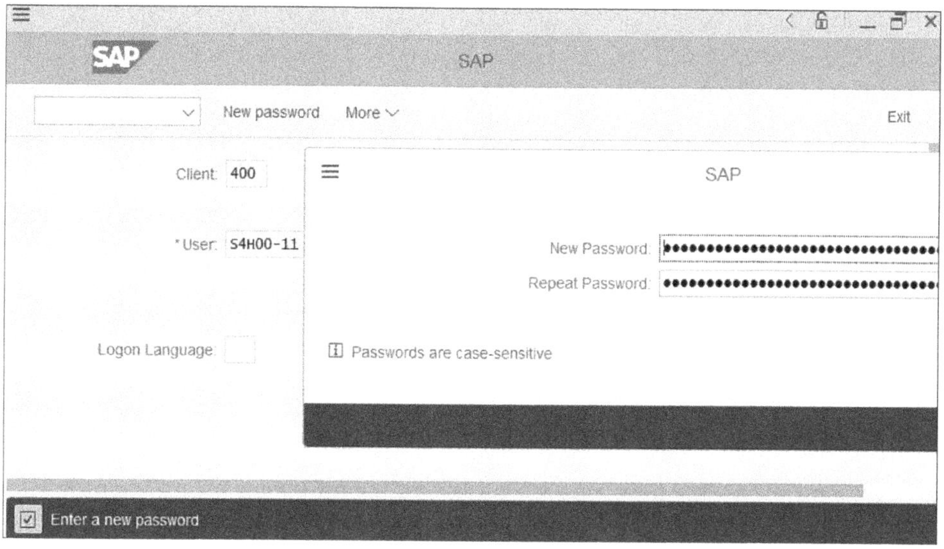

Figure 1.6 Changing a Password

Password Forgotten?

Been on vacation and forgot your password? It happens to the best of us. Ask your SAP admin, but even the most resourceful colleagues can't make your password visible. That would be a huge security risk.

However, an admin can assign a new *initial password* to your user as a replacement for your forgotten password. You must then change this password immediately the next time you log in, just as you would for an initial login.

1.1.3   No Multiple Logons

You can start programs such as Microsoft Excel or Word as often as you like on a PC; but the same isn't true with SAP GUI. Just try it out! Try to log in a second time according to the last instruction. Then, after confirming the logon screen, a message box for multiple logons will appear, as shown in Figure 1.7.

You have the following options in the **License Information for Multiple Logins** window:

- **Continue with this logon and end any other logons in the system**
  Caution: Selecting this option will automatically close the existing login window. After that, only the new user login is active. Unsaved data from the first login will be lost!

- **Continue with this logon, without ending any other logons in the system**
  This sounds good, but beware: this way, you log in a second time without closing existing logins. This second login can be costly because SAP charges for SAP S/4HANA

47

licenses per active user. In many systems, this second option is therefore always hidden. If you want to use the SAP system in several windows at the same time, you can—but not by logging on again.

- **Terminate this logon**
  This option cancels the new logon and leaves the old logon as it is. In this case, you use the Microsoft Windows task manager to switch to the existing SAP window and continue from there. Of course, this doesn't affect the license fees.

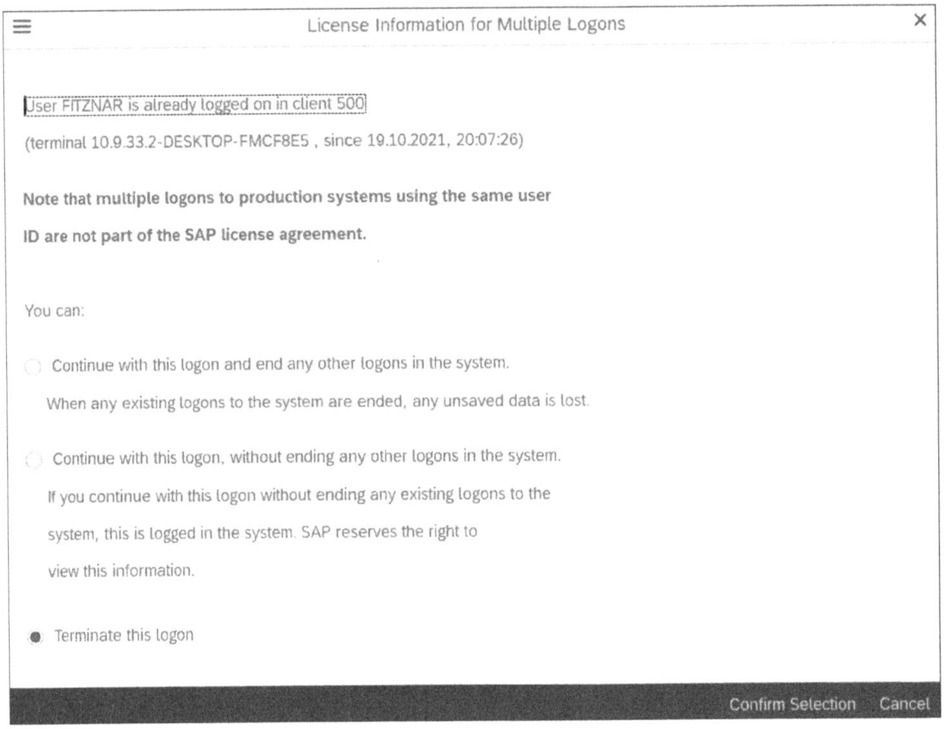

| ≡ | License Information for Multiple Logons | × |

User FITZNAR is already logged on in client 500

(terminal 10.9.33.2-DESKTOP-FMCF8E5 , since 19.10.2021, 20:07:26)

Note that multiple logons to production systems using the same user
ID are not part of the SAP license agreement.

You can:

○ Continue with this logon and end any other logons in the system.

   When any existing logons to the system are ended, any unsaved data is lost.

○ Continue with this logon, without ending any other logons in the system.

   If you continue with this logon without ending any existing logons to the

   system, this is logged in the system. SAP reserves the right to

   view this information.

● Terminate this logon

                                              Confirm Selection   Cancel

Figure 1.7 Window after a Multiple Login

[!]

Data Loss

Consider this example of the first option, **Continue with this logon and end any other logons in the system**: You're logged in at your workstation, you've half-entered a large order there, and you haven't yet saved it. Now you go into a meeting. There, you want to show a colleague an SAP S/4HANA report and also log on to the SAP system with your username using his laptop. As described, the multiple logon window appears. There, you select **Continue with this logon and end any other logons in the system** and bye-bye! The data entered at your workstation is gone; the SAP window there has been closed.

1.1.4   Log Off

One of the most important functions is, of course, the after-work function—logging off. There are several shutdown buttons when only one SAP GUI window is open:

- The ✕ (**Close**) icon in the upper-right corner of the window
- **Exit** button on the far right of the **SAP Easy Access** window
- Key combination [Alt]+[F4] from Windows

As you can see in Figure 1.8, it can be even more complicated: Click the **More** button and open the **System** submenu. There you choose the lowest menu option **Log Off**.

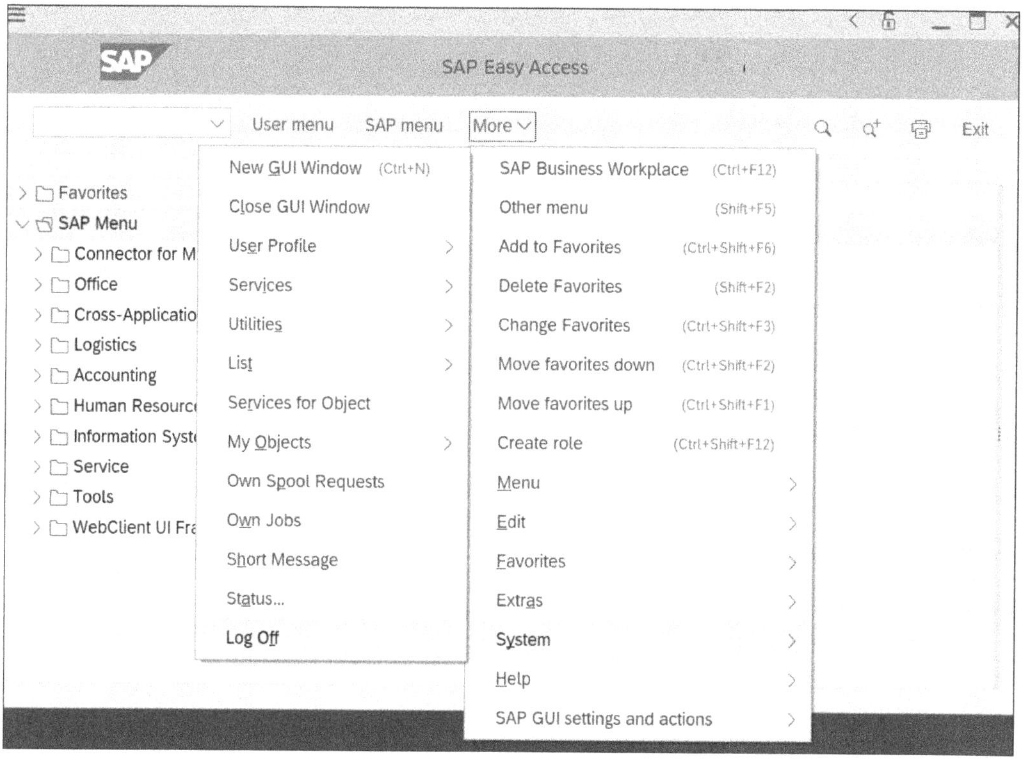

Figure 1.8 System Submenu with the Log Off Option

Help! Where are the More and Exit Buttons?

If you can't find these buttons due to an outdated theme setting, select the **System** menu command in the menu bar and then the **Log Off** menu option. For information on how to set your current theme, see Section 1.2.

No matter which method you use, SAP S/4HANA will ask again if you really want to log off in the **Log Off** window via the message, **Unsaved data will be lost. Do you want to log off?** (see Figure 1.9).

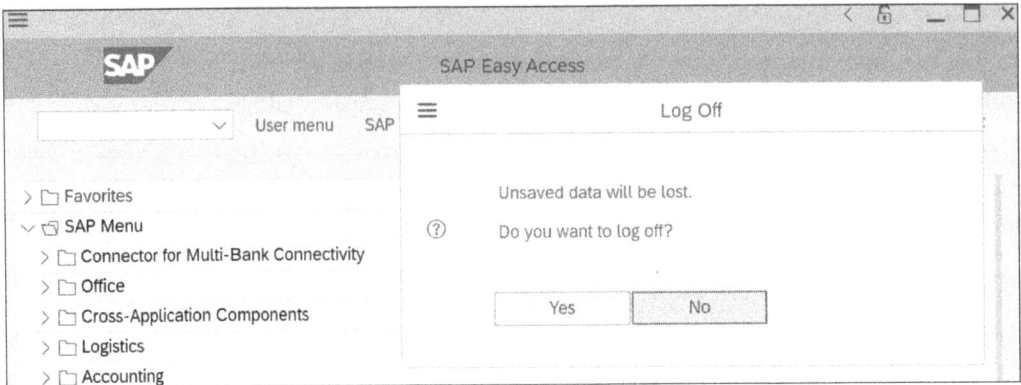

Figure 1.9 Question before Final Logoff

Here you may answer with **Yes** or **No**. Don't be surprised by this question just because you don't have any data to save in our case; unlike Microsoft Word or Microsoft Excel, the SAP system always asks this question, regardless of whether you've changed data or not.

So, now you know how to get in and out of the system. The next section will be fun for you because you get to choose a theme and set your favorite color.

1.2   Select Display Themes and Colors

Belize is more than a country in Central America; it's also the name of a *theme* (or *screen layout*) in SAP S/4HANA. Do you want to see the same functions in the SAP GUI as in this book? Not a bad idea! To do this, set the Belize theme now. You'll also learn how to change the color setting.

We use the Belize theme because it's the default setting in SAP GUI 7.60. Of the various themes in SAP GUI, it's most similar to the SAP Fiori user interface. If you use both SAP GUI and SAP Fiori, then Belize is often the best choice.

If you logged off toward the end of the previous section, log back on now. We'll start in the **SAP Easy Access** initial screen. From here, follow these steps:

1. Select **More** • **SAP GUI Settings and Actions** • **Options**, as shown in Figure 1.10.

> **Help! I Still Don't Have a More Button**
>
> If you don't have this button due to an outdated theme setting, click on the ▣ icon (**Customize Local Layout**) in the upper-right corner. In the menu that opens, select the top command **Options**. After that, you'll also find yourself in the **SAP GUI Options**. After this procedure, we promise you'll also have the **More** button.

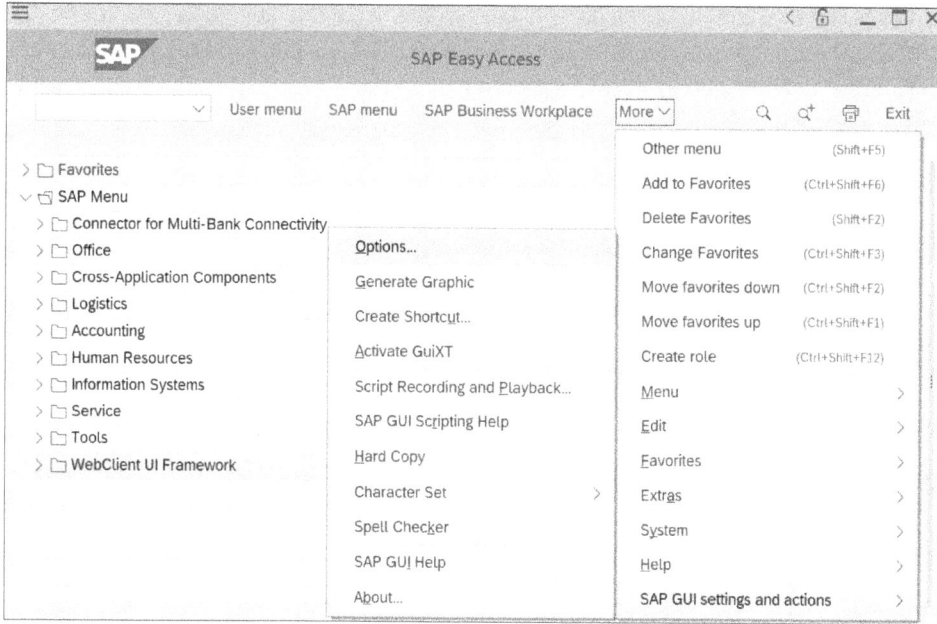

Figure 1.10 Accessing SAP GUI Options

2. Open the **Visual Design** folder in the upper-left corner, and click **Theme Settings**, as shown in Figure 1.11.

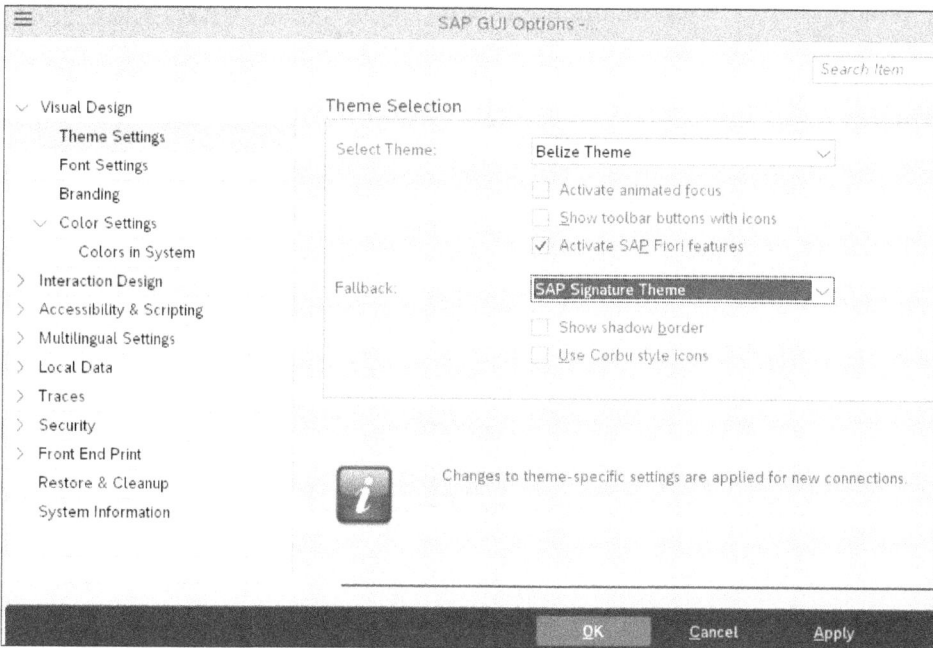

Figure 1.11 Theme Selection

3. In the **Select Theme** field, click ☑ to open the list of themes, and select the **Belize Theme** setting. Deselect the **Show Toolbar Buttons with Icons** and **Activate SAP Fiori Features** checkboxes, as shown in Figure 1.11.

[»]

Is the Belize Theme Missing?

If you can't find the Belize theme, you have an outdated SAP GUI version. In this case, in SAP GUI version 7.50, in the **Theme** field on the top left, select the **SAP Signature Theme** setting, and in the right half of the window, select **Accept Belize Theme**. However, you'll only see buttons with explanatory texts instead of icons. The icons for the Belize theme are available only from SAP GUI version 7.60.

4. While you're here, you may as well set your favorite color. To do this, in the **SAP GUI Options** window in the left part of the screen, choose **Visual Design • Color Settings • Colors in System**. Now you can see the range of colors for the Belize theme in the right part of the window, as shown in Figure 1.12.

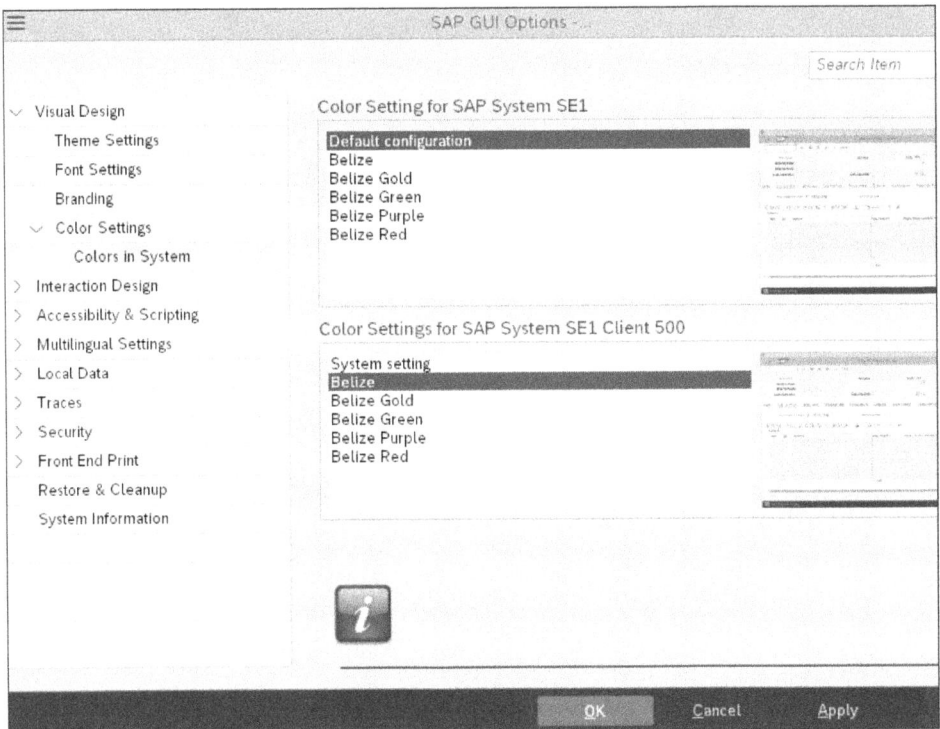

Figure 1.12 Color Settings

5. In the lower-right part of the window, select the color for your client. After each click on a color setting such as **Belize Green**, you'll see a preview of the color to the right.

6. Click the **OK** button to close the **SAP GUI Options** window.

7. Everything still the same? Same here! Press ⌈Ctrl⌉+⌈+⌉ to open a new window; only in a newly opened window will you see your new color setting.

It Doesn't Have to Be Belize

Belize has no menu bar and fewer icons than the other themes, which is why it's clearer and better suited for a soft start. The price for this is that the paths are sometimes longer than with the other themes.

If you later find that you need the **More** button very often, you should try the following three other themes: SAP Signature, Corbu, or Blue Crystal. These themes have a different screen layout and other screen elements. This often saves users, especially SAP GUI power users, many clicks. Figure 1.13 shows an example of the **SAP Easy Access** screen in the Corbu theme.

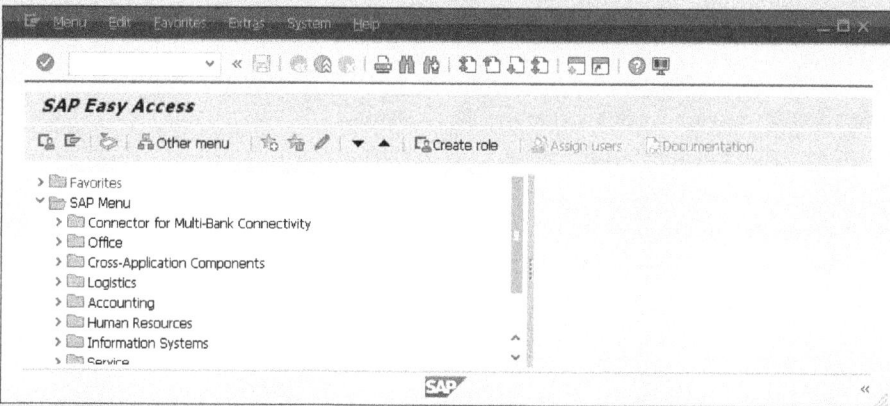

Figure 1.13 SAP Easy Access Window in the Corbu Theme

These themes take you directly from a menu bar to the menu commands more quickly, and additional icons take you more quickly to other functions. In Belize, you often need to use the **More** button to do this.

Now that you've set the basic layout and colors, explore the details of the SAP GUI window in the next section.

1.3   Explore the SAP GUI Screen Elements

A prerequisite for targeted navigation with SAP GUI is knowledge of the screen layout. We'll now discuss this using the **SAP Easy Access** window, as shown in Figure 1.14, which is also the welcome screen after logging on and your central cockpit in SAP GUI. From here, you access all SAP GUI applications in your task area.

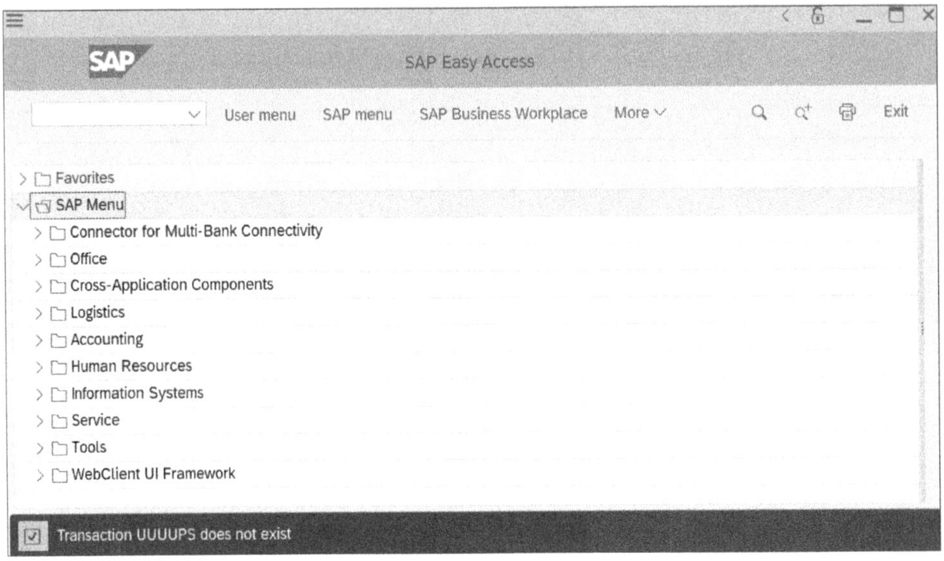

Figure 1.14 SAP Easy Access Window in the Belize Theme

We'll only give you an overview in this section, so you don't get lost in details. We'll discuss individual functions in more detail in later sections precisely when you need them. Now let's get an overview from top to bottom.

1.3.1   Top Bar

In the top bar shown in Figure 1.15, you'll find the ▭ ▭ × window icons known from Windows, which have the same meaning in the SAP window. To the left of this is a system information area, which you close with the ⟩ icon and display with ⟨. You'll rarely need this area in practice. At the top left, there is also a window menu that you open by clicking on the ☰ icon.

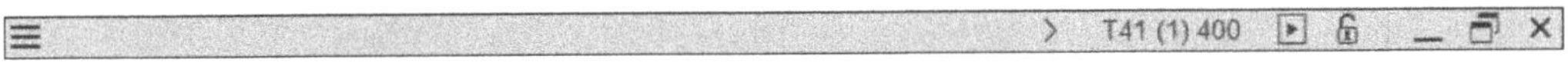

Figure 1.15 Top Bar

1.3.2   Title Bar

As the second uppermost bar, the title bar acts like a navigation device, showing you at which position and in which window you're currently located (see Figure 1.16). Each window has its own title or heading, which is displayed here.

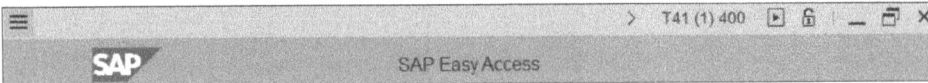

Figure 1.16 Top Bar and Title Bar in the SAP Easy Access Window

1.3.3   Header Line

Below the title bar, you'll see the header line, as shown in Figure 1.17. This is the line you'll need most often in practice. On the far left, you'll find the *command field* for entering commands.

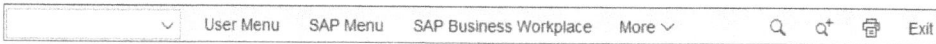

| | | User Menu | SAP Menu | SAP Business Workplace | More ⌄ | | Q | Q⁺ | 🖶 | Exit |

Figure 1.17 Header Line in the SAP Easy Access Window

Further to the right up to the **More** button, you'll find buttons and icons that you need only in the current window. These buttons and icons thus change from window to window. For example, if you're maintaining vendor master data using the **Maintain Business Partner** function, the header line is populated with different buttons and icons as if you were currently in the **SAP Easy Access** window (see Figure 1.18).

☰			>	T41 (1) 400	▶ 🔒	— 🗗 ×
< **SAP**		Maintain Business Partner				
	⌄	Locator On/Off Person Organization Group Open BP More ⌄				Exit

Figure 1.18 Top Bar, Title Bar, and Header Line in the Maintain Business Partner Window

However, many icons are only displayed if you've activated the **Show Toolbar Buttons with Icons** preset for the Belize theme in the **Select Theme** area. In SAP GUI version 7.50, there is no display of icons for the Belize theme yet.

Further to the right, in the header line, you can see buttons and icons that have their fixed place here throughout the system and which are summarized in Table 1.1.

Icon/Button	Short Name	Meaning
More ⌄	More	Accesses special functions, each belonging specifically to the current window, and other general functions
Q	Find	Searches, for example, in a list of orders for a supplier
Q⁺	Find Next	Continues a search after the first hit has been found
🖶	Print	Prints a list or a form
Exit	Exit, Log Off	Terminates the current transaction; in the **SAP Easy Access**: **Log Off**

Table 1.1 Buttons and Icons in the Header Line

[»]

Why Don't I See All the Functions in the Header Line?

There are two possible reasons why your header line doesn't display everything we describe here. First, Q (**Find**), Q⁺ (**Find next**), and 🖶 (**Print**) are only displayed in the windows where there is something to search or print.

Second, the window is too narrow. Try to drag the **SAP Easy Access** window to make it narrower. The buttons and icons "disappear" into the **More** menu.

1.3.4   More Menu in the Header Line

The functions offered in the **More** menu change according to the window width or screen resolution and the window you're currently in. For example, according to Figure 1.19, there are commands in the **More** menu specifically in the **SAP Easy Access** window—and only here—for editing your favorites (left side).

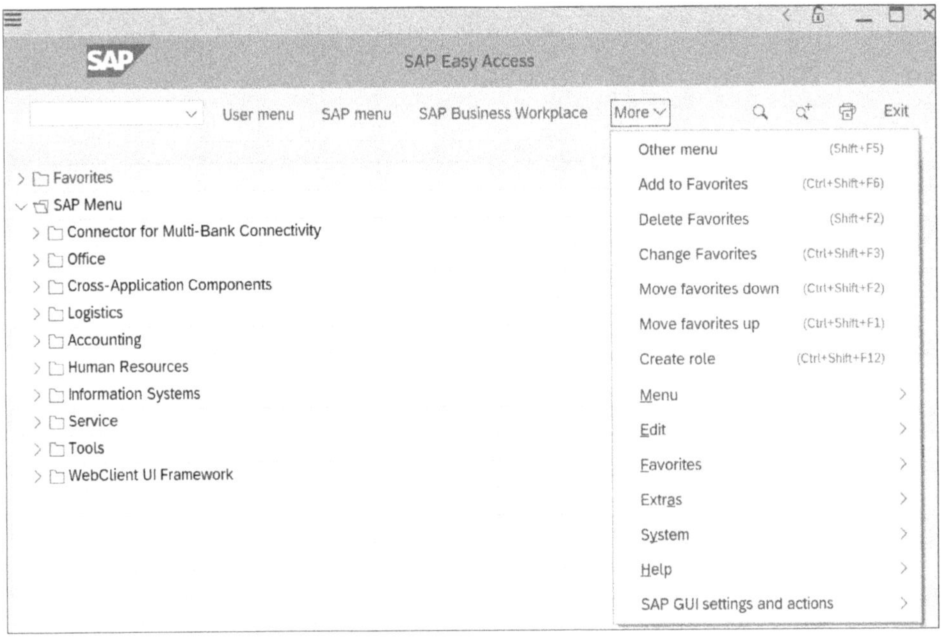

Figure 1.19 More Menu in the Right Part of the SAP Easy Access Window

This rule is confirmed by the following three exceptions: The last three menu items, **System**, **Help**, and **SAP GUI Settings and Actions**, always remain the same, no matter which window you're in. This is because there are menu commands there that can be useful in any window:

- **System**
  This submenu contains various general functions, for example, for sending short messages or for logging out.

- **Help**
  This submenu takes you to different online guides.

- **SAP GUI Settings and Actions**
  This submenu contains a colorful mixture of general commands. You had already used the **Options** command to set the correct preferences for the Belize theme.

1.3.5   Screen Area: Favorites and SAP Menu

The screen area is the area between the header line and the footer line, which is the last screen line. In the **SAP Easy Access** window, this is where you'll find the **SAP Menu** and, above it, the **Favorites**.

The **SAP Menu** with its thousands of entries is probably one of the largest menus available. You open a subfolder ☐ by clicking on ❯ in front of it or by double-clicking on its name, for example, **Office**, **Logistics**, **Accounting**, or **Human Resources**.

As shown in Figure 1.20, you open the **Office** subfolder in the **SAP Menu**. Here, in addition to other subfolders, there are also entries marked with the ⚙ icon. These are the applications that are called transactions in SAP S/4HANA. Sometimes you have to open 10 or more folders in this tree structure to finally get to the transaction you want.

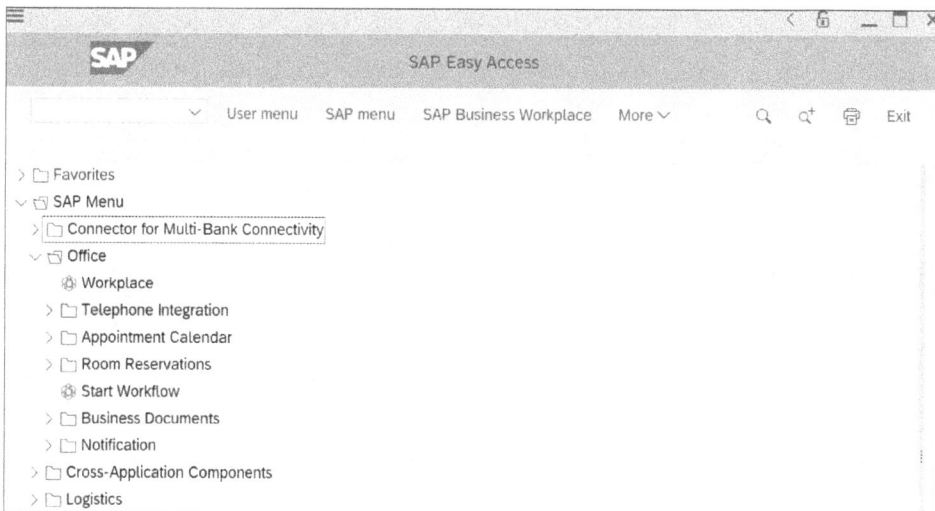

Figure 1.20 SAP Menu with Open Office Subfolder

Open and Close Folders Faster

There are other places in SAP S/4HANA besides the **SAP Menu** where you open and close many folders. This often works faster with the keyboard: "Climb" into the tree to the desired folder, down with `ArrowDown` or up with `ArrowUp`. You open the branches and twigs (i.e., the folders) with `ArrowRight` and close them with `ArrowLeft`. By the way, this trick also works in Windows Explorer.

Double-click on the name to start a transaction. In this example, double-click on **Work-place** in the **Office** folder, which you can use, for example, to send messages to other SAP users. What happens? The screen area with the **SAP Menu** disappears completely and is replaced by the first screen of the transaction, as shown in Figure 1.21. The header line and the **More** menu contain other functions, and the title bar text changes as well.

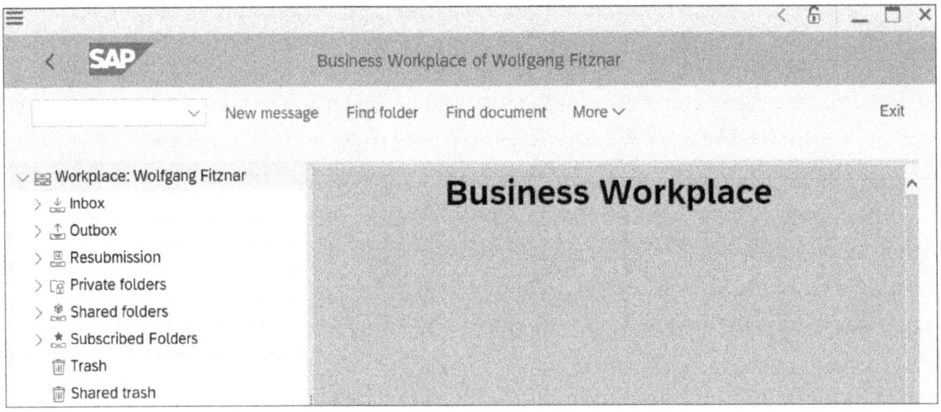

Figure 1.21 Initial Screen of Transaction Workplace from the Office Subfolder

But stop! We're still in the overview, after all. Leave the transaction either by clicking ◁ (**Back**) from the title bar on the top left or by clicking the **Exit** button on the top right.

But do you really have to open all folders individually until you get to the desired transaction? No, there's a much easier way! You can use the **Favorites** folder above the **SAP Menu** to collect the transactions you need there. We'll show you how to do this in Section 1.5.

In the **SAP Easy Access** window, the menu structure is separated from the right part of the window by a vertical separator bar (see Figure 1.22), which you can select with the mouse and drag to the right or left. If you drag to the right, then you'll see more of the left **SAP Menu**. The separator bar, dragged all the way to the right, hides the right part of the window completely.

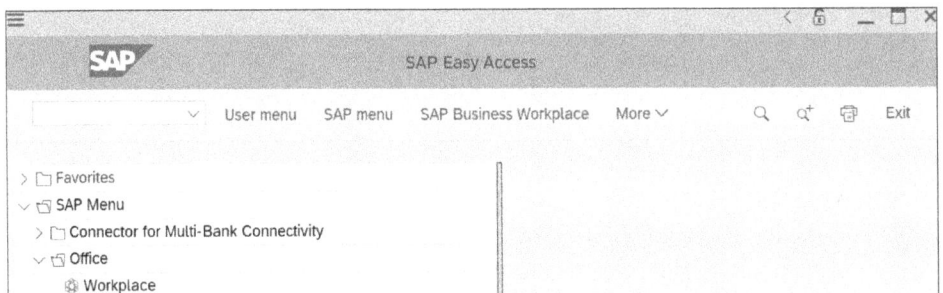

Figure 1.22 SAP Easy Access Window with Separator Bar between the Two Halves of the Screen

1.3.6   Footer Line

The bottom line of the window is called the *footer line*. You probably see nothing here but a black bar and wonder what there is to talk about, but the left part is reserved for displaying messages. Enter "Hello SAP" in the command field at the top left, and confirm this entry with the key. You'll then see a message in the bottom left of the footer line, as shown in Figure 1.23.

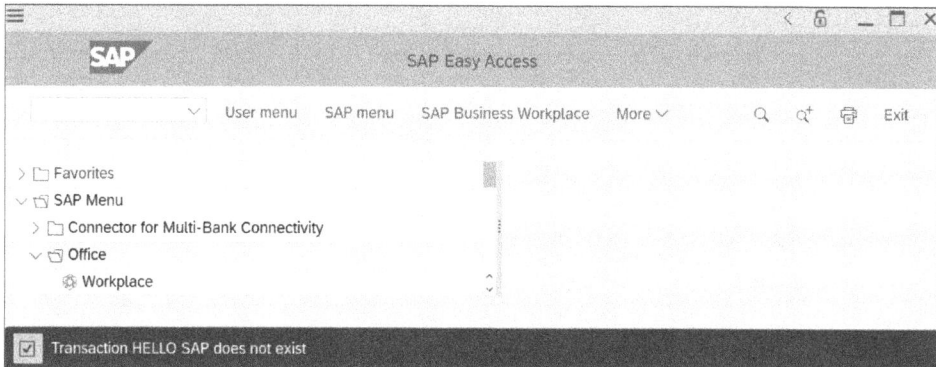

Figure 1.23 SAP Easy Access Window with a Message at the Bottom Left of the Footer Line

In addition, there are buttons such as **Save** or **Continue** on the right side of the footer line in many windows, but not in the **SAP Easy Access** window.

Where Are My Functions in the Belize Theme?

If the Belize theme is new to you, and you've worked with an SAP system before, you're probably a bit irritated because you're missing the former menu bar as well as the system function bar with the overlapping icons.

With the Belize theme, the menu bar functions have been integrated into the **More** menu. The icons from the system toolbar have either been deleted without replacement or moved to various places as described in Table 1.2.

Icon	Short Name	Where in Belize?
	Enter	**Continue** button in the footer line, displayed only when needed
	Close Command Field	Deleted without replacement
	Save	**Save** button in the footer line, displayed only when needed
	Back	Icon in the title bar, except in the **SAP Easy Access** window

Table 1.2 Icons from the Former System Toolbar and Corresponding Functions in the Belize Theme

Icon	Short Name	Where in Belize?
	Exit	**Exit** button in the header line
	Cancel	**Cancel** button in the footer line, displayed only when needed
	Print	The icon in the header line, displayed only when needed
	Find	The icon in the header line, displayed only when needed
	Find Next	The icon in the header line, displayed only when needed
	First Page	Deleted without replacement, use the scroll bar
	Previous Page	Deleted without replacement, use the scroll bar
	Next Page	Deleted without replacement, use the scroll bar
	Last Page	Deleted without replacement, use the scroll bar.
	New GUI Window	Deleted without replacement, but new window can still be opened with the key combination Ctrl + +
	Generate Shortcut	Deleted as icon without replacement, but shortcuts can be created via **More • SAP GUI Settings and Actions • Create Shortcut**
	Help	Deleted as icon without replacement, but field help can still be called with the F1 key
	Customize Local (Layout Menu)	Deleted as icon without replacement, but layout menu accessible via **More • SAP GUI Settings and Actions • Options**

Table 1.2 Icons from the Former System Toolbar and Corresponding Functions in the Belize Theme (Cont.)

This was our first tour of the complete SAP screen in the Belize theme. In the next sections, we'll get more specific, and you'll make practical and useful default settings, such as displaying transaction codes.

1.4   Display Transaction Codes

Lunch conversation in the cafeteria: "You won't believe it! Someone with the username HGW752 changed one of my customer master records. Does anyone know how I can find out who that was?" — "Simple, use SU01D!" (We'll translate the answer for you: "Start the User Display transaction with the SU01D transaction code!")

Each transaction has a short name called a *transaction code*. Examples include Transaction VA01 (Create Sales Order) or Transaction FB03 (Display Document). In this section, you'll learn how to display the transaction codes in the **SAP Menu** and in the header line.

> **What Do I Need Technical Transaction Codes For?**
>
> You need these codes if you want to work professionally, for example, to quick-start transactions or to create favorites. Even in a job interview, you may be asked for transaction codes to verify your SAP experience.

1.4.1   Display Transaction Codes in the SAP Menu

For practical purposes, we recommend that you preset according to the following procedure, so that in the **SAP Menu** and in the **Favorites** list, these transaction codes are displayed in front of the transaction names. And this is how it works:

1. In the **SAP Easy Access** window shown in Figure 1.24, open the **Office** folder. Here, you can already see from the first **Workplace** entry whether Transaction SBWP (SAP Business Workplace) is displayed in addition to this transaction name.

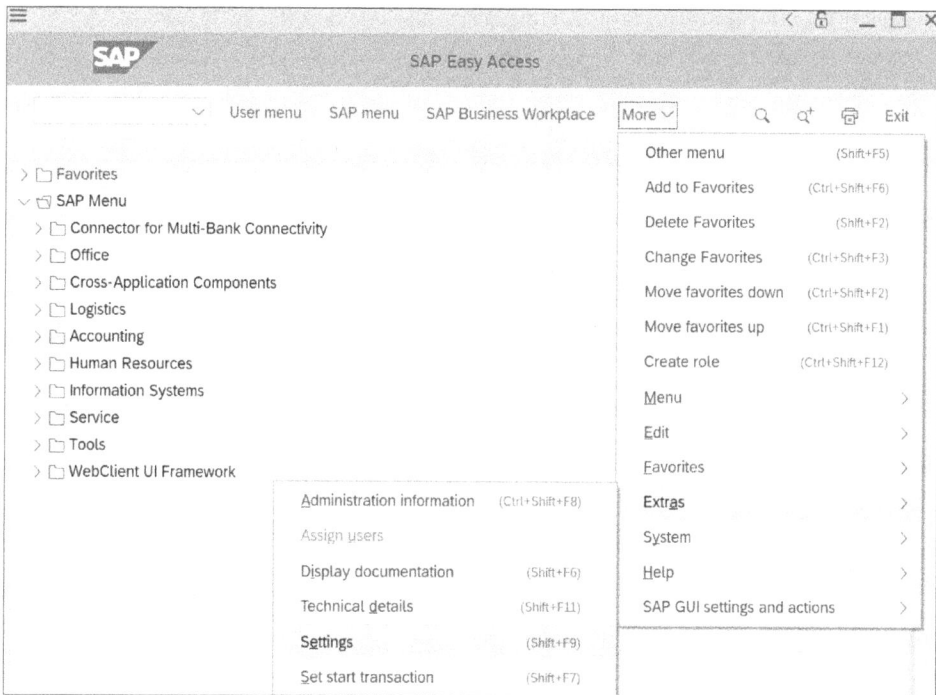

Figure 1.24 Office Folder and More Menu with the Extras and Settings Options

2. If yes, all is well. If not, in the **SAP Easy Access** window, choose **More • Extras • Settings**. The **Settings** window appears, as shown in Figure 1.25.

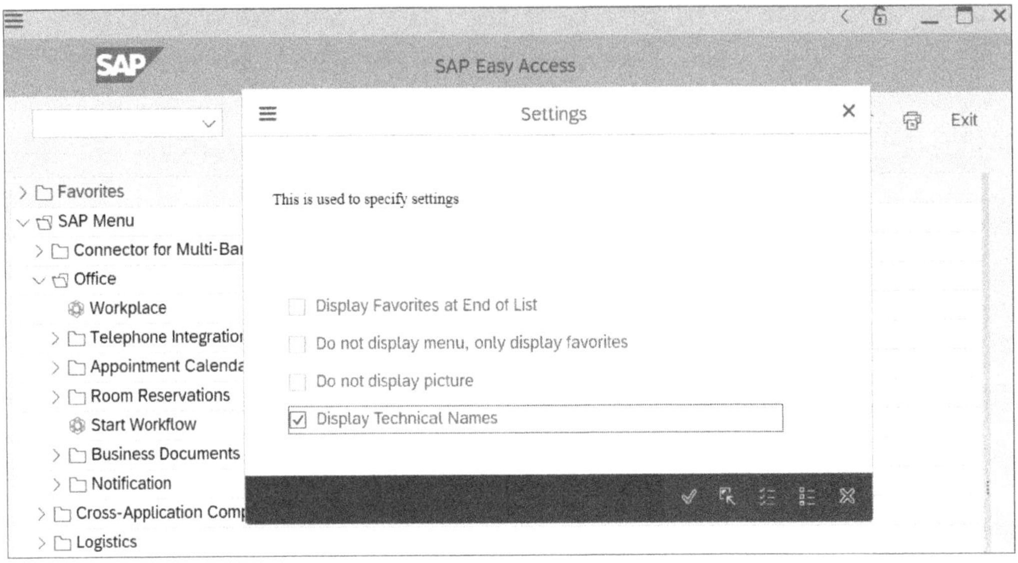

Figure 1.25 Settings Window

3. Select **Display Technical Names**. Transaction codes are referred to here as technical names. Press ⌨Enter⌨ or click ✔ (**Continue**).

Now SAP GUI also shows the transaction codes such as SBWP in the tree structure of the **SAP Menu**. You can see this if you open the **Office** folder again, as shown in Figure 1.26.

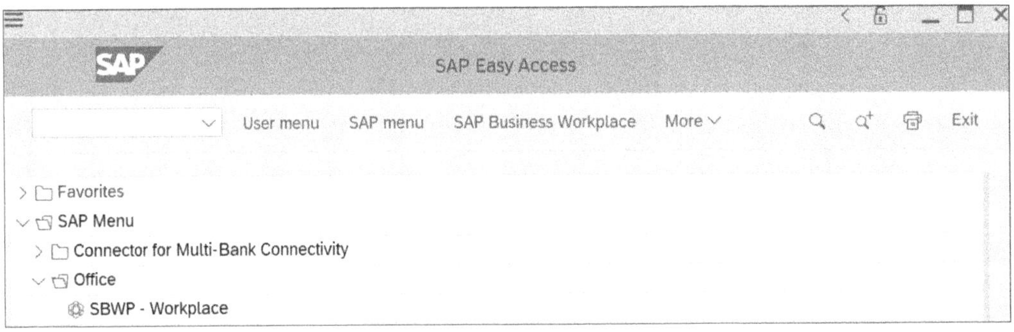

Figure 1.26 Office Folder Showing Transaction Code SBWP

1.4.2   Display the Current Transaction Code

It's often helpful to see the transaction code of the currently called transaction, for example, if you need help with a transaction and are on the phone with support. Use the display in the *system information area* in the top line of the window, and follow these steps:

1. In the **SAP Easy Access** screen, in the **Office** folder, start **Transaction SBWP - Workplace** by double-clicking it.

2. At the top of the system information area, to the right of the ▷ icon, click on the **T41 (1) 400** entry, which may have a different name depending on the system (T41 for us), window number (1 for you and 1 for us), and client (400 for us). As in Figure 1.27, a list opens in which you can see the code of the currently called transaction, in our example, **SBWP**.

Is the Icon Missing?

If you don't have the ▷ icon, the system information area is hidden from you. Click the ◁ icon in the upper-right corner to make this area visible.

---

3. Click the transaction code in the list.

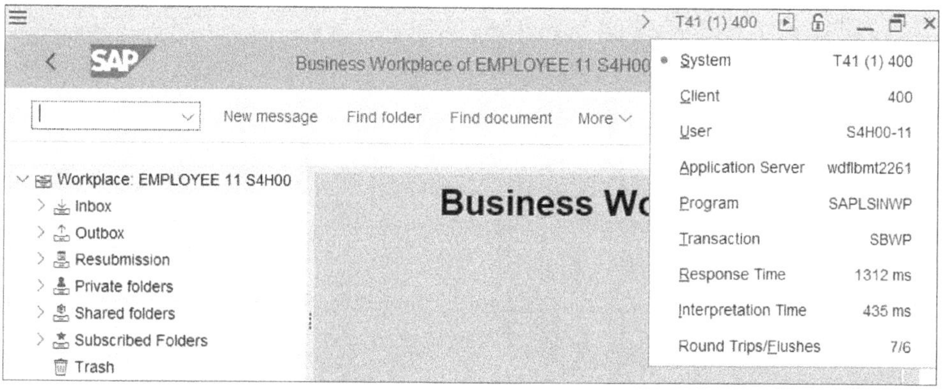

Figure 1.27 Display the Current Transaction in the System Information Area

From now on, you can see the code of the currently called transaction at any time in the upper-right corner of the system information area: ▷ SBWP ▶.

For those who log in with a PKI card and thus without a username, the information about your username is given in the third line, **User**, because you don't see this in the login screen.

To return to the **SAP Easy Access** screen, click ◁ (**Back**) in the upper-left corner.

1.5   Create a Favorites Menu

Now you can get creative and make your own menu. If you use several transactions, you can collect them in a separate **Favorites** menu. This saves you all the work and time of opening and closing folders in the tree structure of **SAP Menu**.

Unlike the **SAP Menu**, the **Favorites** menu is your personal menu. In other words, it's a menu that you can freely design and set up as you would your living room. Almost all SAP users add their transactions to the **SAP Easy Access** window as favorites, and you should too.

You can enter not only transactions but also links to files, subfolders, and web addresses. Do you use SAP GUI and SAP Fiori apps at the same time? With web addresses, you can also integrate links to apps in your favorites.

[+]   **Import Ready-to-Use Favorites Menu**

Before you create your own **Favorites** menu, here's the first tip: You don't need to reinvent the wheel. Have you perhaps already seen a wonderful **Favorites** menu from a nice colleague set up just the way you need it? Great! Your colleague can download the **Favorites** menu as a file in the **SAP Easy Access** by choosing **More • Favorites • Download to PC** and then copy it to a USB stick or e-mail it to you. You then select **More • Favorites • Upload from PC** at your workstation, and you have the same favorites as your colleague.

By default, the **Favorites** folder is empty. Now we'll enter some general transactions as examples, which are briefly introduced here:

- **Transaction SU3 (Maintain Users Own Data)**
  You use this transaction to specify your personal information, such as default values. The abbreviation SU stands for *system* and *user*.

- **Transaction SU01D (User Display)**
  In many places, SAP S/4HANA often only shows a cryptic username instead of a real name, for example, when SAP S/4HANA locks the processing of data for you because another user is currently processing it. You then ask yourself: Which user is hiding behind the cryptic username? With Transaction SU01D (User Display), you can find the matching real name to the SAP username.

- **Transaction SBWP (SAP Business Workplace)**
  In the SAP system, you may receive emails and other messages. With this transaction, you display messages or compose SAP system-internal messages to other SAP users.

The codes we've discussed in this section are the technical abbreviations for transactions that you'll use in the next section.

1.5.1   If You Know the Transaction Code

In this section, we'll cover a quick method for creating favorites that works only if you know the transaction code.

Follow these steps:

1. If necessary, switch to the **SAP Easy Access** window.
2. Now for an elegant right-left combination: right-click on the topmost **Favorites** folder, and select **Insert Transaction** from the context menu (see Figure 1.28). The SAP system will ask for the transaction code in a new window.

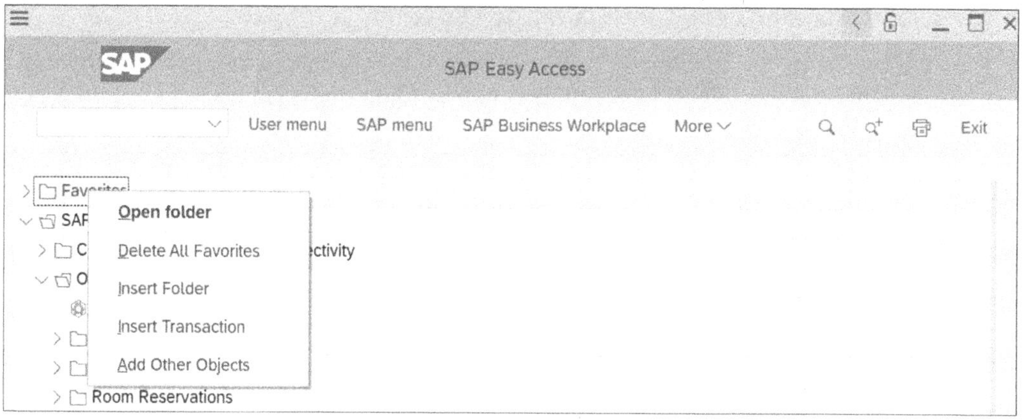

Figure 1.28 Favorites Context Menu

3. Enter the transaction code here, for example, "SU3", as shown in Figure 1.29, and press the [Enter] key. The window will now close, and the new entry will immediately appear in the **Favorites** menu.

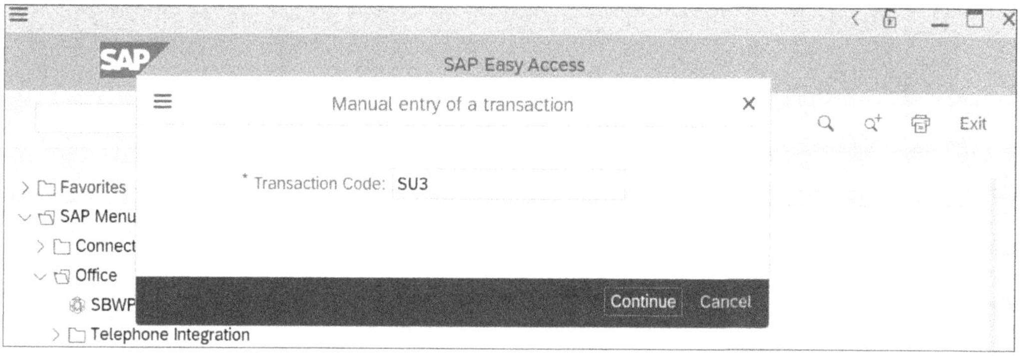

Figure 1.29 Entering the Transaction Code

4. Repeat the previous three steps for Transaction SU01D (User Display) and Transaction SBWP (SAP Business Workplace). The screen will look like Figure 1.30.

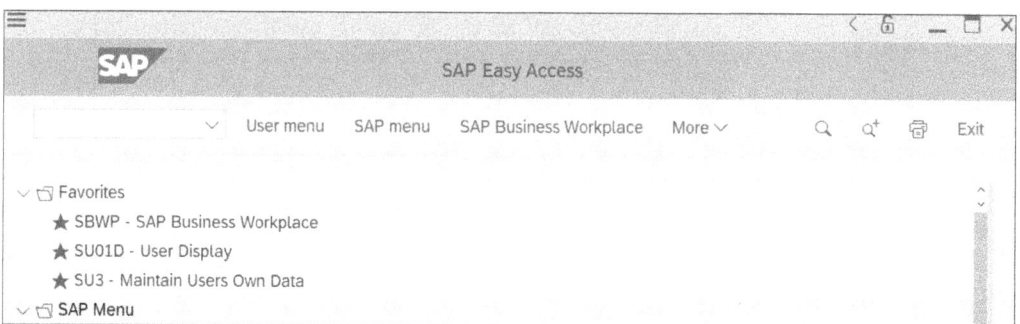

Figure 1.30 Favorites List

1.5.2 If You Don't Know the Transaction Code

Don't know the transaction code? Just click through the folders in the **SAP Menu** until you can see the desired transaction. You can also add this transaction to the **Favorites** menu with the drag and drop method, which follows these steps:

1. For our example, open the **Office** and **Appointment Calendar** folders in the **SAP Menu**. Here you can see Transaction SSC1 (Owner), as shown in Figure 1.31.

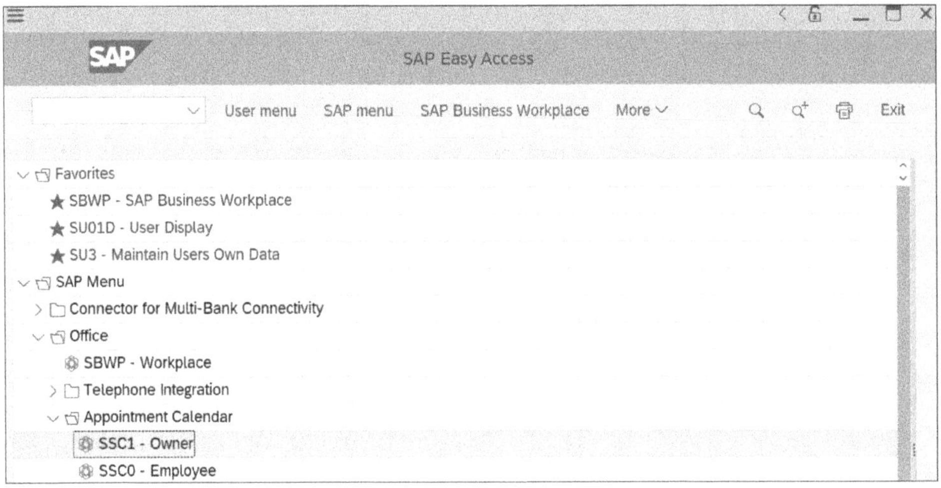

Figure 1.31 Dragging and Dropping to Favorites

2. Click (once, not twice) on **SSC1 - Owner**, and hold down the left mouse button.
3. While holding down the left mouse button, drag the **SSC1 - Owner** entry to the top of the **Favorites** folder, and release the mouse button.

[+] **Even More Options**

In our case, the drag-and-drop path was very short. If it's longer and more cumbersome, there is a better alternative: right-click on the transaction entry in the **SAP Menu**, and then select the **Add to Favorites** menu option.

In Chapter 5, Section 5.1, you'll learn how to create an optimal **Favorites** menu; there's still a lot to discover!

1.6 Set User Data Appropriately

As the user, personal user data is stored for you that includes your name as well as other information. In the following, you'll learn how to set the login language, the decimal and date display, your output device, and other settings for the printout.

Before you start making changes with the following instructions, you should know that depending on the system settings, fields may be locked against input or changes.

Follow these steps:

1. Select **More • System • User Profile • User Data**, or start the transaction by double-clicking on the favorite **SU3 – Maintain Users Own Data**, which you created in the previous section. The **Maintain User Profile** window appears, as shown in Figure 1.32.

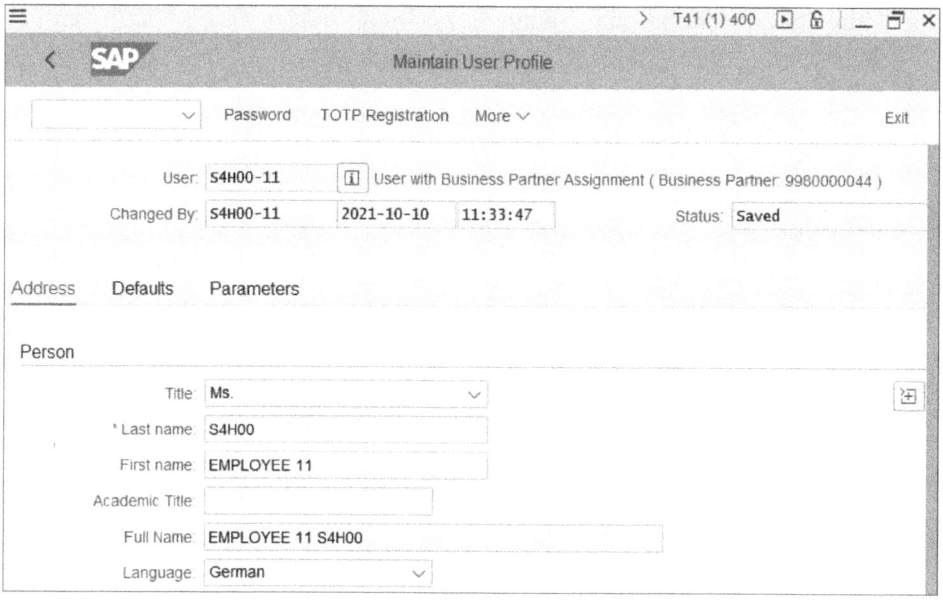

Figure 1.32 Maintain User Profile: Address Tab

2. Depending on the system configuration, you can enter your first and last name and other personal information in the corresponding fields. However, these fields are locked in many systems so that only the administrator can change this information.

    SAP S/4HANA also transfers the name entered here to forms such as purchase orders or sales orders. If set differently and possible, choose **English** in the **Language** drop-down field ☑ because this field determines in which language certain texts are displayed and printed.

3. Click the **Defaults** tab to check some basic presets there, as shown in Figure 1.33.

4. Note the following fields for your first access:

    – **Start Menu**: This field should be empty for our examples or have the entry S000 so that you can see the complete **SAP Menu** and not just any special submenu when you start the program, that is, immediately after logging in. Sometimes there are also reduced company- and department-specific menus.

    – **Logon Language**: Enter "EN" for "English," this specification saves you having to make an entry on the login screen.

- **Decimal Notation**: Use the ▽ icon to select the display **1,234,567.89** from the list, so that you get the comma as a thousand separator and the dot as a decimal separator.
- **Date Format**: Use the ▽ icon to select the international date format **YYYY-MM-DD** or the American date format **MM-DD-YYYY** from the dropdown list.
- **Output Device**: Enter "LOCL" for the normal Windows printer. Be sure to use the letter "O", not zero.
- **Print Now**: Check this box; otherwise, you won't receive an automatic printout after a print command.
- **Delete After Output**: Don't check this box; otherwise, you won't be able to make a reprint, as the printout won't be saved in the SAP S/4HANA system.

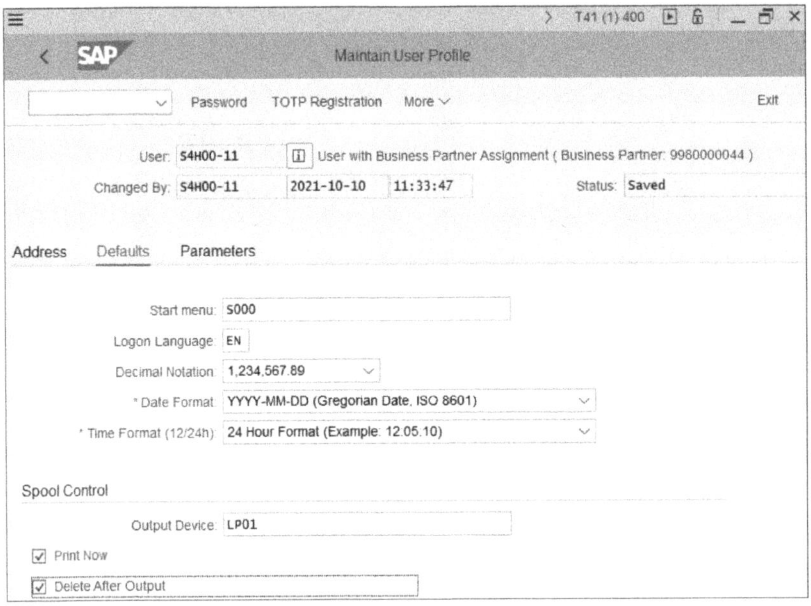

Figure 1.33 Maintain User Profile: Defaults Tab

5. Save your entries by clicking the **Save** button at the bottom right of the footer line. You're then sent back to the **SAP Easy Access** window.

Some new settings, such as the login language and decimal and date display, only take effect the next time you log in to the SAP S/4HANA system.

1.7 Summary

We've come to the end of the first chapter, and you now know how to log in and out, know the screen layout of SAP GUI, can display transaction codes, and have set your important user data. Great!

Next, we'll discuss how to use transactions to display and maintain data.

Chapter 2

Displaying and Maintaining Data
with SAP GUI Transactions

The first SAP GUI chapter was only the prelude. Now it's time to pay attention because from this point on, the productive work begins. In this chapter, you'll display, enter, and change your data elegantly and efficiently using SAP transactions.

For this purpose, we'll first show you another procedure for starting transactions, which, in many cases, is even more convenient and faster than starting from the **SAP Easy Access** window.

Then we'll cover exciting functions available to you for navigating the transactions and entering data. Finally, we'll provide some self-help assistance by showing you how to locate the specific record you need among the often millions of records and how to find out the meaning of fields.

What You'll Learn

- How to start transactions faster and to navigate in transactions safely and purposefully
- How to efficiently enter and change data
- How to find input values, documents, and master data
- How to get support via various SAP help sources
- How to work with multiple windows

We won't always use the most commonly used transactions and apps throughout Part I and Part II of the book. Instead, we've selected simple and clear examples for a smooth introduction. Another criterion for the selection was that the rules of the transactions and apps should be simultaneously representative of as many others as possible. Based on this knowledge, in Part III of the book, we'll switch to the frequently used applications with which the standard processes are carried out in the system.

2.1   Start a Transaction

You already know from the previous chapter that you can start a transaction by double-clicking in the **SAP Menu** or by selecting an entry in the **Favorites** list. In this section, you'll learn a particularly fast and elegant method that is popular with professionals and power users: entering the transaction codes in the command field.

2.1.1   Transactions and Transaction Codes

For this method, you should know the transaction codes by heart, which won't be difficult because you'll soon dream about these codes, as you'll encounter them every day in the **SAP Menu**. To make it even easier for you to remember the codes, you can find some of them in Table 2.1, and then we'll use them to derive the corresponding rules for you.

Transaction Code	Name	Description
FI01	Create Bank	Create bank master data
FI02	Change Bank	Correct bank master data
FI03	Display Bank	Display bank master data without the option to modify
MM01	Create Material	Create material master data
VF02	Change Billing Document	Change an outgoing invoice
KS03	Display Cost Center	Display the data of a cost center in cost accounting
PA20	Display HR Master Data	Display data such as addresses or salaries
FSP0	G/L Acct Master Record in Chrt/Accts	Display, change, and enter general ledger accounts in financial accounting
SU3	Maintain Users Own Data	Maintain personal data such as names or default values
SBWP	SAP Business Workplace	Display and create, for example, SAP-internal messages to other SAP users

Table 2.1 Transaction Codes from Different Modules

Did you notice anything while reviewing the previous table? The codes from the table are almost all four-digit codes. But there are also shorter and longer codes. And you can *often* read the main module from the first character and *sometimes* the component from the second character. In the example of Transaction VF02, this means that V stands for the sales module (in German, *Vertrieb*) and F for the invoicing component (in German, *Faktura*). Or in the example of Transaction PA20, P stands for the personnel module and A for the personnel administration component.

The last two characters stand for the executed action: 01 usually means create data, 02 usually denotes a change transaction, and 03 usually refers to a display transaction. So, according to these rules, what is the code of the Display Material transaction in the materials management module? That would be MM03!

However, there are some exceptions to this rule, as the words *often* and *sometimes* have already indicated. For example, Transaction FSP0 (G/L Acct Master Record in Chrt/Accts) performs all three actions within one transaction: creating, changing, and entering an account in the financial accounting module. One aspect that you'll notice is that the SAP system is very diverse.

Company-Specific Transactions

SAP provides its customers with the ABAP programming language. With this, companies develop their own transactions that are missing in the scope of delivery of SAP software. You can recognize these home-grown transactions very quickly because, for technical reasons, their transaction codes usually begin with the letter "Y" or "Z".

2.1.2   Quick Start with the Commands /n or /o

Now you'll experience how to jump directly into a transaction without any detour by using one of the commands /n or /o and entering a transaction code.

Follow these steps:

1. Click in the command box at the top left of the header line.

2. As shown in Figure 2.1, enter "/n" before Transaction SU3. Entering "/nSU3" is your command to the system to start a new transaction ("n" as in *new*).

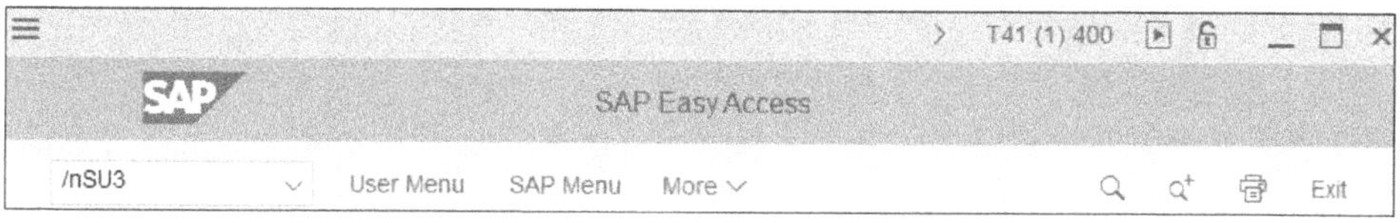

Figure 2.1 Starting a New Transaction with the Command Box

3. Confirm with [Enter] so that the SAP system can execute the command. You're now in Transaction SU3 (Maintain Users Own Data) using the *same* window.

4. Enter "/oSBWP" in the command field, as shown in Figure 2.2, and confirm with the [Enter] key. Prefixing the command with "/o" opens the transaction, in this case, Transaction SBWP (Workplace), in an *additional* window ("o" as in *open*).

Figure 2.2 Starting a New Transaction in an Additional Window with the Command Box

5. Now enter "/n", as shown in Figure 2.3, and confirm this with ⎡Enter⎤. This command ends a transaction and takes you directly to the **SAP Easy Access** window. The /n command is great if you get lost or stuck.

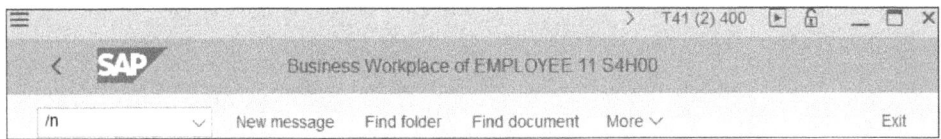

Figure 2.3 End Transaction with the Command Box

6. Use the task bar to switch to the **Maintain User Profile** window, which you saw earlier in Figure 2.2. Close this window by clicking on ⊠ (**Close**) in the upper-right corner of the window.

7. In the **SAP Easy Access** window, enter only the transaction code "SU3" and press ⎡Enter⎤. This will take you to Transaction SU3 (Maintain Users Own Data) without opening an additional window (see Figure 2.4). So, a transaction code entry works without the "/n" addition, but only in the **SAP Easy Access** window.

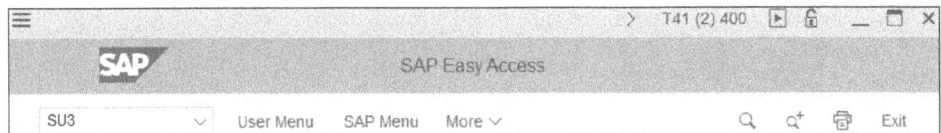

Figure 2.4 Entering Transaction Code SU3

[!]

Avoid Data Loss

Have you changed data and not yet saved it? Save the data before you call a new transaction with the /n command because, by doing so, you end the previous transaction without the SAP system alerting you of a possible loss of data.

Table 2.2 provides an overview of the commands covered so far, where "XXXX" stands for any transaction code.

Command Field Input	Description
/nXXXX	Start a transaction in the same window. Attention: unsaved entries will be lost without warning!
XXXX	As with /n: Start a transaction in the same window. Omitting "/n" only works in the **SAP Easy Access** window, however.
/oXXXX	Start a transaction in an additional window.
/n	Cancel a transaction in the same window and return to the **SAP Easy Access** window. Caution: Unsaved entries are lost without warning!

Table 2.2 Commands for Starting and Canceling Transactions

Start Transactions in the Command Field without Typing

You don't like typing? Neither do we! That's why we use the ☒ icon to the right of the command field instead of manual command entry, as shown in Figure 2.5.

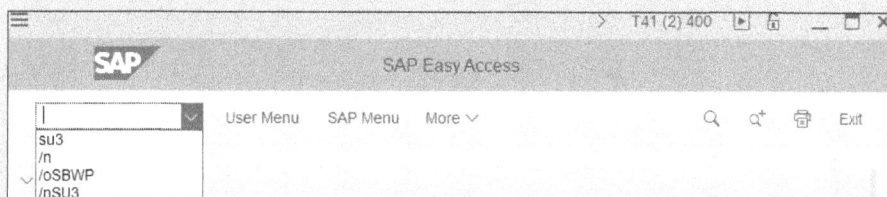

Figure 2.5 Call Up Recently Used Commands

Here you'll find the 15 recently used commands, and you can transfer a command to the command field with one click.

Which Transaction Startup Method is Best?

You now know several ways to start transactions. Depending on where you are, different ways are faster:

- If you're in the **SAP Easy Access** window, we favor starting from the **Favorites**.
- If you're *not* in the **SAP Easy Access** window, calling a transaction by entering the transaction code in the command field is often faster because the command field is a fixed part of the header line and is therefore available everywhere.

2.2   Navigate in a Transaction

The basic flow of transactions is often the same, as many transactions for data display and maintenance usually consist of two screens: the *initial screen* and the subsequent

73

detail screen. You'll learn how to perform change transactions in the following subsections.

2.2.1   Perform a Change Transaction

As an example transaction, we'll use Transaction FIO2 (Change Bank) from finance because it's very clear and has the basic rules that you can apply to many of the other transactions. We'll change the address of a bank in the following instructions.

Follow these steps:

1. Click in the command field, and start Transaction FIO2 (Change Bank) by entering "/nFIO2", as shown in Figure 2.6. Press Enter. The initial screen will then appear.

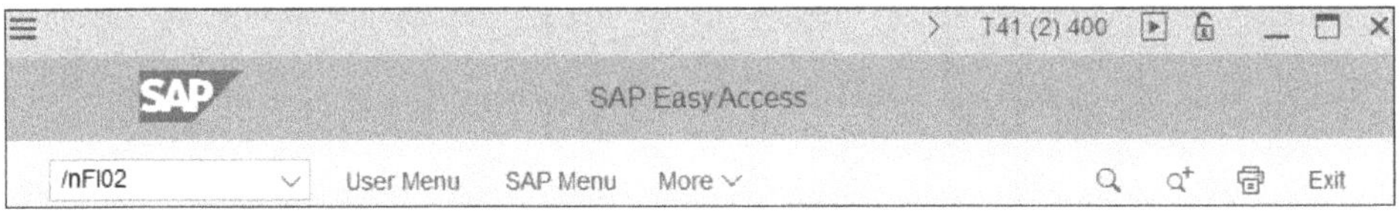

Figure 2.6 Entering the Transaction Code

2. On the **Change Bank: Initial Screen**, as shown in Figure 2.7, enter "US" in the **Bank Country** field. Enter a valid bank code in the **Bank Key** field, and press Enter to confirm these entries. The bank code ABNAUS33XXX works in the SAP Live Access training system.

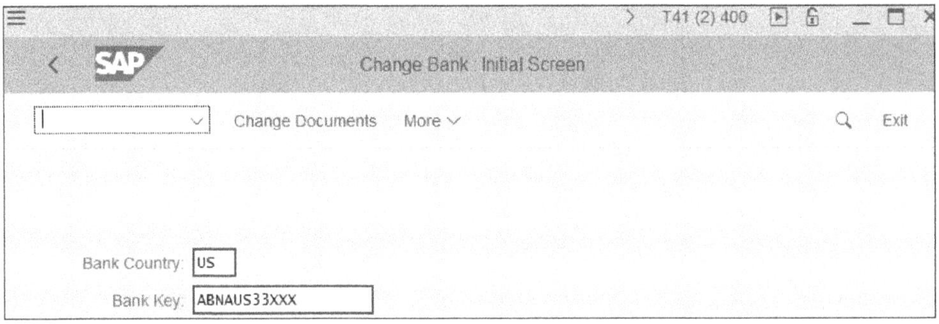

Figure 2.7 Initial Screen

3. After confirming with Enter, the **Change Bank: Detail Screen** will appear, as shown in Figure 2.8. In this screen, you can change any data of the address. By pressing Enter, you can optionally have your changes checked.

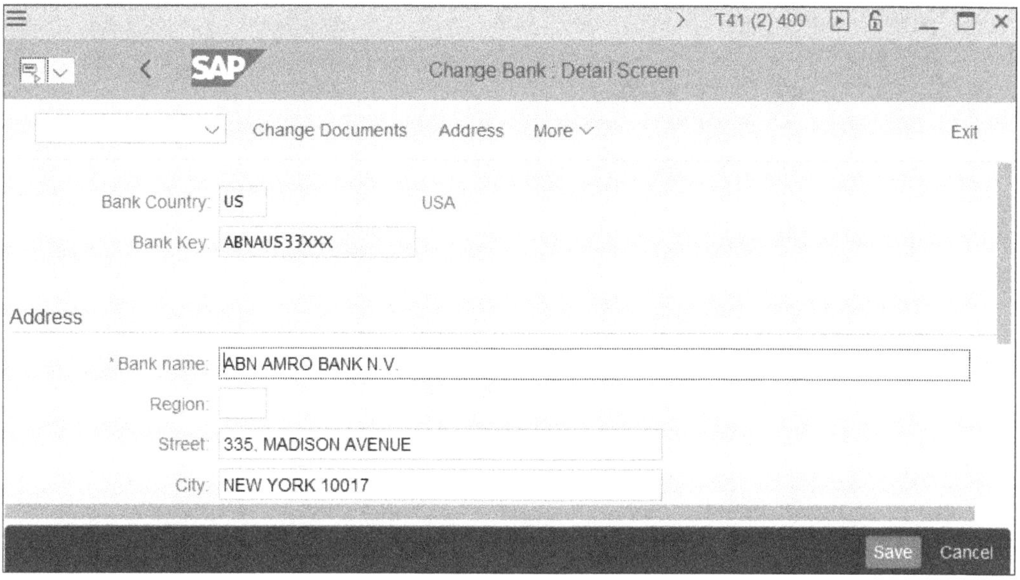

Figure 2.8 Detail Screen

4. After the change, you have several options in the **Change Bank: Detail Screen**, which you can find in Table 2.3.

Function in Detail Screen	Description
(Back) from the title bar, **Cancel** button from the footer line	Discard your change and go back to the initial screen.
Exit button from the header line	Discard your change and immediately return to the **SAP Easy Access** window.
Save button from the footer line	Save the changes and return to the initial screen.

Table 2.3 Functions in the Detail Screen

5. Click the **Save** button at the bottom right.

6. SAP assumes that after a change has been made, another change is to be made to another data record. Therefore, you'll return to the **Change Bank: Initial Screen**, as shown in Figure 2.9. Click (Back) from the header line. This will take you back to the starting point of the excursion: the **SAP Easy Access** window.

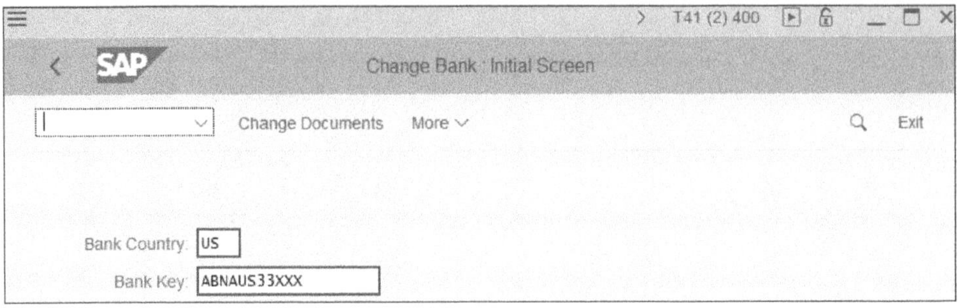

Figure 2.9 Change Bank: Initial Screen

Embrace this sample instruction because you'll use it to perform many transactions successfully. Display transactions also work according to the same pattern, except there is nothing to save in the latter, of course.

Note that there may be differences from transaction to transaction. For example, the **Exit** button should generally end a transaction and immediately return to the **SAP Easy Access** window. However, it often doesn't, which is just one example of the diversity of the SAP GUI transactions.

2.2.2 Lock Records

One more thing you should know: If you change a data record—such as a bank in our example—you're in *edit mode*. If you're just displaying the data in the SAP system, you're in *display mode*.

If you edit a record in edit mode, as you just did, no other user can edit this record at the same time. The record is locked for all colleagues. If another person tries to call this master record, an error message appears in the footer line, as shown in Figure 2.10, which usually also shows the username of the locking user.

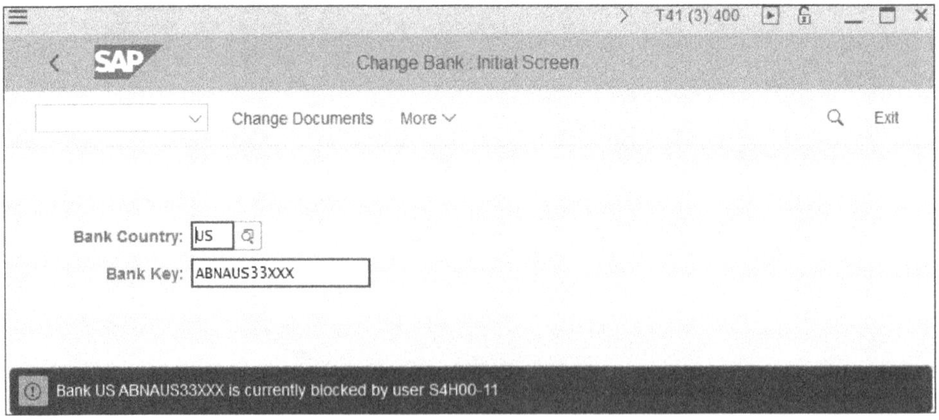

Figure 2.10 Error Message When Trying to Call an Already Locked Record for Modification

This lock is necessary because otherwise a colleague might override your changes. If a data record is in edit mode with you, no one can make changes. And they won't be able to do so until you leave the locked record by saving it or leaving editing without saving. However, the record that is being edited can be viewed by another user.

2.2.3   Find the Locking User

And now? According to the error message from Figure 2.10, the SAP system only displays the username and not the real name of the person causing the block. So, who is this user? With Transaction SU01D (User Display), you can find out the real name very quickly to notify the colleague. Follow these steps:

1. Start Transaction SU01D (User Display) by entering "/nSU01D" in the command field, as shown in Figure 2.11, and confirming with Enter. The initial screen will appear.

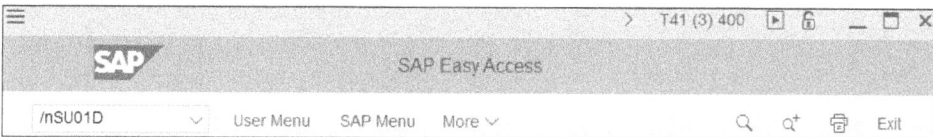

> T41 (3) 400
SAP   SAP Easy Access
/nSU01D   ⌄   User Menu   SAP Menu   More ⌄   Q Q⁺ 🗊 Exit

Figure 2.11 Starting Transaction SU01D

2. On the initial screen, enter a username in the **User** field, as shown in Figure 2.12. You can also use your own username as an alternative.

3. Click on **Display** 6ὸ in the header bar or on the corresponding **Display** button, because the Enter key doesn't work here. (Did we already mention that the SAP system is very diverse?)

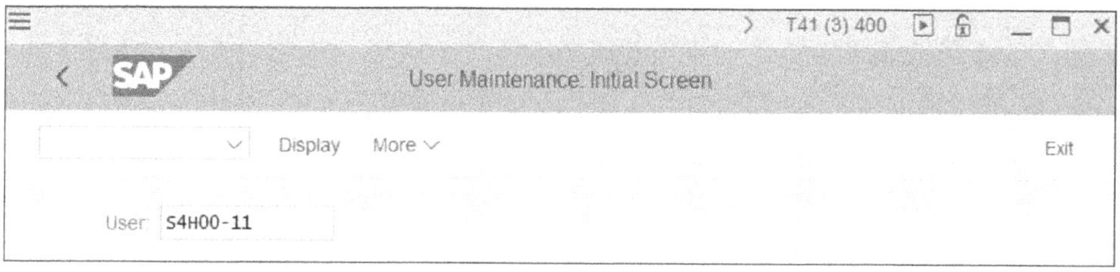

> T41 (3) 400
‹ SAP   User Maintenance: Initial Screen
                ⌄   Display   More ⌄   Exit

User S4H00-11

Figure 2.12 Entering a Username

4. In the detail screen, as shown in Figure 2.13, you'll find not only the name of the user but also the associated building address, premises, or telephone number further down.

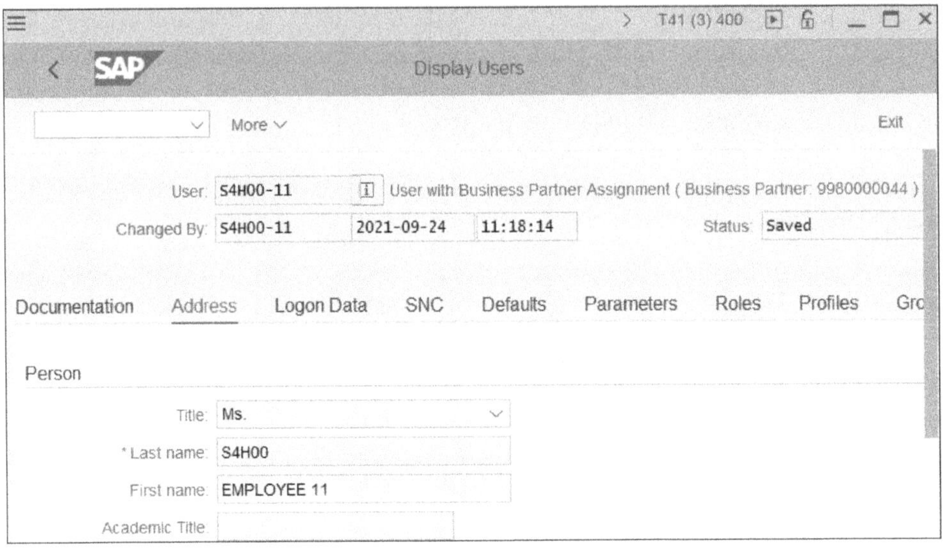

Figure 2.13 Detail Screen Containing User Information

5. After clicking the **Exit** button in the header line twice, you'll be back in the **SAP Easy Access** window.

2.3   Record Data

In this section, you'll enter a new data record using a bank as an example. Data entry can be done in a cumbersome or efficient way. Of course, you'll learn how this can be done efficiently.

2.3.1   Keys for Data Entry and Modification

After reading the previous section, you know how to move from screen to screen. But what is the fastest way to move from field to field within a screen? Of course, you can click the mouse in a field to enter or change data there. But by using the key functions and shortcuts in Table 2.4, you'll often move much faster.

Key/Key Combination	Meaning
Tab	Jump to the next field
Shift + Tab	Jump to the previous field

Table 2.4 Keyboard Functions for Data Acquisition and Modification

Key/Key Combination	Meaning
ArrowUp	Jump to the field in the line above
ArrowDown	Jump to the field in the line below
Insert	Switch between override and insert mode

Table 2.4 Keyboard Functions for Data Acquisition and Modification (Cont.)

Try out these helpful keyboard operations in the next section.

[+]

Override or Insert Mode?

Use the Insert key to switch between overtype and insert mode. Just give it a try! To see the effect, enter a few digits at the beginning of the **Bank Key** field in the following instructions. Then enter more characters at the beginning of the field. In insert mode, the rest of the input is shifted to the right, that is, inserted.

As a rule, set the insert mode. The overtype mode is better only in one exceptional case: when characters have to be replaced, that is, typed over, when you have a lot of data.

2.3.2   Perform a Recording Transaction

You'll record a new bank in the following instructions. You should pay close attention to these instructions and the methods because they are also representative of many other processes from your field of work.

Follow these steps:

1. Start Transaction FI01 (Create Bank) by entering "/nFI01" in the command field, as shown in Figure 2.14, and confirming with Enter. The initial screen will appear.

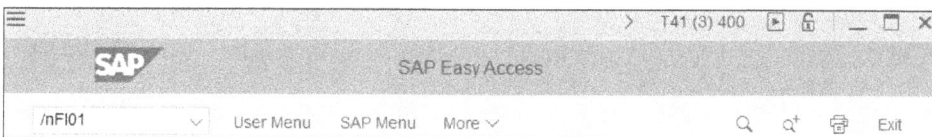

Figure 2.14 Starting Transaction FI01 (Create Bank) in the Command Field

2. In the initial screen, enter only the first letter "U" in the **Bank Country** field. With the appropriate default settings, a dropdown list of all country abbreviations and countries beginning with "U" will automatically appear, as shown in Figure 2.15. After entering further characters, the search result is reduced accordingly.

Figure 2.15 Entering the Bank Country

3. You can now click on the desired entry in the value list. In our case, however, it's faster to enter the second letter "S" and then immediately switch to the next field **Bank Key** with the ⟨Tab⟩ key.

[»]

Automatic Value Lists

The automatic supply of value lists is also called automatic enhanced search or *type-ahead input help* in the SAP system. If it's activated in your system, it doesn't exist in all fields, but only in frequently used fields such as **Country, Account Number, Vendor**, or **Material**. As you've seen with the **Bank Country** field, it searches not only the key field but also several fields, in our case, also the **Name** field. When the enhanced search is available for a field, (Search) appears instead of the normal input help icon.

4. In the **Bank Key** field, as shown in Figure 2.16, enter a bank code such as "87654321", and confirm with ⟨Enter⟩.

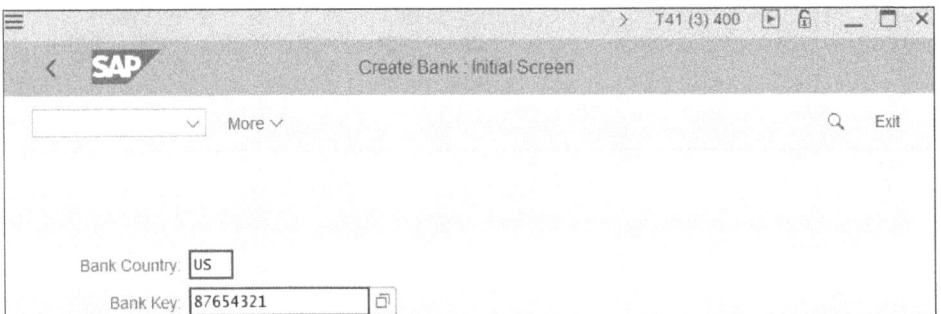

Figure 2.16 Entering the Bank Key

5. On the detail screen, as shown in Figure 2.17, enter only the address using any data. Enter any street and any city, but don't yet enter a name for the **Bank Name** field.

6. Now, we'll let you stumble here because this can also happen to you in practice, but don't worry, we'll get you back on your feet afterwards. So, click on the **Save** button now. As you can see in Figure 2.17, an error message appears in the footer line. The SAP system makes sure that you enter your data completely and correctly, and that's a good thing. You can find out what the error message is all about in the upcoming text box.

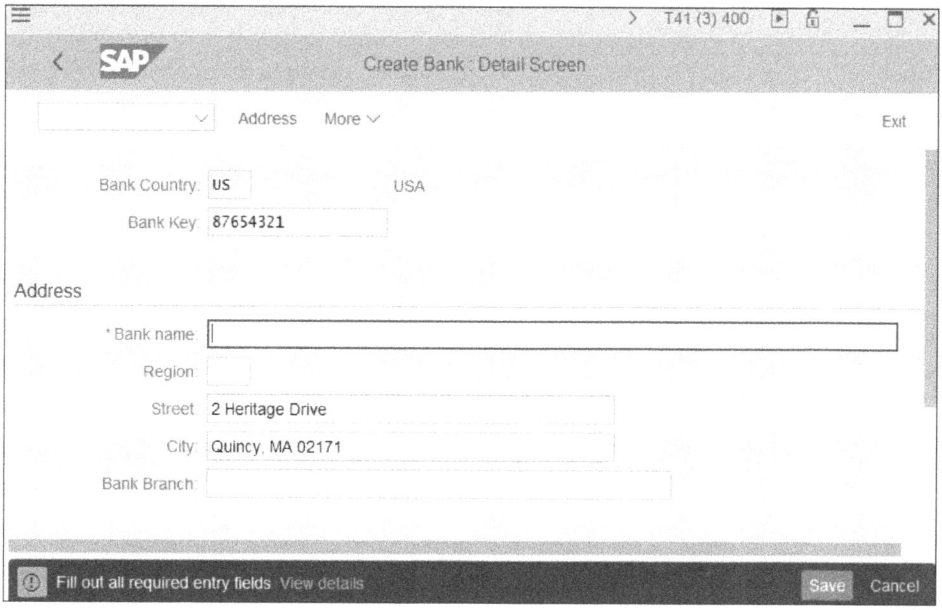

Figure 2.17 Enter Address Data and Error Message in the Footer Line

---

Required and Optional Fields

The **Bank Name** field is a *required field* and should have been filled in before saving. Required fields are often—not always—marked with a red asterisk in front of the field name. Without filling in all required fields, the SAP system refuses to save a record, resulting in an error message in the footer line. Use the **View Details** link to the right of the message to see a sometimes more and sometimes less helpful explanation of the error message.

Also note that SAP refers to fields that may be left blank as optional fields.

---

7. Enter the name "XYZ Bank" in the **Bank Name** field, and click the **Save** button again. If the success message appears as shown in Figure 2.18, you've created your first master record in the SAP system!

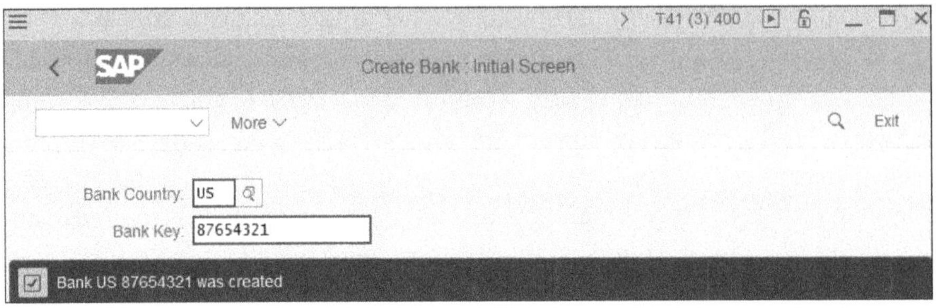

Figure 2.18 Success Message

In addition to the required fields, there are other data entry rules that you should follow:

- For most transactions, you'll return to the initial screen after saving because the SAP system assumes that you'll want to enter more data records afterwards.
- In the detail screen, you can also cancel the entry—just like in a change transaction—with the **Cancel** button from the footer line or with the **Exit** button from the header line.
- Some fields are protected against entries, for example, the **Bank Country** and **Bank Key** fields in the detail screen. If there are incorrect entries here, you have only one option: cancel and start again on the initial screen.

2.3.3   Interpret Messages Correctly

In the previous section, you saw two types of messages: a red *error message* due to a missing entry in a required field (see Figure 2.19) and a friendly green *success message* after saving (see Figure 2.20).

Figure 2.19 Error Message

Figure 2.20 Success Message

A third variant, yellow *warning messages*, warn of a possible incorrect entry, as shown in Figure 2.21. You must confirm such a warning message, or you won't be able to continue. You're forced by the SAP system to check again whether there is an incorrect entry or whether you've made this entry intentionally. If the entry is correct, confirm with [Enter].

Figure 2.21 Warning Message

[+]

2

How to Prevent Tendonitis

Often, you've already entered something once that you'll need again. There is a very helpful function that helps you avoid an enormous amount of superfluous typing.

Using a function called *input history*, just press the space bar or the [Backspace] key at the beginning of an empty field, and you'll get a list of all your entries from the past three months, as shown in Figure 2.22. With one click, you can apply the desired value.

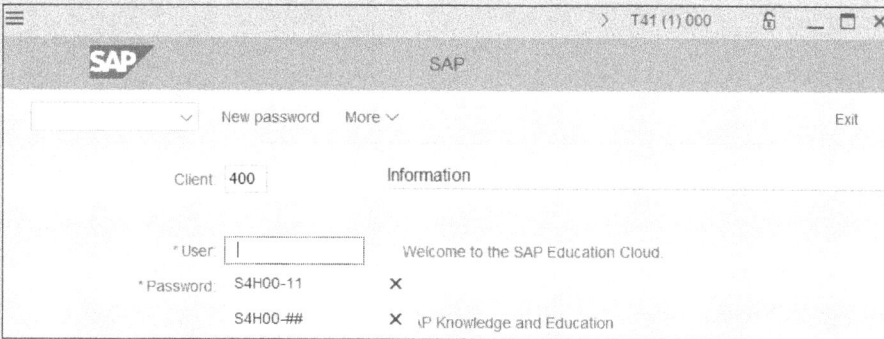

Figure 2.22 Input History for the User Field on the Login Screen Showing a List of Your Entries

Try it, for example, the next time you log in on the login screen, where after pressing the space bar, the username is displayed in the **User** field, so you no longer have to enter it manually every time you log in.

2.4   Use Value Help to Show Input Options

In the bank master data, you entered the abbreviation for the bank country and the bank key in the initial screen. You may have asked yourself what you can do if you don't have this information, for example, if you only know the bank name and not the bank key. When this issue arises, the SAP system offers you powerful *value and search aids*. We'll discuss these aids in this section in the context of input options, and in Section 2.5, in the context of documents and master data. Specifically, we'll cover the value help in this section, including how to transfer, search for, sort, and perform other functions with value lists.

2.4.1   Transfer a Value from the Value List

Value lists exist, for example, for countries, currencies, or company codes. In contrast to master data (e.g., materials) or documents (e.g., journal entries), value lists are input fields with a limited number of possible input values.

For example, you want to check the data of a foreign bank and are looking for an input value for the **Bank Country** field from a value list because you don't know the SAP system's country code for this country. To do this, you use the *value help*, which is also known as *F4 help* because it can also be called by using the F4 key.

Follow these steps:

1. In the command field, start Transaction FI03 (Display Bank) by entering "/nFI03", as shown in Figure 2.23, and confirm with Enter.

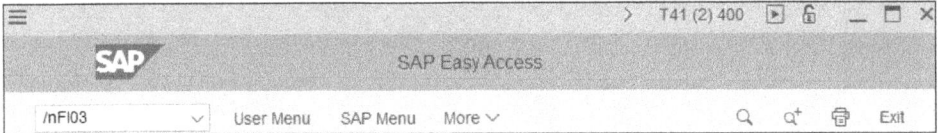

Figure 2.23 Starting Transaction FI03

2. In the initial screen, the cursor is already in the **Bank Country** field. Have you ever noticed the ⃞ icon (**Value Help**) to the right of the **Bank Country** input field? This is what you're testing now. This icon exists for many fields, but you'll see it only when the cursor is in the respective field. Now click on ⃞ (**Value Help**), or press the corresponding function key F4.

3. The SAP system then displays the value list in a separate window, as shown in Figure 2.24.

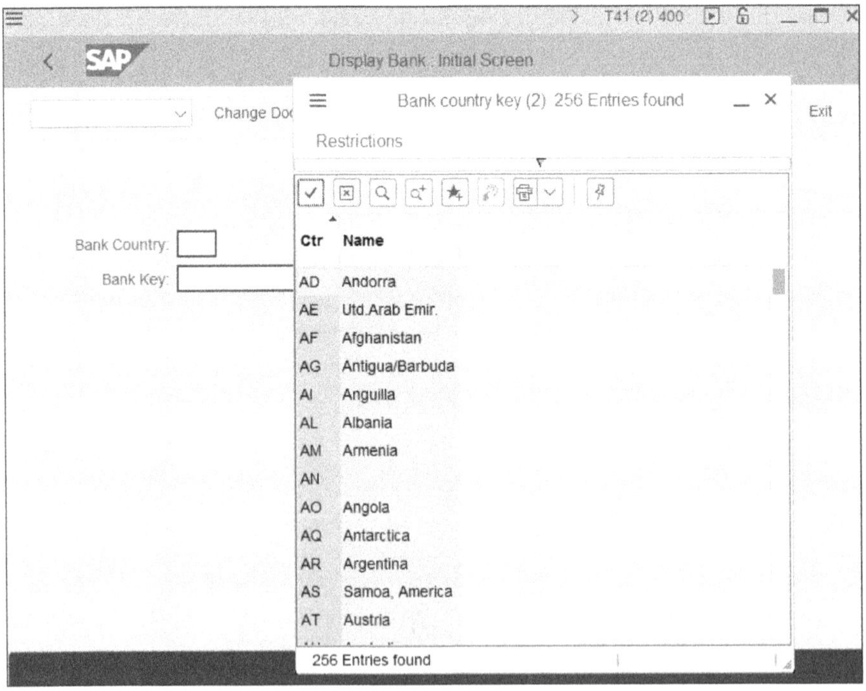

Figure 2.24 Initial Screen with Value Help

Here, you can see all available country abbreviations and, to the right, the countries in plain text. For example, double-click on the **AR (Argentina)** line in the list. This double-click transfers the value—in our case, the country abbreviation "AR"—into the field and closes the window with the value list.

4. Click again on ⌕ (**Value Help**) to the right of the **Bank Country** field because now we want to show you some of the many functions that exist in the value list.

2.4.2   Search the Value List

Suppose you're looking for the country key for the country Vietnam in the value list. What is the fastest way to get there? Looking through all the countries takes too long because, after all, there are more than 200 entries.

You could sort the list of values alphabetically (see the next section). But you could also search for a term in the opened list of values by using ⌕. After clicking on the icon, the small **Find** window will appear where you enter the search term "Vietnam" and confirm it with ⌐Enter⌐. As a result, the line for Vietnam will immediately appear in the list, as shown in Figure 2.25.

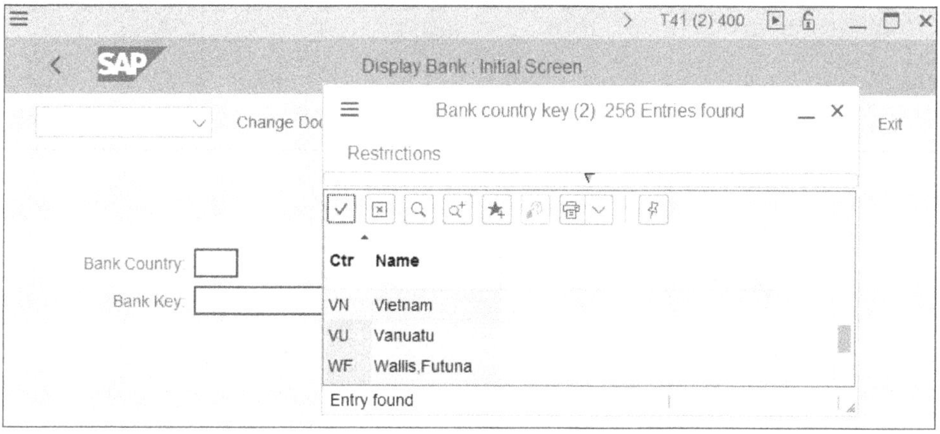

Figure 2.25 List of Values for Countries with the Result of the Search

2.4.3   Sort the Value List

Sorting in the opened list of values is easier to perform than in Microsoft Excel: you only need to click on one of the column headings, in our case, **Ctr** or **Name**, and the list will be sorted in ascending alphabetical order. If you click again on the same heading, it will be sorted in descending order.

2.4.4   Other Functions in the Value List

Windows with a value list have their own icon bar. In Table 2.5, you'll find the most important functions and their effects.

Icon	Name	Key(s)	Effect
✓	Copy	`Enter`	Transfers the selected value into the input field; also possible by double-clicking
✗	Close	`Esc` or `F12`	Closes the window without applying a value
🔍	Find	`Ctrl`+`F`	Searches for values using search terms
🔍⁺	Find Again	`Ctrl`+`G`	Continues a search if there are several entries with the search term in the list
★	Insert in Personal List	–	Copies the selected value to a personal value list
📖	Personal Value List	–	Displays the personal value list
🖶	Print	`Ctrl`+`P`	Provides an immediate printout of the list of values on the local printer on which you also print your Microsoft Word texts or Microsoft Excel spreadsheets by default
📌	Keep	–	Keeps the value list displayed until you exit the transaction (value list is normally closed automatically after double-clicking on a value)

Table 2.5 Functions in the Value List

[+]

Create Personal Value Lists

Value lists are often very long and contain unnecessary baggage. From large value lists with several hundred entries, you often only need one or two dozen entries! In these cases, create personal value lists with exactly the entries you need. You'll learn how to do this—and it's fast—in Chapter 5, Section 5.3.

2.5   Use Search Help to Find Documents and Master Data

The value help from the previous section immediately shows a list with all values. This would be more than impractical for master data (e.g., materials) or documents (e.g., journal entries), as there are often tens of thousands of values here. For these cases, *search help* is provided. Search help is a form in which you first enter search specifications before a list is displayed. In this section, you'll see how to use the search help to find data sets, and you'll explore techniques such as using wildcards and making multiple selections.

2.5.1   Find a Data Set with the Search Help

In the following example, you're looking for a domestic receivables account in financial accounting for which you only know a piece of text, but not its number. To do this, use Transaction FSP0 (GL Acct Master Record in Chrt/Accts).

Follow these steps:

1. Start Transaction FSP0 (GL Acct Master Record in Chrt/Accts) by entering "/nFSP0" in the command field, as shown in Figure 2.26. Confirm by pressing [Enter].

Figure 2.26 Starting Transaction FSP0

2. In the **Chart of Accts** field, enter "YCOA", as shown in Figure 2.27.

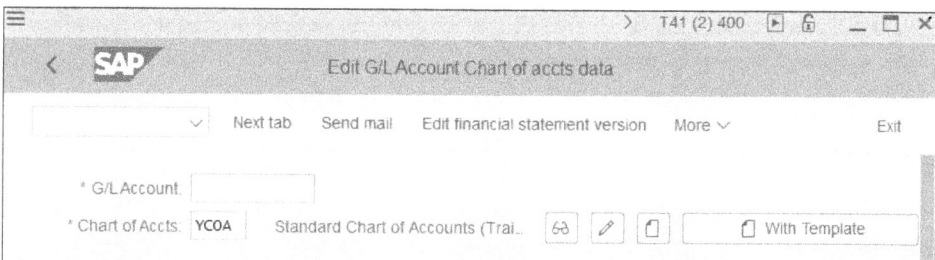

Figure 2.27 Chart of Accounts Field

3. Click in the **G/L Account** field and then on [Q] (**Value Help**) to the right of this field, or press the corresponding [F4] function key. The search help form opens, as shown in Figure 2.28.

4. In the search window, the first search form **G/L Account No. in Chart of Accounts** is open, as shown in Figure 2.28. You want to enter a part of the account description as search text in this search help, but, unfortunately, there is no field for this here.

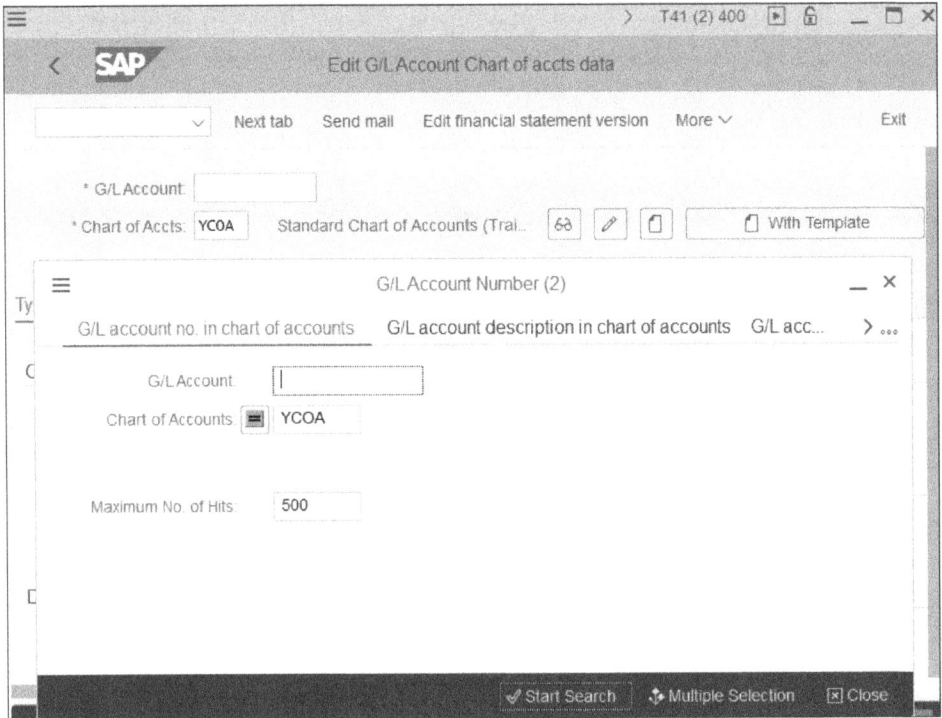

Figure 2.28 Search Help Window: G/L Account No. in Chart of Accounts Tab

5. Don't give up too soon! There are several search forms in this search window. This is the case with almost all master data and documents. Click accordingly on the second search form **G/L Account Description in Chart of Accounts**.

6. In the **G/L Long Text** field, enter the search term "*Receivable*", as shown in Figure 2.29, and press ⌈Enter⌉. You use the * character to get all accounts with the term "Receivable", no matter if this term is in the beginning, middle, or end of the account name.

[+]   **Find and Change Search Form Quickly**

Sometimes there are 20 or more search forms in a search help. You can display more form titles in the row with the form titles using ⟩ (**Forward**) and ⟨ (**Back**). However, it's smarter to click on ⊙⊙⊙ on the far right because then you'll see all the forms in a list, as shown in Figure 2.29, and you can immediately click on the appropriate form here. The SAP system remembers the last-used search form for the next search.

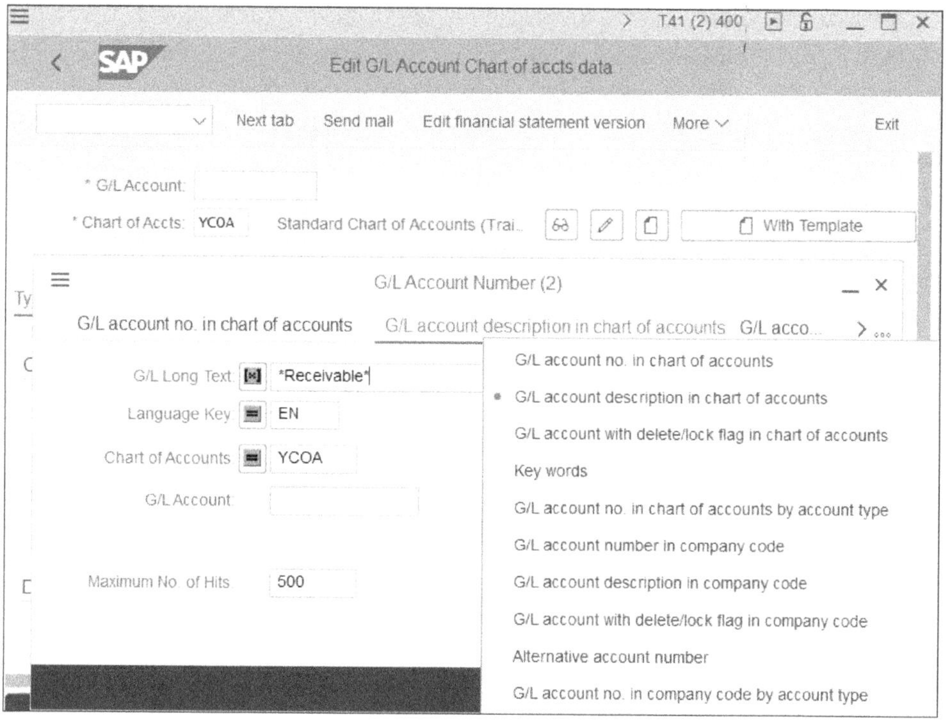

Figure 2.29 Search for Receivable with G/L Account Description in Chart of Accounts Tab

7. Now you can see the *hit list*, as shown in Figure 2.30. Here you have all the functions at your fingertips that you already learned about in the previous section about the value list.

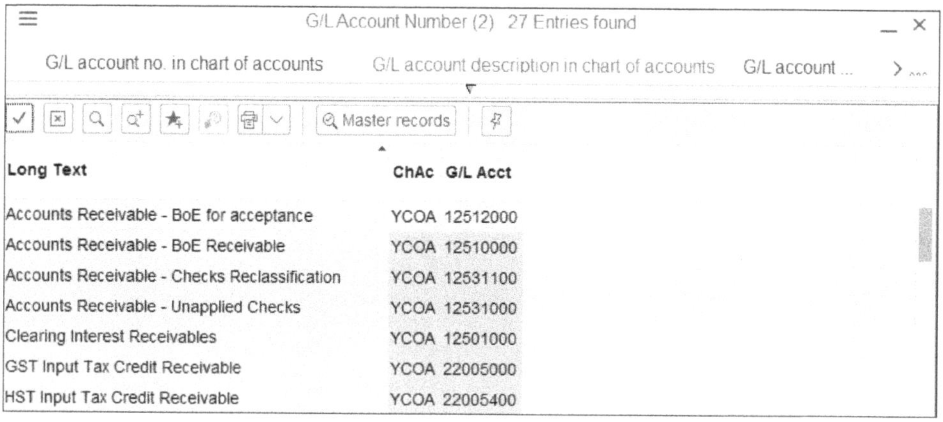

Figure 2.30 Hit List

8. At the top of the title bar of the hit list, you can see the number of lines found. This is too many hits for you, so you narrow down the search. Many users close the window

here with ❌ (**Close**) and call it up again. But not you! You're clever and click in the middle of the third line of the window.

9. Now you can see the search help again with the original search specifications and can narrow down the search further. To do this, add your entry in the **G/L Long Text** field, and now enter "*Receivable*Domestic*", as shown in Figure 2.31. Press the ⌨ Enter key to get a hit list again.

≡	G/L Account Number (2) 27 Entries found		— ✕

| G/L account no. in chart of accounts | G/L account description in chart of accounts | G/L account wi... | ❯ ... |

G/L Long Text ⦿ *Receivable*Domestic*

Language Key ≡ EN

Chart of Accounts: ≡ YCOA

Long Text	ChAc G/L Acct
Accounts Receivable - BoE for acceptance	YCOA 12512000
Accounts Receivable - BoE Receivable	YCOA 12510000

Figure 2.31 Narrowing Down the Search

10. The receivables domestic account you're looking for is now listed in the hit list, as shown in Figure 2.32. As with the value list, double-click on the line with the long text ***Receivables*Domestic*** to transfer the account number to the input field. (And here's one more tip for those of you who are interested in accounts in finance: by confirming with ⌨ Enter, you'll see the data of the account.)

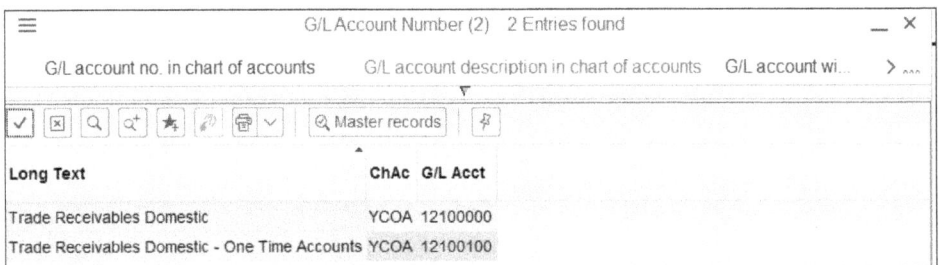

≡	G/L Account Number (2) 2 Entries found		— ✕

| G/L account no. in chart of accounts | G/L account description in chart of accounts | G/L account wi... | ❯ ... |

Long Text	ChAc G/L Acct
Trade Receivables Domestic	YCOA 12100000
Trade Receivables Domestic - One Time Accounts	YCOA 12100100

Figure 2.32 Hit List with New Values

11. At the same time, the hit list is hidden again because it has now served its purpose. Click the **Exit** button from the header line, and you'll return home, that is, to the **SAP Easy Access** window.

You now know the procedure for picking out a targeted data set even from very large data sets, such as from 100,000 or more customers or orders.

2.5.2   Wildcards

You know wildcards from the card game: With them you replace one or more desired cards. There are two different wildcards available in the SAP GUI search help: * and +.

The * wildcard, which is often used in practice, replaces any number of characters. As in the example of the previous section: If you enter the search term "*Receivable*" in the **G/L Long Text**, you'll get all accounts with this text in the hit list. Without using the wildcards, you would have had no hits, because there is no account with the stand-alone name "Receivable".

The + wildcard is used less frequently. It replaces exactly one character at a fixed position in a string.

> **Type Less, Use the Wildcard**
>
> Have we mentioned before that we're very lazy about typing? That's exactly why we like to use wildcards. For example, before we enter the words "hexagon head screw" in a material search, we enter "hex*". You too can use the * wildcard to reduce the amount of typing you have to do when searching.

2.5.3   Multiple Selection

Now you want to search for the receivables account in English and in German. To do this, you specify several language abbreviations as search defaults. This is called *multiple selection*. Follow these steps:

1. You're still in the initial screen of Transaction FSP0 (G/L Acct Master Record in Chrt/ Acct), as shown in Figure 2.33. Click again on [🔍] (**Value Help**) to the right of the **G/L Account** field. The search help will then appear again.

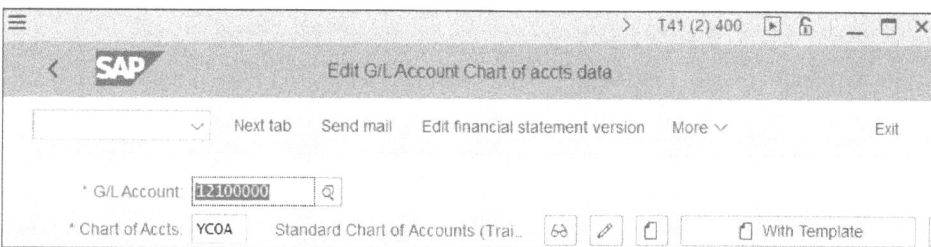

Figure 2.33 Transaction FSP0 Initial Screen

2. Enter the number "*12100000" in the **G/L Account** field, as shown in Figure 2.34. Don't forget the "*" wildcard at the beginning of the number; otherwise, the search in your system may not work. (In the SAP system, leading zeros are stored at the beginning of the account number and some other numbers. In our case, the account number would then be 0012100000, and you need the * wildcard for any leading zeros.)

Two floors above, only one abbreviation can be entered in the **Language Key** field, but two are needed for your search: EN and DE (German). To solve this problem, click in the **Language Key** field, and then click ⚡ (**Multiple Selection**) at the bottom of the window. In other places in SAP GUI, the multiple selection function is hidden behind the ⬚⬈ icon.

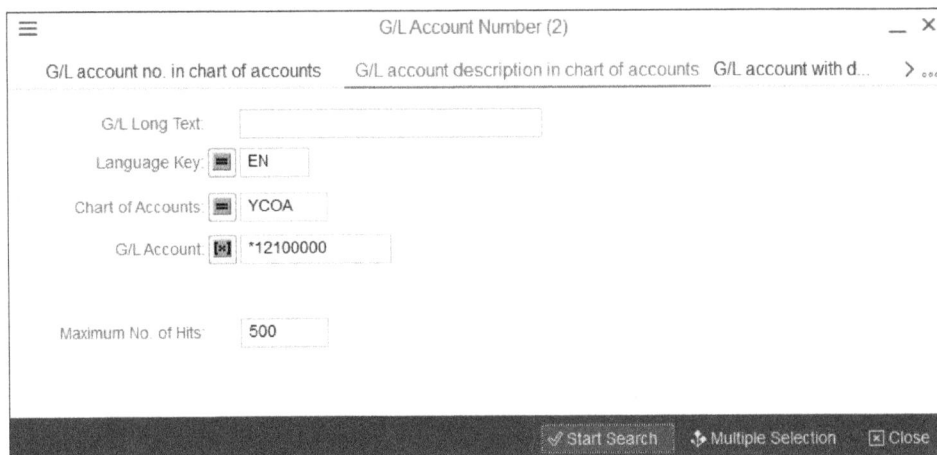

Figure 2.34 G/L Account Search Help

3. Now the **Multiple Selection** window appears, as shown in Figure 2.35, in which the input of several language abbreviations is possible at the top. Enter the abbreviation "DE" under the abbreviation "EN".

4. In the lower part of the window, labeled **Range**, you can also specify intervals, which are often usable in a field for postal codes to specify postal code areas, for example.

Figure 2.35 Multiple Selection for the Language Key

After confirming with Enter , you're back in the search form. Don't be irritated by the fact that only the language key **EN** is still being displayed here.

5. Now start the search in the search help by pressing Enter or the **Start Search** button. The hit list shows that both language keys have been taken into account, as you can see in Figure 2.36.

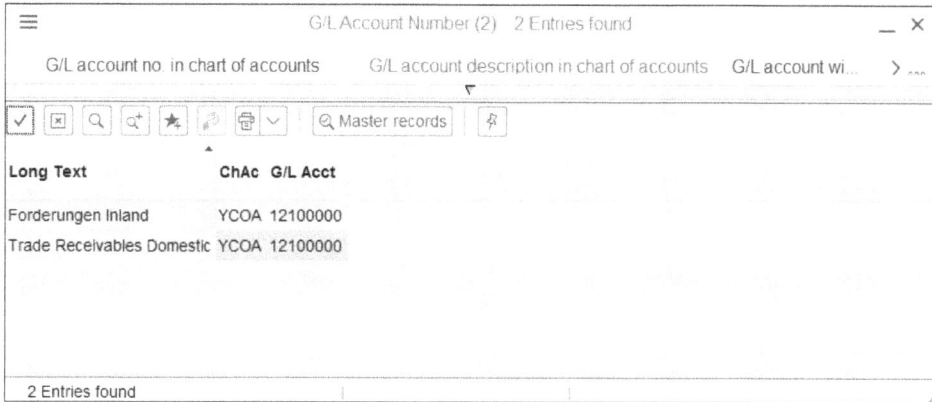

Figure 2.36 Hit List with Multiple Selections

2.6   Use the Field Help

We also often use the SAP software's *help* because it's so extensive that no human being can have everything ready to hand at all times. More than 300,000 different windows and more than 100,000 transactions are too much for one brain!

You may already be familiar with the F1 key from other programs such as Microsoft Excel or Microsoft Word as a first stop for quick help. As you can see in the following sections, F1 also works in many places within the SAP system.

2.6.1   Field Help

You've already dealt with the **Bank Key** field in this chapter. But now you want to know what this term actually means. Let's walk through the steps to use the field help:

1. Start Transaction FIO3 (Display Bank) again by entering "/nFIO3" in the command field, as shown in Figure 2.37, and confirming with Enter .

Figure 2.37 Starting Transaction FIO3

2. On the initial screen, as shown in Figure 2.38, click in the **Bank Key** field, and press F1 . An additional window with the lovely name **Performance Assistant** appears.

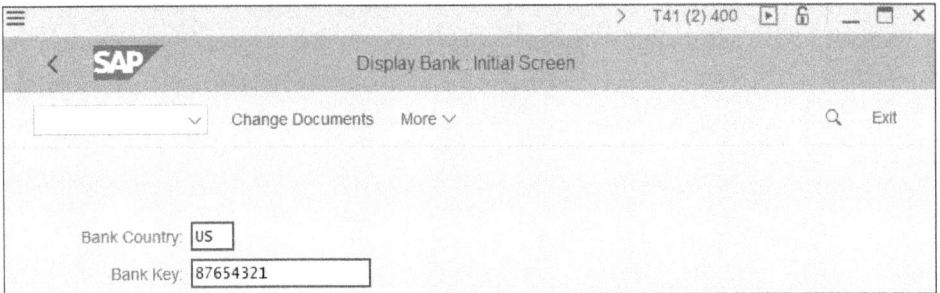

Figure 2.38 Transaction FI03: Display Bank Initial Screen

3. In the **Performance Assistant** window, as shown in Figure 2.39, you'll find sometimes more and sometimes less detailed and comprehensible information. In our case, you'll learn here that the term *bank key* means nothing other than the bank code.

Figure 2.39 Performance Assistant

4. Close the help window by clicking on ✖ (**Close**).

5. Use the **Exit** button from the header line to return to the **SAP Easy Access** window.

2.6.2   Transaction Help

Do you need help with a transaction? First, select the corresponding transaction in the **Favorites** or in the **SAP Menu**. Then, follow these steps:

1. For our example, from the **SAP Menu**, choose **Accounting** • **Financial Accounting** • **Banks** • **Master Data** • **Bank Master Record**, as shown in Figure 2.40.

2. In the **Bank Master Record** folder, click **FI01 - Create** (as an exception, don't double-click!).

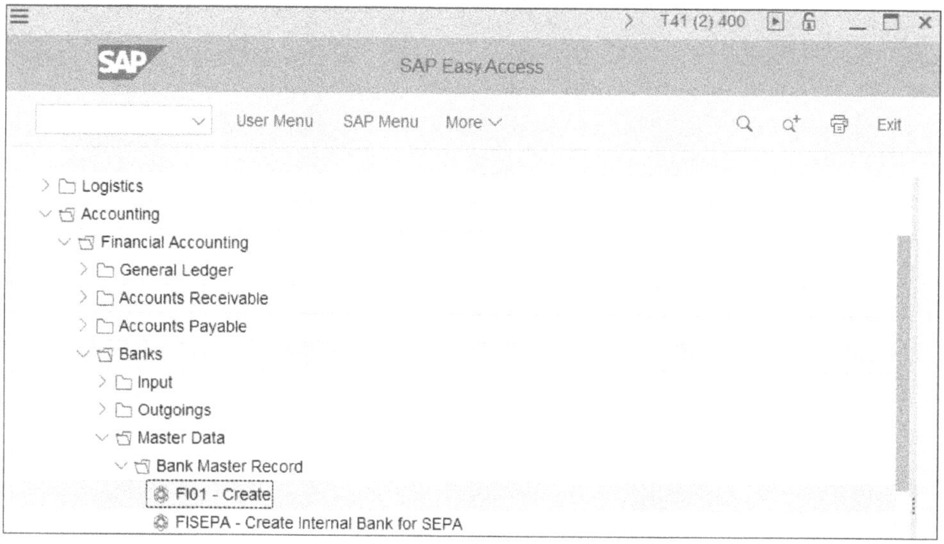

Figure 2.40 Navigating to Transaction FI01 (Create Bank)

3. Press the F1 key. This opens a window in your internet browser, as shown in Figure 2.41. If this help function is configured accordingly in your system, you'll be taken to the SAP Help Portal for SAP S/4HANA in a new window.

4. In the new window, you'll see a table of contents with topics related to banks on the left. Click on the entry **Banks: Create** to see the detailed transaction help on the right.

5. Close the help window by clicking ✖ (**Close**).

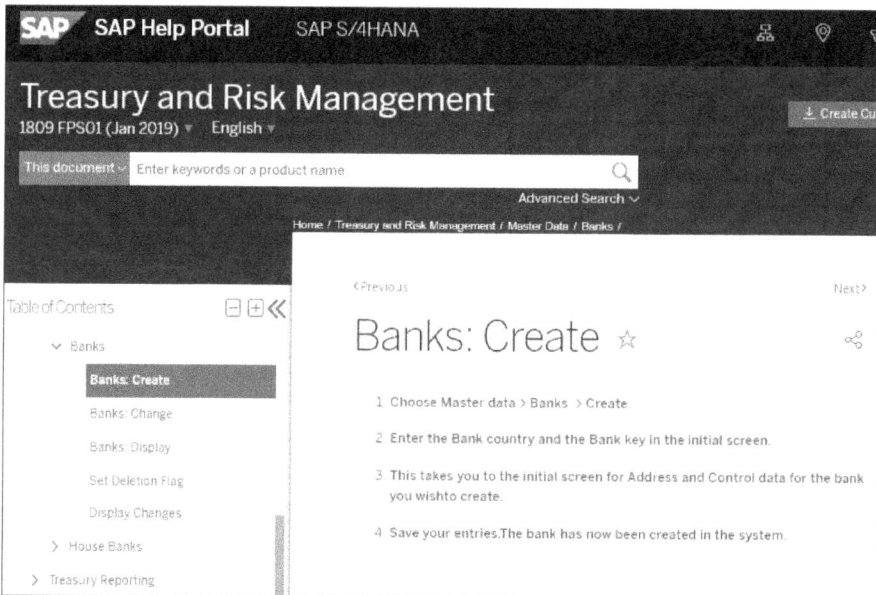

Figure 2.41 SAP Help Portal

[+]   Even More Help for Future Professionals

Even more help is available in the **More** menu. Select **More • Help** in the header line, as shown in Figure 2.42. The **Product Assistance** command in the **Help** submenu corresponds to the transaction help with the [F1] key, but the transaction must have been called beforehand. Below this, you'll find an **SAP Help Portal** link, which leads directly to the initial screen of this portal.

The **SAP Support Portal** option offers SAP specialists various tools for problem-solving. The **Settings** command can be used to perform various presettings for value lists and hit lists, as well as for the [F1] help.

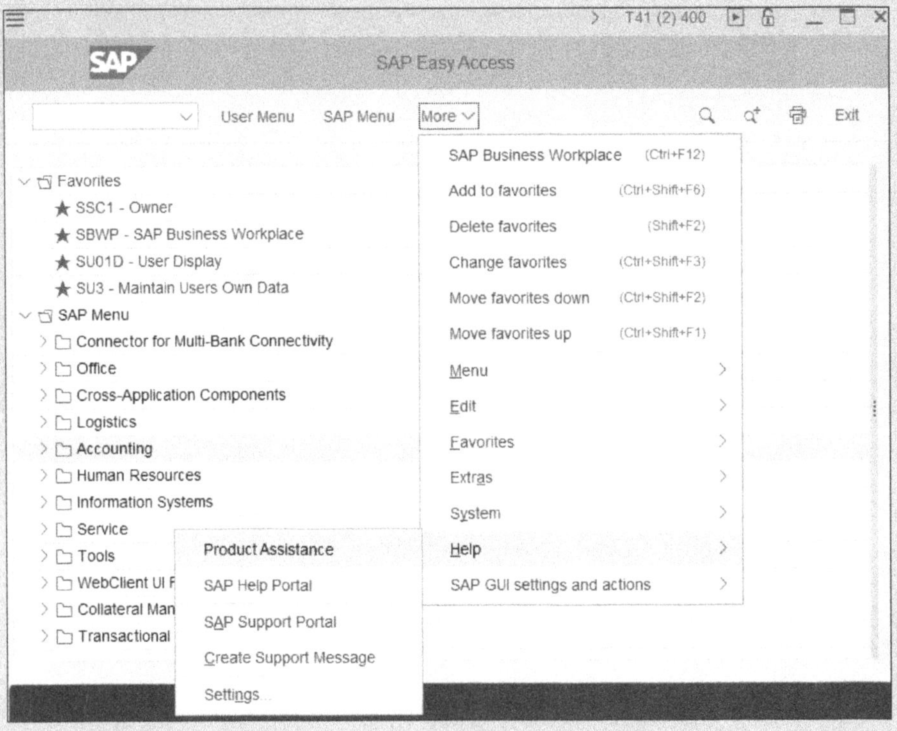

Figure 2.42 Help Area in the More Menu

2.7   Work with Multiple Windows

The phone rings. A colleague is on the phone and needs information from you immediately and urgently, which you have to obtain from the SAP system. You know these kinds of interruptions, and they happen often. What do you do when you are already on the move in a transaction? You open a new window.

In addition, you can also use a second window to bridge waiting times that occur in the first window, for example, due to the creation of an extensive and time-consuming report. We'll walk through the steps in the following sections.

2.7.1 Open a New Window

You may remember the following two methods from previous sections of this book:

- Enter "/o" and the transaction code in the command field and press ⌈Enter⌋ to open a transaction in an additional window.

- Use the key combination ⌈Ctrl⌋ + ⌈+⌋ to open the **SAP Easy Access** view in an additional window.

For our example, open Transaction SBWP (SAP Business Workplace) in a second window by entering "/oSBWP" in the command field and confirming with ⌈Enter⌋, as shown in Figure 2.43.

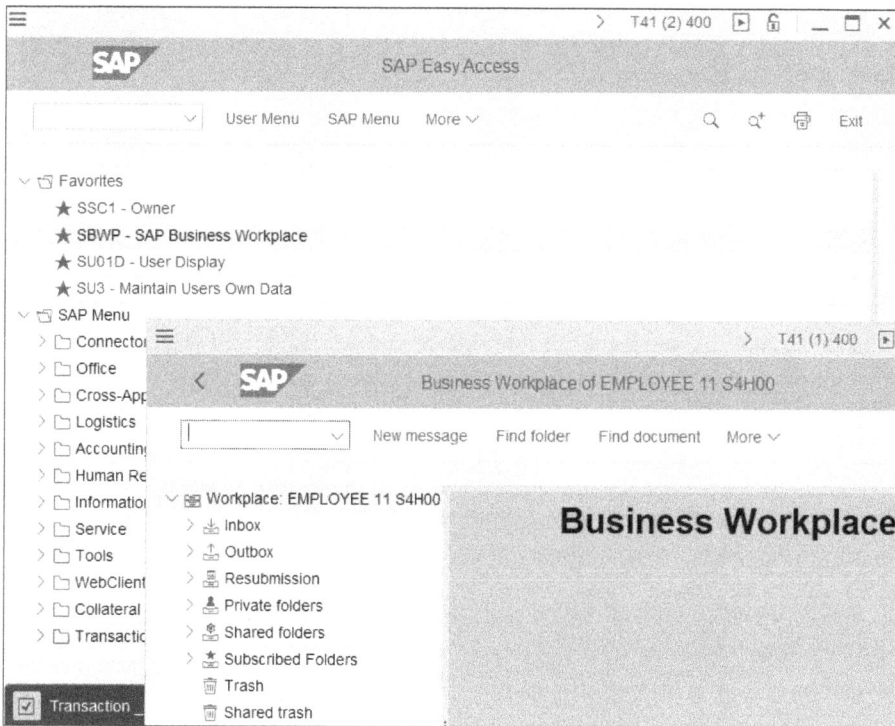

Figure 2.43 Screen with Two Windows

Not Too Many Windows [!]

You're welcome to use many windows, but at some point, the SAP system refuses to open another window. Too many open windows put a strain on performance, that is, the response time of the SAP system. In addition, if you have too many windows open, you may lose track of what's going on. This may also cause you to lock yourself and others during change transactions.

For these reasons, your system administrators set a maximum number of allowed windows. By default, a maximum of six windows is allowed. If you've reached the maximum number of windows that can be opened, you'll be notified by a message. At this point, at the very latest, you should close any windows that aren't urgently needed.

2.7.2 Switch between Windows

You can switch between the individual SAP GUI windows in the same way you do in Windows, for example: by clicking on the nonactivated window when it's visible or by using the task bar.

Another elegant way to switch between (Windows) windows is the key combination [Alt]+[Tab], but only if the SAP window is displayed as an extra Windows screen and not as part of a browser window (the latter is the case, for example, when accessing the SAP Live Access training system).

2.7.3 Close Windows

You have several options for closing an SAP window:

- Clicking the familiar Windows ☒ icon (**Close**) in the upper-right corner of the window
- Using the (hopefully) familiar Windows key combination [Alt]+[F4], but only if the SAP window is displayed as an extra Windows screen and not as part of a browser window

For people who have too much time, there are two more methods:

- Choosing **More • System • Close GUI Window**
- Entering "/i" in the command field and pressing [Enter]

These procedures close only one window at a time, even if you're currently working with multiple windows. However, when the last window is reached, that is, only one window is still open, you log out of the SAP system at the same time by closing the last window.

When the end of the working day is in sight, and you want to close your work on the SAP system in one go but you still have several windows open, there are two ways to close all windows in one fell swoop:

- Select **More • System • Log Off** from the header line.
- Enter "/nex" in the command field (the command "/n" and "ex" for exit). This causes a logout without further query.

Our Recommendation for Closing Windows

As much as we like to launch commands via the command field, we very rarely use the /nex command, even if we have many windows to close. It could be that there is still data to be saved in one of the windows, and this would then be gone without further prompting.

That is why we prefer to use the key combination [Alt]+[F4] because we can see once again in each window if an action wasn't completely executed.

[+]

2

2.8   Summary

Done! Now, at the end of this chapter, the foundation is in place, and you know the basics as well as some valuable tips on how to perform transactions. Next, let's continue with the exciting topic of reporting.

Chapter 3
Reporting with SAP GUI

Data entry isn't just an end in itself. You also enter data so that exciting evaluations can be made in your company, for example, order value analyses, sales statistics, cost center evaluations, or even control reports for master data. Good reporting is the basis for good decisions, which in turn lead to better business results. In this chapter, you'll first learn the basic reporting techniques.

What You'll Learn

- Where to find suitable reports and how to start them
- Which methods you use to select exactly the data you need
- How to search and filter data within lists
- How to print lists and save them as PDF files
- How to transfer lists via download to Microsoft Excel and other programs

---

3.1 Find and Call Reports

To call reports in SAP GUI, you use *information systems* in the **SAP Menu**. An overview of the reports for your work area can be found in a folder in the **SAP Menu**, which is usually called **Information System** and usually contains further subfolders.

3.1.1 Start a Report from the Information System

Are you looking for specific evaluations? Then simply check the **SAP Menu** in the information systems of your components to see if there are any hidden treasures there! In the following, you'll find some examples of paths or folders in the **SAP Menu** where you might find what you're looking for:

- Accounting • Financial Accounting • Accounts Receivable • Information System
- Accounting • Controlling • Cost Center Accounting • Information System
- Logistics • Sales and Distribution • Sales • Information System
- Human Resources • Personnel Management • Administration • Information System

As an example, we use a small but nice report that clearly lists your customer addresses.

Follow these steps:

1. In the **SAP Menu**, choose **Accounting • Financial Accounting • Accounts Receivable • Information System • Reports for Accounts Receivable Accounting • Master Data** (see Figure 3.1).

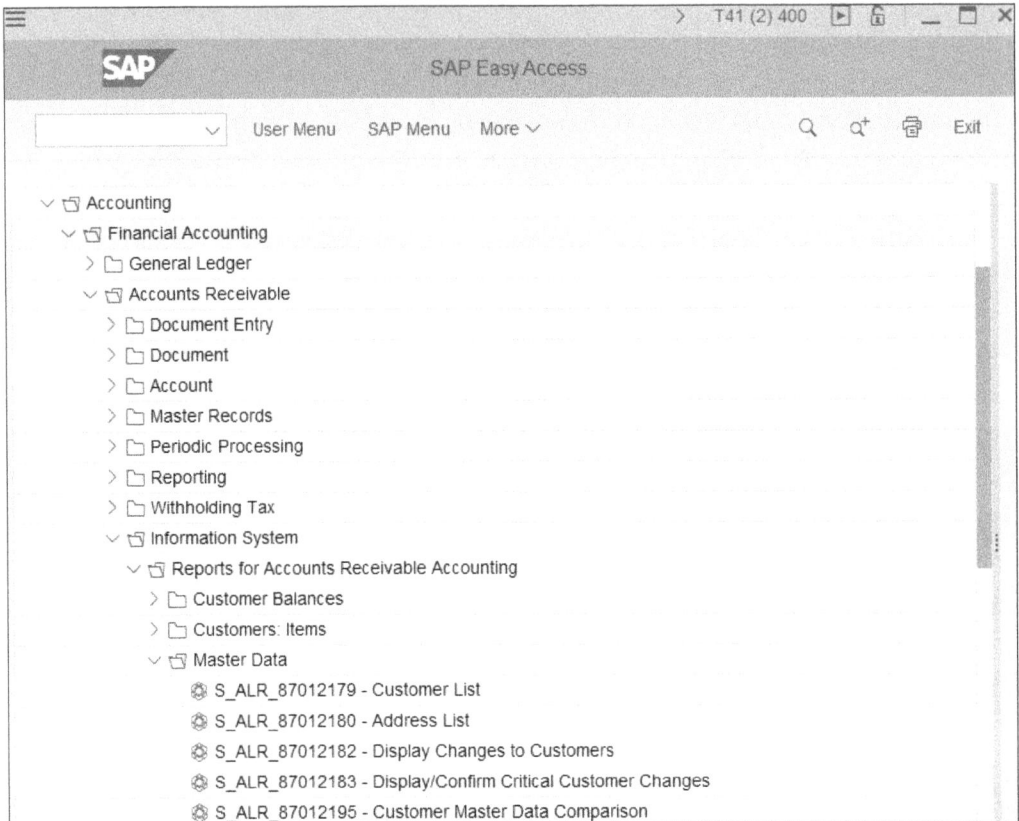

Figure 3.1 Navigating to the Master Data Folder of Accounts Receivable

In this folder, you'll then see the entry **S_ALR_87012180 – Address List**. Yes, S_ALR_87012180 is also a transaction code. Theoretically, you could enter "/n" or "/o" followed by the transaction code in the command field of the system toolbar, and the report would be started. In practice, however, this is too cumbersome because the codes for calling reports are often very long and not very meaningful.

2. Double-click on the **S_ALR_87012180 – Address List** entry. The initial screen of the report appears, which is also called the *selection screen*, as shown in Figure 3.2.

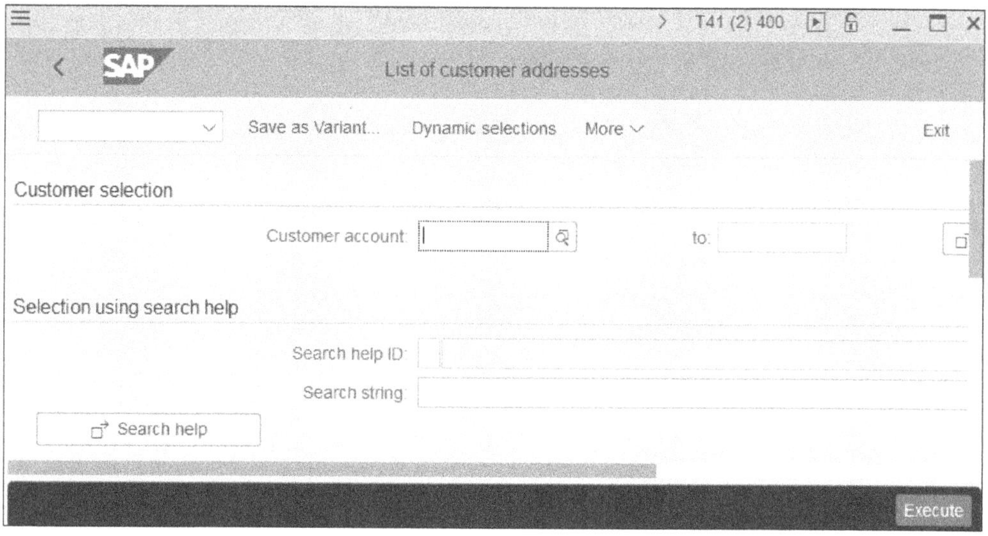

Figure 3.2 Address List: Selection Screen

3. In the selection screen, you determine which data will eventually be displayed in the list. We'll discuss this in detail in Section 3.2; for right now, click the **Execute** button at the bottom right.

4. A popup appears in which the SAP system asks for the maximum number of records to be read, which is preset to "100", as shown in Figure 3.3. Press ⌈Enter⌋ to continue to the list.

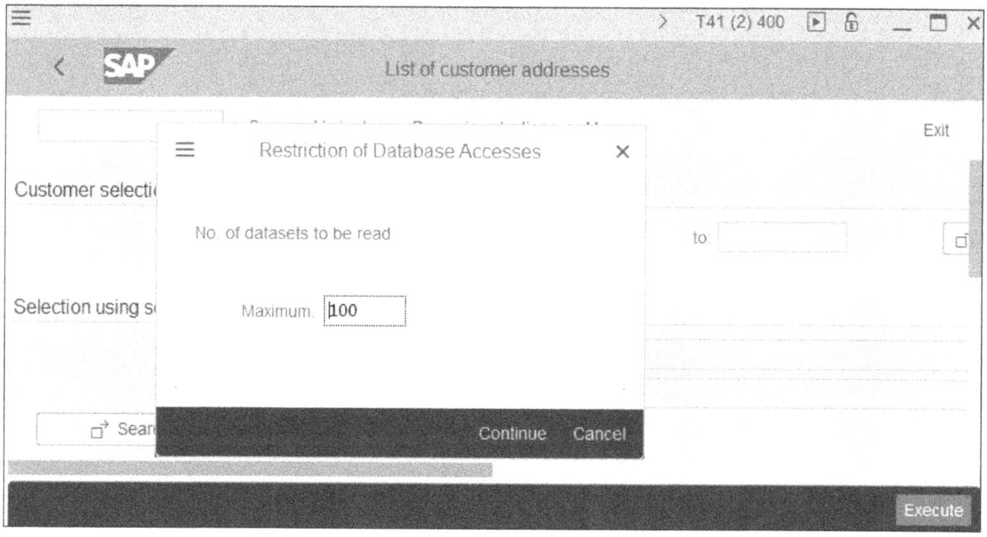

Figure 3.3 Maximum Number of Records

Take a closer look at the list shown in Figure 3.4. Two lines per address? How impractical! Didn't we promise you a clear list? In Section 3.2, you'll create a clear one-line address list.

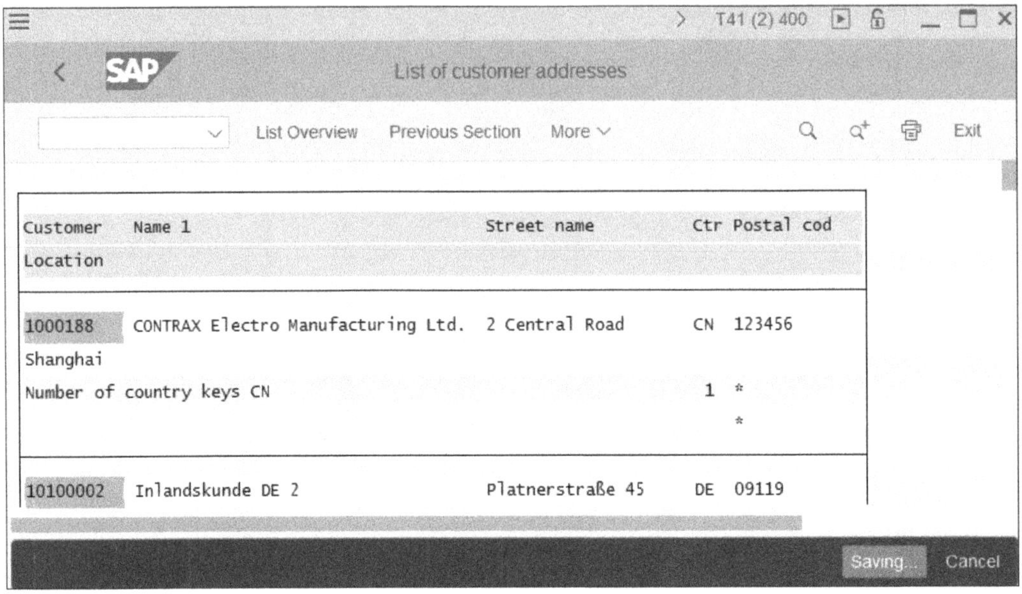

Figure 3.4 Address List with Two Lines per Address

5. Click the **Exit** button to close the list and thus switch back to the **SAP Menu**.

You've seen in these instructions that a report consists of a selection screen and the actual list, and that you can start a report in the **SAP Menu** from an information system. In the next section, we'll show you how it can be done even faster.

3.1.2   Add a Report to the Favorites

Working with the SAP system can be fun. However, it's no fun to first have to open the information system to start a report. Our tip: Simply add the transactions for reports to your **Favorites** list. Follow these steps:

1. Right-click on the transaction entry in the **SAP Menu**, in our case on **S_ALR_87012180 – Address List**.

2. As you can see in Figure 3.5, the context menu opens. Click on the **Add to Favorites** command. From now on, you can easily start this report transaction by double-clicking it from your **Favorites** list.

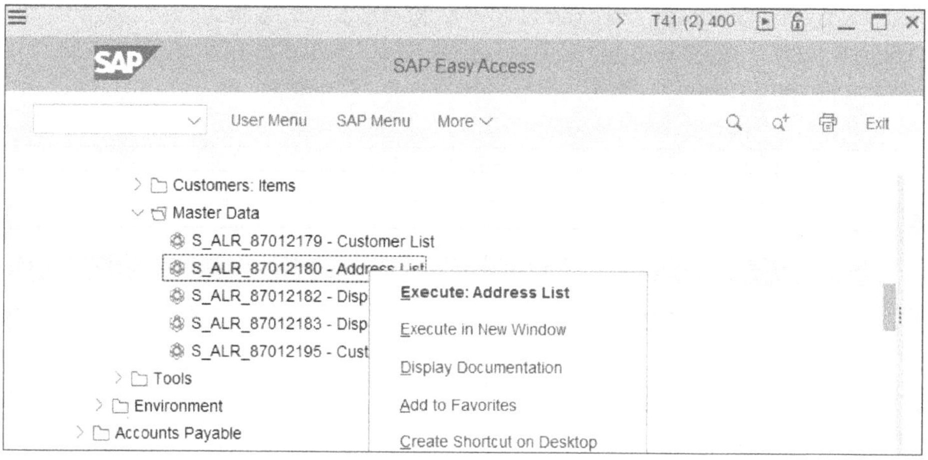

Figure 3.5 Context Menu for the Transaction Entry in the SAP Menu

3.2   Find the Right Data by Selection

In the selection screen, you choose the data that will be displayed in the later list. In addition, we'll make good on the promise of a clear list by showing you how to choose a better output format in the selection screen for our address list, which determines the type of display. We'll discuss these forms, as well as options for dynamic selection, in the following sections.

3.2.1   Selection and Output Formats

In the following instructions, we'll show you how to limit the display to the customer numbers you're interested in by selecting an appropriate range of customer numbers.

Follow these steps:

1. Double-click on the entry **S_ALR_87012180 – Address List** in your **Favorites**, as shown in Figure 3.6.

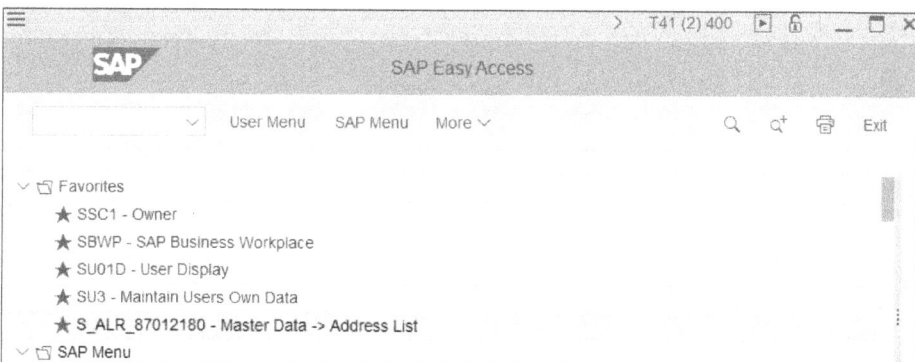

Figure 3.6 Double-Clicking to Access the Address List

2. To select the customers, only the **Customer Account** and **To** fields are available for the selection in the selection screen of the report. Enter a number interval here, for example, the customer number "17100001" in the **Customer Account** field and the customer number "17100005" in the **To** field, as shown in Figure 3.7.

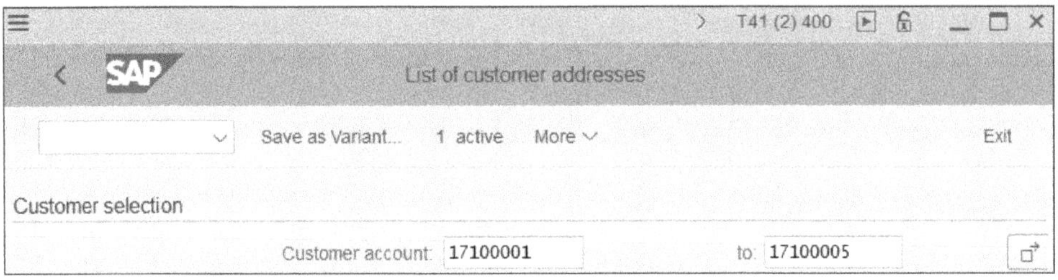

Figure 3.7 Selecting Customers

Alternatively, you could also enter "1710000*" because the * and + wildcards may also be used in the selection mask. You've already become familiar with wildcards in Chapter 2, Section 2.5.2.

[»]

Multiple Selection in Selection Screen

You open the multiple selection feature for a field in our report, as shown in Figure 3.8, by clicking ⬚ (**Multiple Selection**).

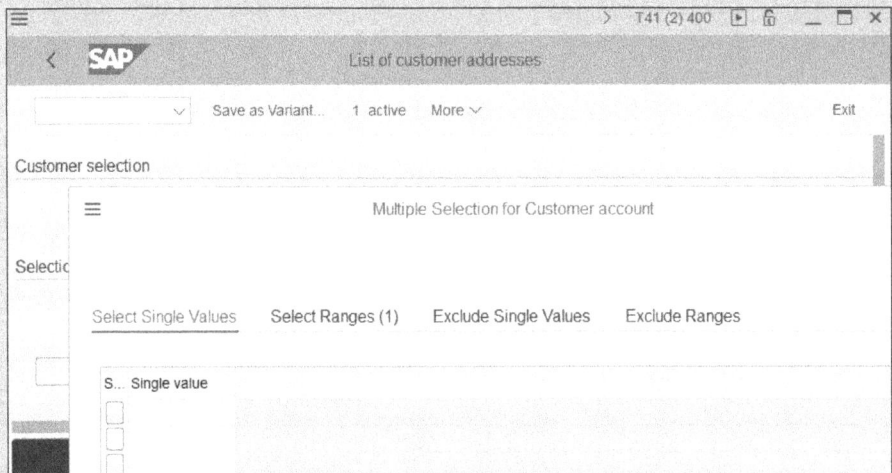

Figure 3.8 Multiple Selection for Customer Account Screen

The SAP system allows you to enter multiple single values on the first **Select Single Values** tab in the **Multiple Selection For …** window that is also displayed. On the second **Select Ranges** tab, you have space for entering several intervals. On the other hand, you can exclude values and intervals on the remaining two tabs. If you click ⬚ (**Cancel**) at the bottom right, this window disappears, and you return to the selection screen.

3. In the selection screen, select the correct **Output Format**, as shown in Figure 3.9. Unfortunately, the obsolete output format **ABAP List** is preset by default. Choose the **SAP List Viewer** instead.

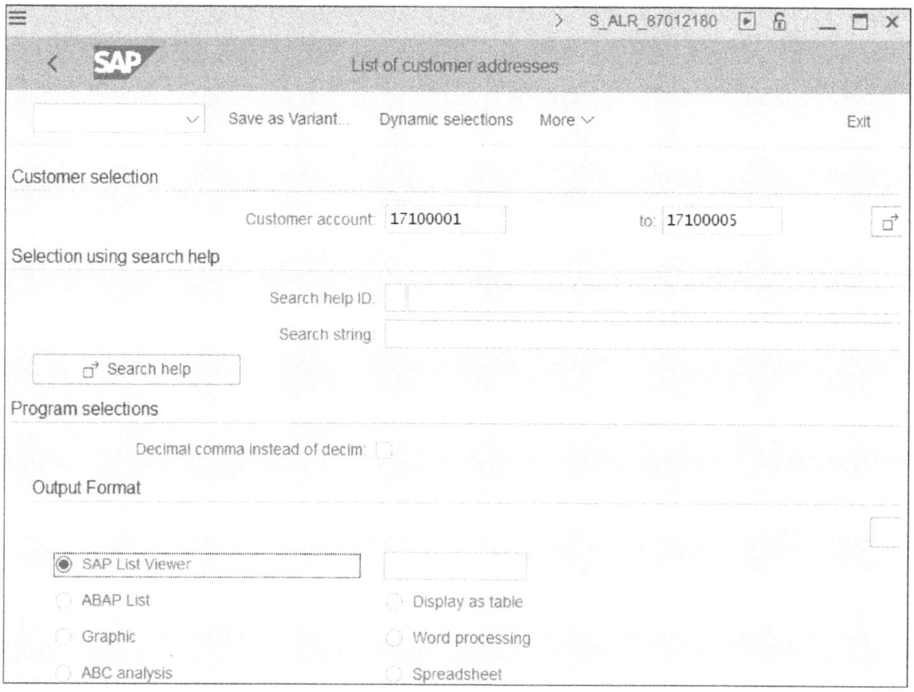
S_ALR_87012180
SAP
List of customer addresses
Save as Variant...
Dynamic selections
More
Exit
Customer selection
Customer account: 17100001
to: 17100005
Selection using search help
Search help ID:
Search string:
Search help
Program selections
Decimal comma instead of decim:
Output Format
SAP List Viewer
ABAP List
Display as table
Graphic
Word processing
ABC analysis
Spreadsheet

Figure 3.9 Selecting the Output Format

4. Click on the **Execute** button in the footer line and possibly confirm with [Enter] for the maximum number of records to be read, and you'll see the previously selected records in the now single-line displayed list, as shown in Figure 3.10. Thanks to the **SAP List Viewer** output format, the display is now much clearer than in our first attempt.

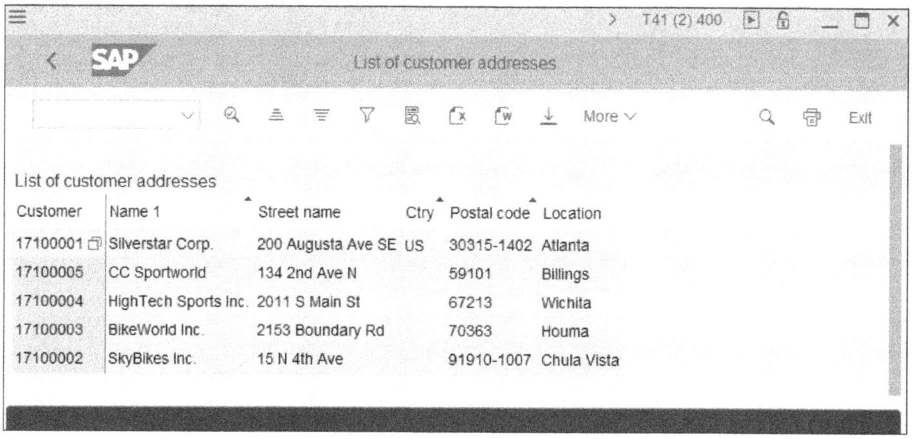
T41 (2) 400
SAP
List of customer addresses
More
Exit
List of customer addresses
Customer
Name 1
Street name
Ctry
Postal code
Location
17100001 Silverstar Corp. 200 Augusta Ave SE US 30315-1402 Atlanta
17100005 CC Sportworld 134 2nd Ave N 59101 Billings
17100004 HighTech Sports Inc. 2011 S Main St 67213 Wichita
17100003 BikeWorld Inc. 2153 Boundary Rd 70363 Houma
17100002 SkyBikes Inc. 15 N 4th Ave 91910-1007 Chula Vista

Figure 3.10 Single-Line Address List

5. Click [<] (**Back**) in the title bar. This takes you back to the selection screen, and this is where we make an additional selection in the following instructions.

[»]

Output Format

When you first called the address list in the previous section, you saw the obsolete output format **ABAP List** preset by SAP, which is also called *ABAP List Viewer (ALV) classic*.

If you have the choice, we recommend the output format **SAP List Viewer** (other term: *ALV grid*) because it usually offers more and easier design options. Many lists already appear automatically in this current representation.

You may wonder why there are two different ALV representations with ALV classic and ALV grid? This has historical reasons. Many lists already existed in ALV classic format before ALV grid was introduced in the late 1990s. After that, SAP provided many lists in this new format, but the old format is also still offered.

Other output formats are needed rather rarely. For the output format **Graphic**, numbers are needed as numeric values, which don't exist in an address list. Even if you have such numbers, it's better not to create graphics in SAP GUI, but with SAP Fiori. SAP Fiori offers much better functions for graphics than SAP GUI.

3.2.2   Use Dynamic Selections

Are you surprised that only the **Customer Account** field appear in the selection screen because you would have liked to select by other fields such as **Country** or **Postal Code**, for example? In many reports, SAP has hidden additional selection fields in the *dynamic selections*. Still other reports offer a button such as **Further Selections** for the same purpose. For our address list and representative for many other reports, we'll now show you the dynamic selections. Follow these steps:

1. In the selection screen of the address list, delete the customer numbers in the **Customer Account** field and in the **To** field.

2. Click the **Dynamic Selections** button in the header line, as shown in Figure 3.11.

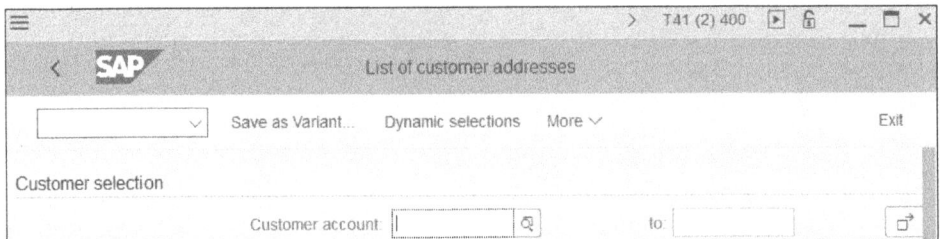

Figure 3.11 Dynamic Selections Button in the Header Line

3. An additional window area with many fields opens at the top, as shown in Figure 3.12. It includes two main areas:

- In the **Dynamic Selections** area on the right, you can see the fields that the SAP system provides by default for entering additional selection values.
- The left area contains many more fields that you can transfer to the right area of the window for selection. The fields that are already available in the right area of the window are highlighted in color in the left area of the window.

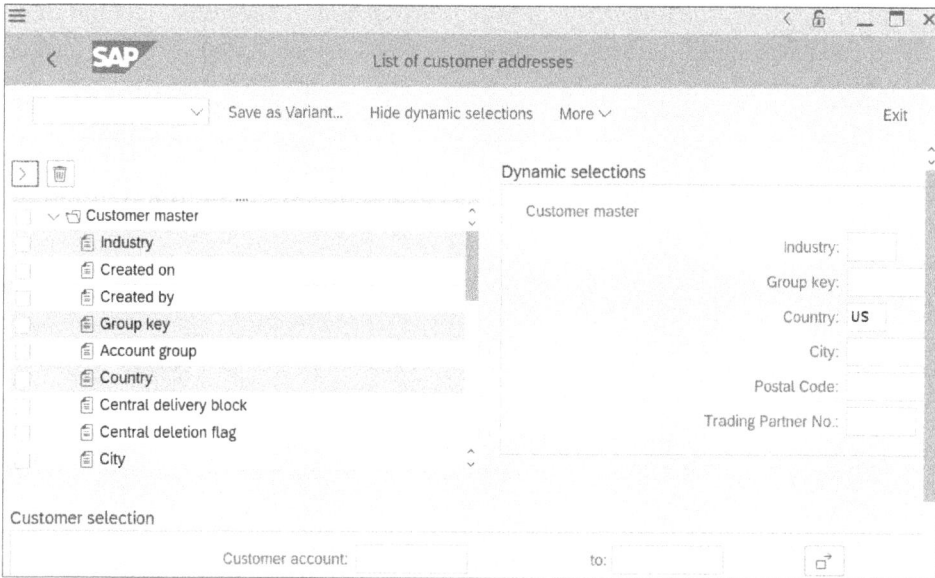

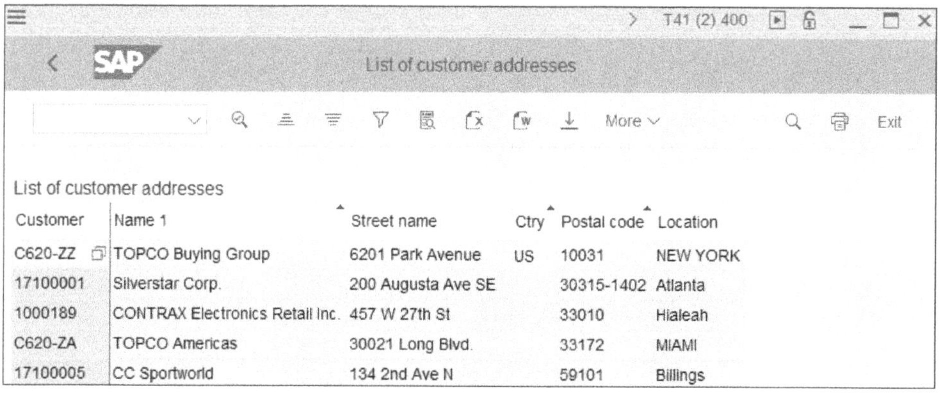

List of customer addresses
Save as Variant...
Hide dynamic selections
More
Exit
Dynamic selections
Customer master
Industry
Created on
Created by
Group key
Account group
Country
Central delivery block
Central deletion flag
City
Customer master
Industry:
Group key:
Country: US
City:
Postal Code:
Trading Partner No.:
Customer selection
Customer account:
to:

Figure 3.12 Window Area with Dynamic Selections

4. Double-click on a field in the left area of the window to transfer it to **Dynamic Selections**.
5. In the **Dynamic Selections** area of the window, enter "US" in the **Country** field.
6. Confirm the selection by clicking the **Execute** button in the footer line. The SAP system now displays the list taking into account your dynamic selections; in our case, this is only the customers from the United States, as shown in Figure 3.13.

T41 (2) 400
List of customer addresses
More
Exit
List of customer addresses
Customer
Name 1
Street name
Ctry
Postal code
Location
C620-ZZ
TOPCO Buying Group
6201 Park Avenue
US
10031
NEW YORK
17100001
Silverstar Corp.
200 Augusta Ave SE
30315-1402
Atlanta
1000189
CONTRAX Electronics Retail Inc.
457 W 27th St
33010
Hialeah
C620-ZA
TOPCO Americas
30021 Long Blvd.
33172
MIAMI
17100005
CC Sportworld
134 2nd Ave N
59101
Billings

Figure 3.13 Address List after Selection in the Country Field

3.2.3 Change and Delete Dynamic Selections

You're still in the list, so now we'll show you how to change or delete dynamic selections. To do so, first return to the selection screen by clicking ◁ (**Back**) in the title bar of the address list. Then, follow these steps:

1. In the selection screen, as shown in Figure 3.14, the dynamic selections are no longer visible. However, you'll see the indication **1 Active** in the header line. This indicates that you have made a dynamic selection specification for one field. Click the button **1 Active** or the ▤ icon to open the dynamic selection part of the window.

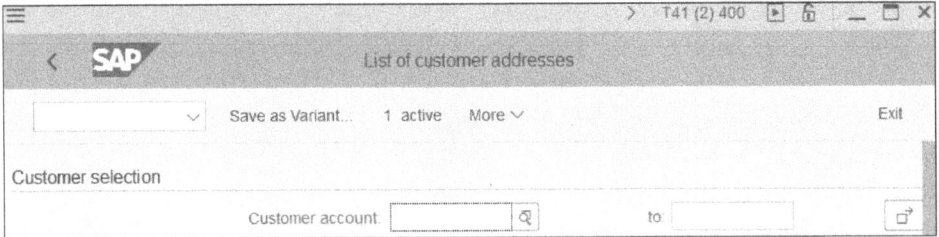

Figure 3.14 Selection Screen with 1 Active Button for Dynamic Selection

2. In the **Country** field (see Figure 3.15), delete the "US" entry, and click the **Execute** button in the footer line.

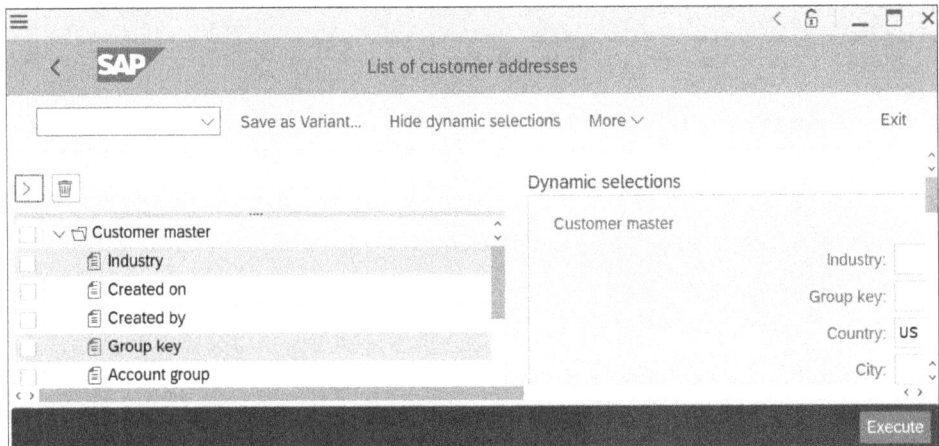

Figure 3.15 Dynamic Selections

Now you'll be shown again the international list with all the customers from the specified number interval.

Instead of deleting the "US" entry, you could also restrict the search here to another country and enter "GB" (United Kingdom) instead of "US", for example.

Leave the list open because the following section will continue with how to make the best use of lists.

[+]

3

Save Selections for Reports

The SAP system forgets your dynamic selections very quickly because the next time you start the transaction, the screen is back to its old state.

Do you often or constantly need additional selection fields or other complex selections? If you're clever, you can create your own report variant for this purpose. You can find out how this works in the next chapter.

3.3   Working with Lists: Find and Filter Data

For the explanations in this section, we assume that your list is already optimally set up, that is, you've sorted the columns and put them in the appropriate order. You'll learn here how to quickly find and filter data. The techniques presented here are especially helpful in large lists.

3.3.1   Find Records

Are you looking for a specific entry in a long list, such as a customer from Wichita in the address list? The fastest way is to use $\mathbb{Q}$ (**Find**). Follow these steps:

1. If you still have the list open, good; if not, create it as described in Section 3.2.2.

2. Click in the first line of the list, as shown in Figure 3.16.

   Why is this necessary? Unfortunately, the search function only searches from top to bottom by default. If your current marker is below the search term you're looking for, the search function won't find this term.

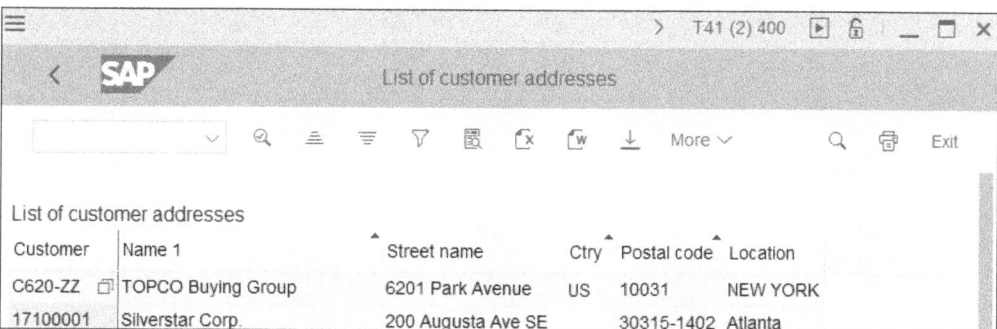

List of customer addresses

Customer	Name 1	Street name	Ctry	Postal code	Location
C620-ZZ	TOPCO Buying Group	6201 Park Avenue	US	10031	NEW YORK
17100001	Silverstar Corp.	200 Augusta Ave SE		30315-1402	Atlanta

Figure 3.16 Selecting the First Line in the Address List

3. Click $\mathbb{Q}$ (**Find**) in the upper-right corner. The **Find** window opens.

4. In the **Find** window, as shown in Figure 3.17, enter the search term, such as "Wichita". It's also sufficient if you enter only part of the term, such as "Wichi".

5. Select the **Display Number of Hits** option and confirm with $\boxed{\text{Enter}}$.

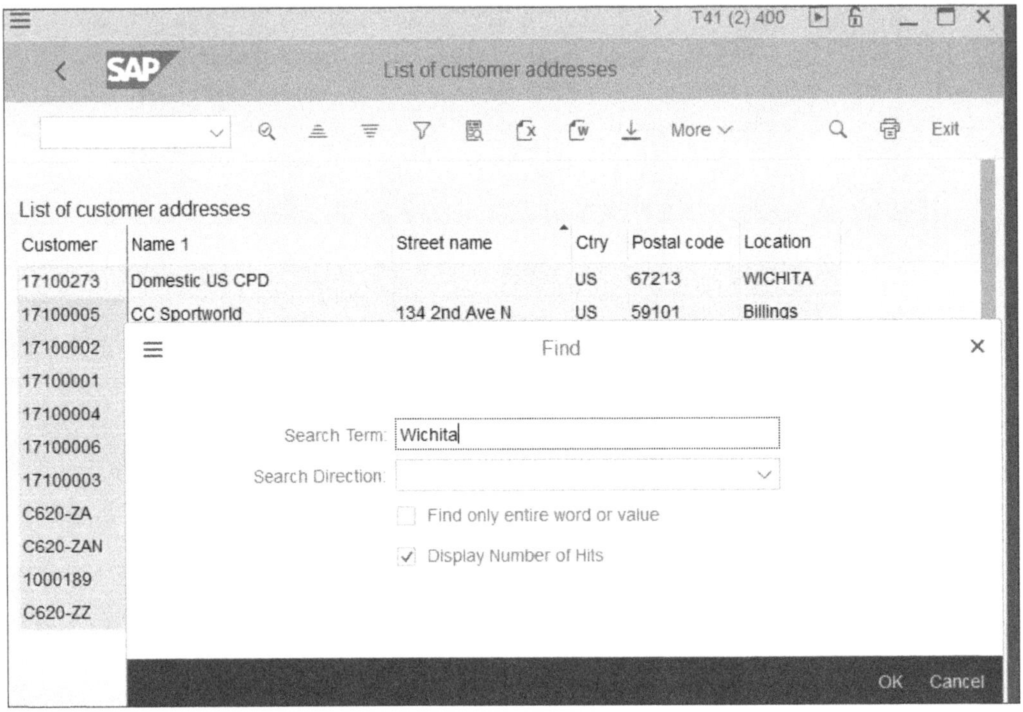

Figure 3.17 Find Window

6. The first line of the list containing the search term moves to the beginning of the list.
The **Find** window shows you the number of hits at the bottom next to [i] based on
the corresponding preset.

Pressing [Enter] will take you to the next hit, and clicking [X] (**Close**) in the upper-right
corner or pressing [Esc] will close the **Find** window.

3.3.2   Filter Records

The filter function allows you to select directly in the list. This saves you the detour via
the selection screen and the determination of the desired data records there. Another
and important advantage of filtering in the list is that you can filter your data by *any*
field. In the selection screen and in the dynamic selections, there are often fields missing that you might want to select by. In other words, you can make some selections
only in the list with filters.

Are you motivated? Great! Then, on to our example. Only the lines with the entry
"Wichita" should be displayed in the list. We'll first show you the "official" way and then
a little trick. Follow these steps:

1. Click on the desired column heading, which, in our case, is **Location**, as shown in Figure 3.18.

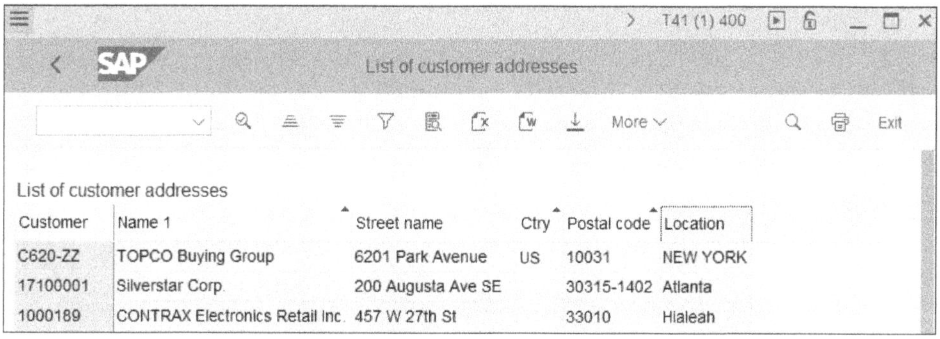

Figure 3.18 Selecting the Location Column

Select Multiple Columns

If you want to filter several columns at once, you can also select several columns at once. To do this, first hold down the ⌜Ctrl⌝ key and then select the corresponding column headers with individual clicks.

---

2. Click ▽ (**Set Filter**) in the header line. The **Determine Values for Filter Criteria** window appears, as shown in Figure 3.19.

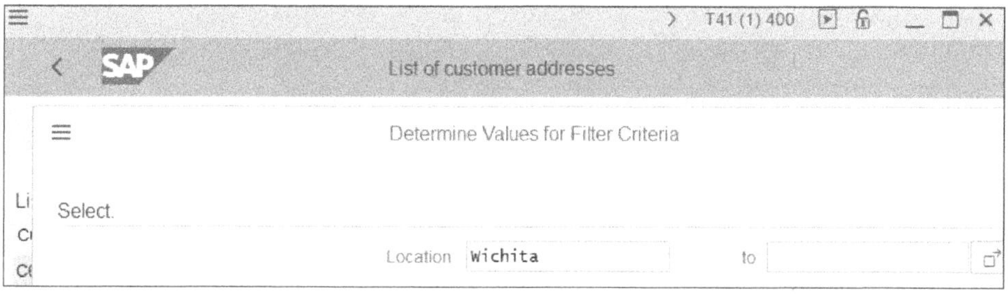

Figure 3.19 Determine Values for Filter Criteria Screen

3. Specify the value for the selected column; in our case, this is "Wichita", for the **Location** column. Use the wildcards * or +, or the multiple selection feature with a click ⬚⁂ (**Multiple Selection**), if you need more complex selections.

4. Confirm your selection with the ⌜Enter⌝ key. You'll immediately receive the filtered list.

5. Click ▽ icon (**Set Filter**) in the header line to see the whole list again.

To the left and right of the ▽ icon (**Set Filter**), there are many more icons in the header line, with which you can sort or change the layout, for example. You'll appreciate these icons in the next chapter of the book because there you'll learn, among other things, how to optimally design SAP List Viewer (ALV) lists.

[+]

Filter with Three Quick Clicks

If the value you want to filter by is on the screen, you can save yourself the column marking and manual entry of the value. It's faster like this:

1. In the list, right-click on the value you want to filter by, in our case **Wichita**. A context menu appears, as shown in Figure 3.20.

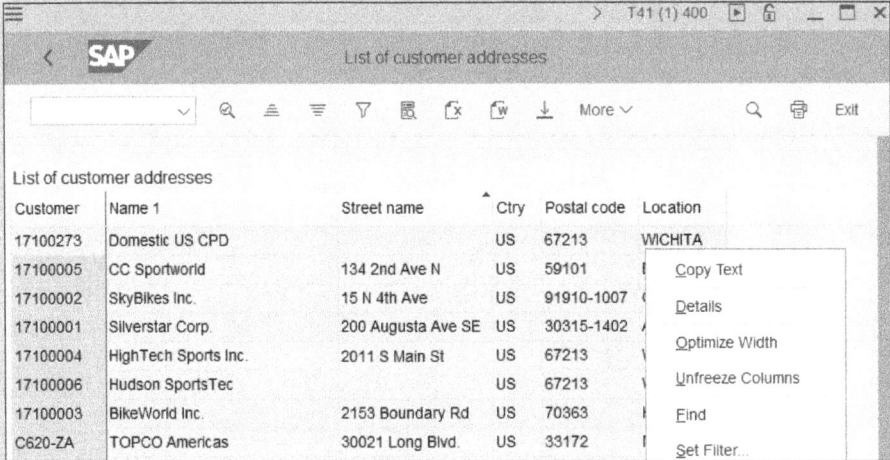

Figure 3.20 Opening the Context Menu

2. In the context menu, select the **Set Filter** command. The **Determine Values for Filter Criteria** window opens, as shown in Figure 3.21.

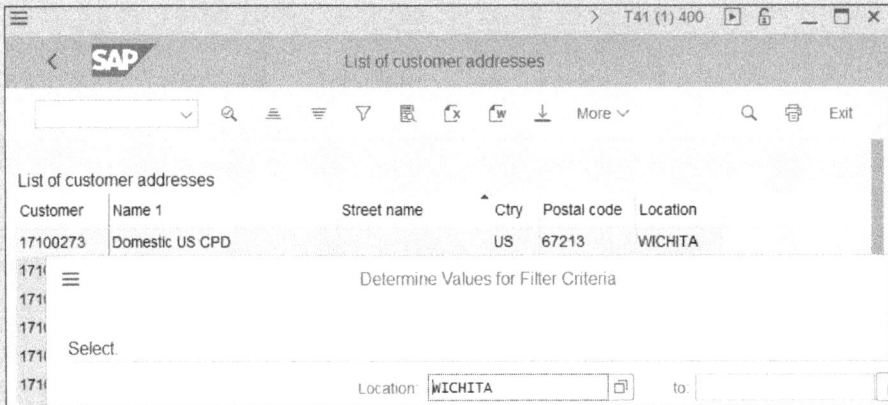

Figure 3.21 Setting Values for Filter Criteria

3. As you can see, the value **Wichita** you clicked on was automatically taken over into the **Location** field. You just press the ⌜Enter⌟ key here, and you've the filtered list.

3.4   Print Lists or Save as PDF Files

The paperless office hasn't yet become a reality. In this section, you'll learn how printing with the SAP system works and how to make the necessary settings.

In the following instructions, you'll learn how to initiate an immediate printout or have the list output as a PDF document:

1. Is the address list of customers still open, as shown in Figure 3.22? If not, call it up as shown in Section 3.2.2.

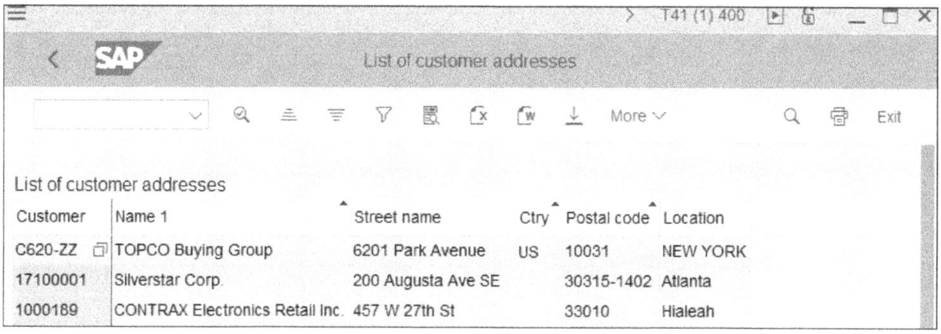

Figure 3.22 Address List

2. Click 🖶 (**Print**) on the far right of the header line. The **Print ALV List** window appears, as shown in Figure 3.23. In this window, you can make the following specifications:
   - **Output Device**
     Select whether to output to a printer or to a PDF file. If you want to print on your normal Windows printer, enter "LOCL" as the output device.

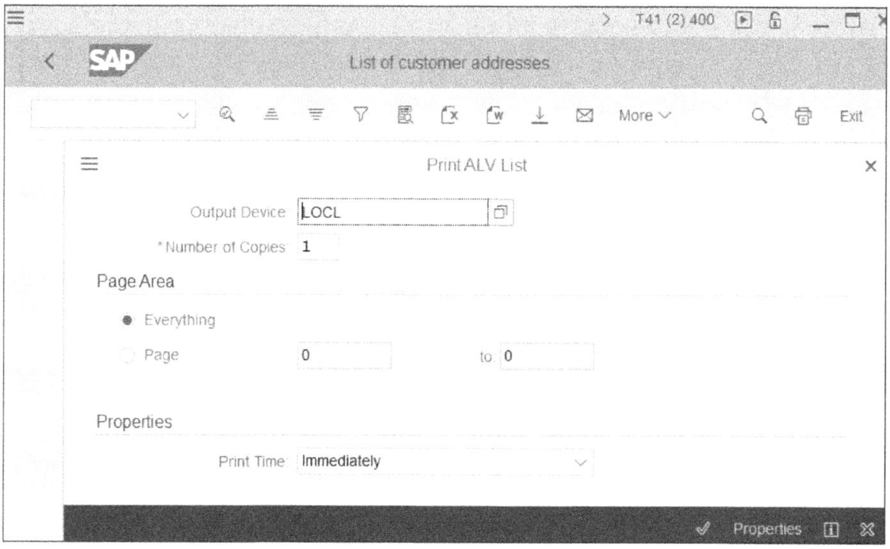

Figure 3.23 Print ALV List Settings

- **Number of Copies**
  Specify the number of printouts.
- **Page Area**
  Specify whether you want to print all pages, single pages, or a page interval.
- **Print Time**
  This field is particularly interesting. As a rule, you select the **Immediately** setting here so that the list is printed. With the alternative **SAP Spool Only for Now**, you create the possibility of triggering an output yourself at a later point.
3. Click on the green checkmark icon in the lower-right corner to initiate the printout.

   Pay attention to the footer line, as shown in Figure 3.24. Here you'll now see the message that a spool request has been sent to the SAP printer.

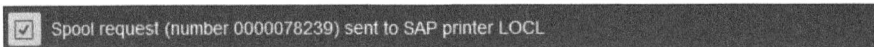

☑ Spool request (number 0000078239) sent to SAP printer LOCL

Figure 3.24 Success Message

4. If you entered a PDF device in the **Output Device** field, you'll be prompted for a file name. Otherwise, the Windows **Print** window will now appear. Click **OK**, and the printout is created.

Are you still in the address list? Leave it open because this is where we'll continue in the next section to download lists.

3.5   Download Lists

Why do we authors love SAP GUI? Because it is really entertaining and varied. The download functions are very good examples of this variation. But let's start with a practical case first. A colleague who doesn't have SAP access needs the address list as a Microsoft Excel file. SAP has hidden the download functions in different places depending on the list. We'll find and explain these functions in the following sections.

3.5.1   Download Functions in the Context Menu

The following instructions will show you how to perform a download from the address list. If you're not still in the list of customers, you can find your way to this list thanks to Section 3.2.2.

Now you want to start the download to a Microsoft Excel file. Follow these steps:

1. Right-click anywhere in the list, and the context menu appears, as shown in Figure 3.25.

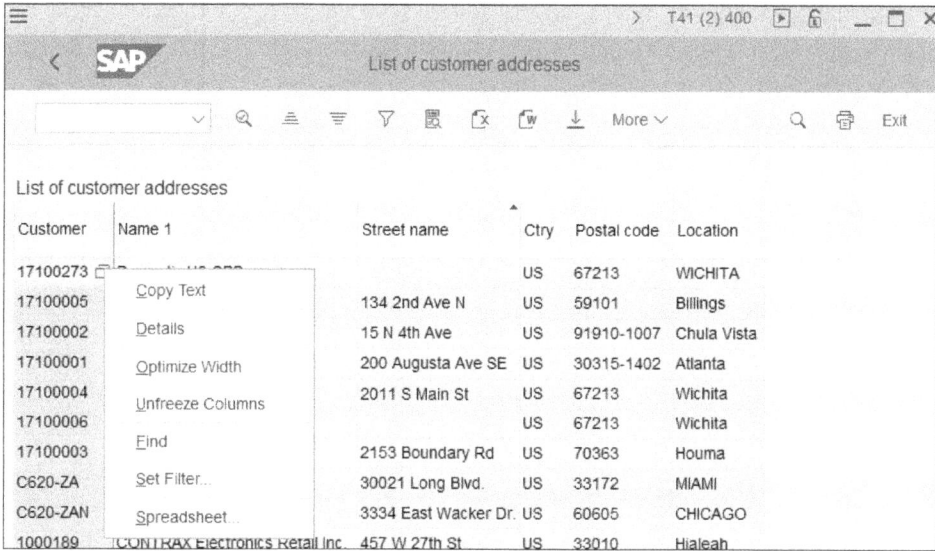

Figure 3.25 Opening the Context Menu

2. Here's a download function for Microsoft Excel (it doesn't always exist in the context menus). Click on the last menu option, **Spreadsheet**. A dialog box appears, as shown in Figure 3.26, where you can select the desired format.

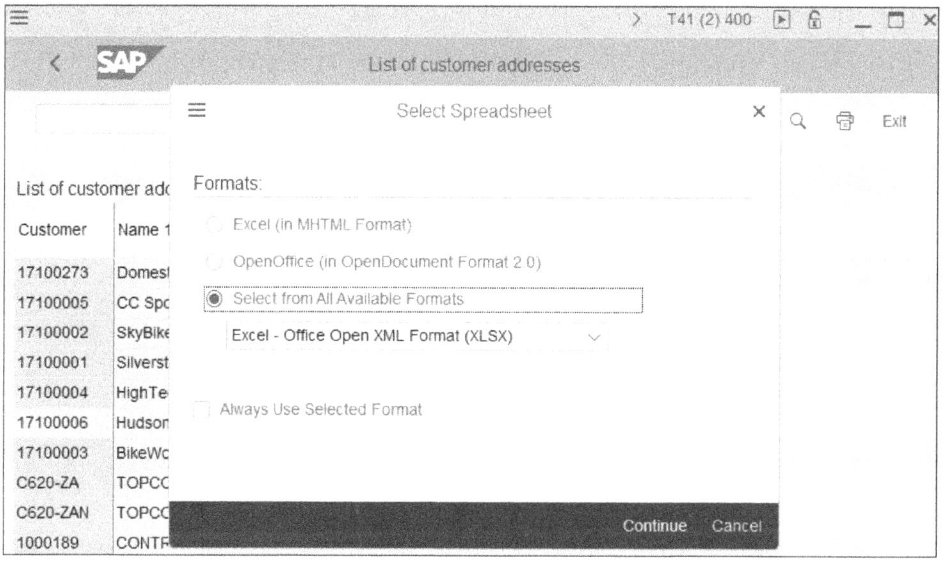

Figure 3.26 Selecting the Microsoft Excel Format

3. Keep the default **Excel – Office Open XML Format (XLSX)** in the **Select from All Available Formats** list box. Press the ⌐Enter⌐ key to continue.

4. In the **Save As** window, select the location on your PC. To do this, click **Desktop** on the left, for example. This allows you to conveniently open the file from the Windows desktop with a double-click after the process is complete. The file name **EXPORT.XLSX** is already preset (see Figure 3.27), but it can be changed, of course.

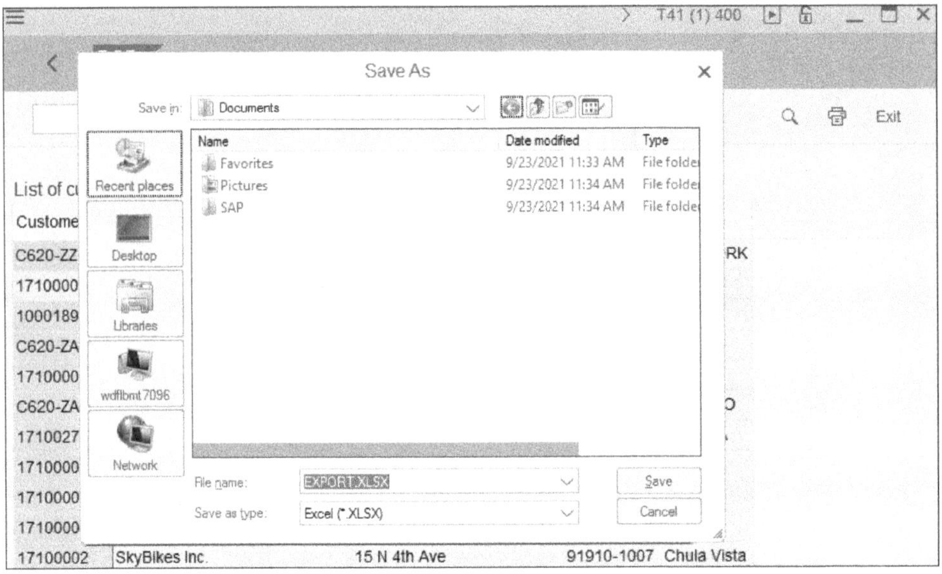

Figure 3.27 Save As Window

5. Now click the **Save** button in the lower-right corner of the window.

6. If the **SAP GUI Security** window appears, select the **Remember My Decision** setting here, and then click the **Allow** button.

If necessary, confirm again, and the Microsoft Excel file is displayed in a window, as shown in Figure 3.28. Because the file is saved (on the desktop), you may close the Microsoft Excel window again.

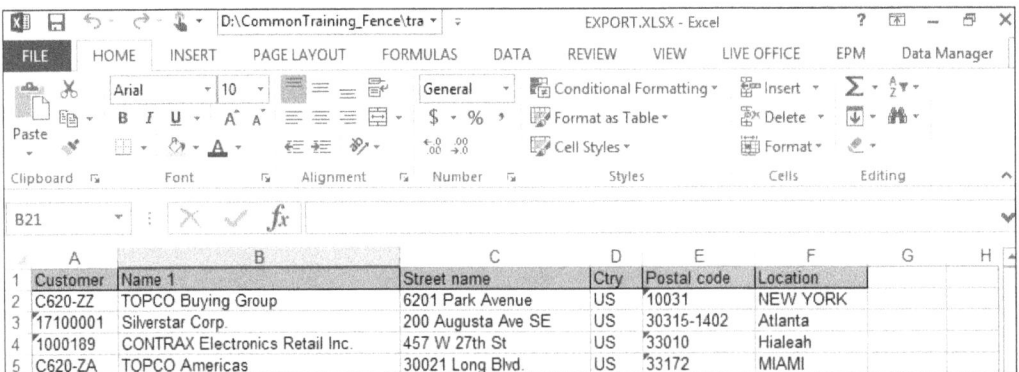

Figure 3.28 Microsoft Excel Spreadsheet Displayed

[+]

3

Select and Copy Table Ranges

Often you don't need the whole ALV table downloaded, but only a table range or a single field content. Simply use the Windows clipboard for this purpose.

There's a problem, however, if you want to select a subrange in the list with the left mouse button pressed because, in SAP List Viewer, you can only select one line by holding down the left mouse button. Here is the solution to this problem:

1. Press the key combination [Ctrl]+[Y].
2. With the left mouse button pressed, select the part of the table on the screen that is to be copied.
3. Now you can copy the selected table content to the clipboard with the key combination [Ctrl]+[C] and paste it elsewhere with [Ctrl]+[V].

Often, the SAP system offers you one or both icons from Table 3.1 in the header line for download.

Icon	Function
⬇	The **Local File** icon transfers the records from a PC to a local file. You can choose the file format such as *.xls* (Microsoft Excel) or *.txt* (text file). However, not all file formats are available in every report.
⌐x	The **Microsoft Excel** or **Spreadsheet** icon transfers the data from a PC directly into Microsoft Excel .

Table 3.1 Download Icons

3.5.2 Download Functions in the More Menu

In some reports, there is no download function either in the header line or in the context menu. Well, what to do? There is one more menu we haven't tried yet.

Select the following menu path in the address list: **More • List • Export • Spreadsheet**, as shown in Figure 3.29. You'll see that this works!

And if even this menu command doesn't work, there's one very last arrow in our quiver: choose **More • System • List • Save • Save**. This menu option is not found in every list.

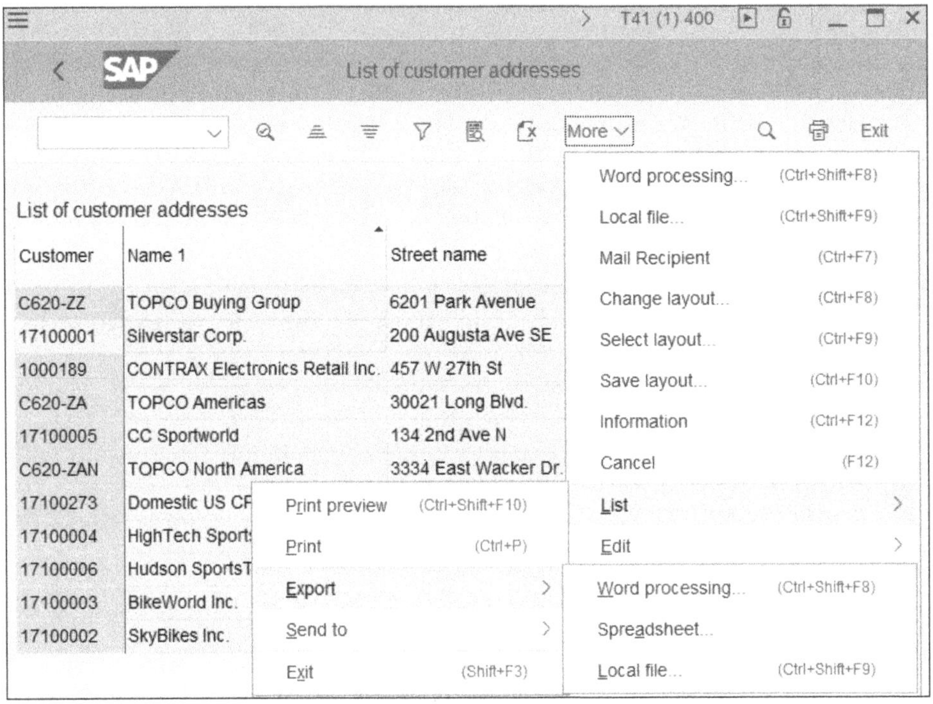

Figure 3.29 Download Menu Command in the More Menu

[+]

Never Give Up Too Early!

One explanation for the fact that SAP GUI offers so many possibilities is that some transactions were programmed in the past millennium, others 10 years ago, and still others only recently. Depending on when a transaction was created, different guidelines applied to the developers. And because there are now more than 100,000 transactions and more than 300,000 different windows, SAP has never harmonized the operation of the transactions.

But what does that mean for you? If you're looking for a function in SAP GUI and can't find it, you should never give up too soon. It may—as you've seen in the download—simply hide elsewhere. If you search, you will find!

We recommend the following sequence to search for a function:

- First search in the header or footer line. You can find out the meaning of the icons by hovering the mouse cursor over the icon to see the meaning displayed in a tooltip.

- Very often, you'll find useful functions in the context menus that you don't see in the header or footer line. Note here that a right-click gives different results depending on where it's placed. If you click on a column header in the address list, you get different menu commands than if you click in the middle of the list or on an empty space above the list.

- As a third chance to find a function, you're left with the **More** menu.

The fact that not all functions can always be accessed in the same place may be a bit annoying at times, but it doesn't matter so much when you look at the tremendous variety of functions and customizability on the other side. To learn more about this, you should definitely read the next chapters. Maybe you'll just do like we do: we sometimes grumble, but we just take the SAP system as it is.

3.6   Summary

You've learned about key SAP GUI reporting features, including finding, filtering, printing, and downloading tables. Let's continue with the steps to customize SAP GUI reports.

Chapter 4
Changing SAP GUI Reports

Remember the address list from the previous chapter? There you used a ready-made standard report. However, standard reports are like off-the-rack clothes: sometimes they fit perfectly, but often they don't. If not, the SAP system offers you many options to create a customized report from a standard report with little effort, which is another example of the freedom you have to set up SAP GUI just like you want it.

Like many users, you probably use the same reports every day or every week. This chapter is especially worthwhile in this case because you'll learn how to permanently save your settings for your reports and how to retrieve them with just a few clicks.

What You'll Learn

- How to save default values for the selection screen in a report variant
- How to quickly activate your default values for the selection screen by calling up the corresponding report variant
- How to change and delete your report variants
- How to arrange the layout of the list clearly
- How to save a layout so you don't lose your settings
- How to retrieve and use ready-made layouts
- How to maintain your layouts

4.1 Save Default Values in a Report Variant

You store the default values for reports in *report variants* (also called *selection variants*). The use of report variants has several advantages:

- You save yourself from having to reenter the selection criteria each time you start the reports.
- Precise selections in the report variants result in lists containing only the necessary data. This makes the lists shorter, clearer, and faster to create. Shorter lists in turn have the side effect that the SAP system is less burdened, making system response times shorter as well. Your colleagues will thank you!

- Often, the most important advantage is that you always get the right results in your lists and evaluations after you've saved the appropriate defaults in report variants and then use them consistently.

Only three steps are necessary to create a report variant:

1. Start the report.
2. Enter your selection in the selection screen.
3. Save this selection as a report variant.

In the following sections, we'll show you the process for creating and using a report variant by using the address list from the previous chapter as an example.

4.1.1   Start the Report and Define a Selection

A report variant always contains your complete selection specifications from the selection screen. The preliminary work therefore consists of first making your specifications in the selection screen after calling up the report. In our example, you select all customers from the United States for the address list.

Follow these steps:

1. Start in your **Favorites** list by double-clicking on **Transaction S_ALR_87012180**. (If you can't find this transaction in your **Favorites**, call it in the **SAP Menu** by choosing **Accounting • Financial Accounting • Accounts Receivable • Information System • Reports for Accounts Receivable Accounting • Master Data**.)

   After starting, the selection screen appears. You always create a report variant by starting from the selection screen of the report.

2. In the selection screen shown in Figure 4.1, click **Dynamic Selections** to specify the country code.

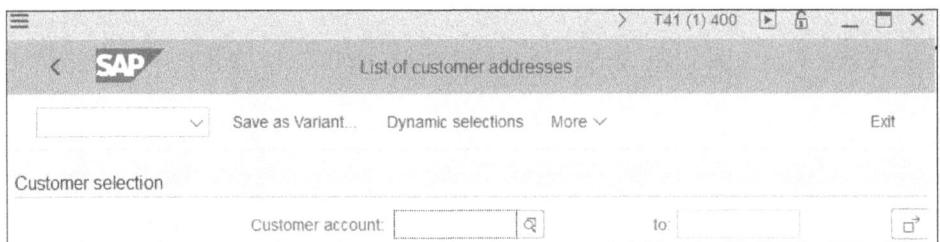

Figure 4.1 Selection Screen

3. In the **Dynamic selections** area of the window, as shown in Figure 4.2, enter "US" in the **Country** field. Further down, select the **SAP List Viewer** output form if it's not already selected from a previous use of the list.

4. Test the selection by clicking the **Execute** button in the footer line.

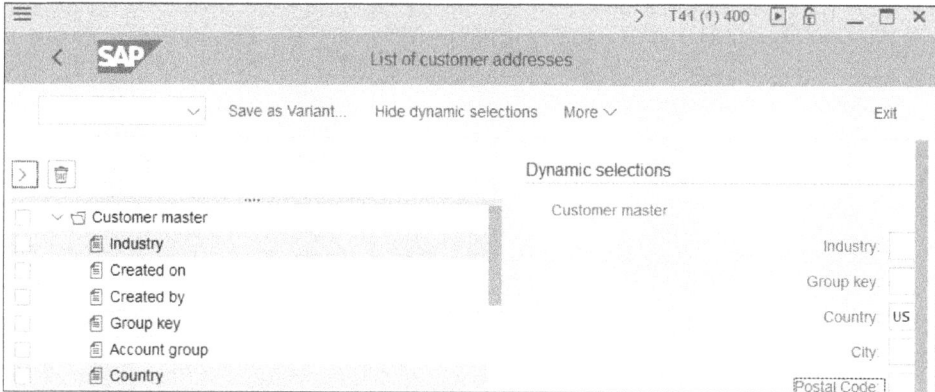

Figure 4.2 Dynamic Selections

5. Press Enter when you're asked about the number of records to be read.

6. If everything turns out to your satisfaction (see the list in Figure 4.3), switch from the list back to the selection screen by clicking ◁ (**Back**).

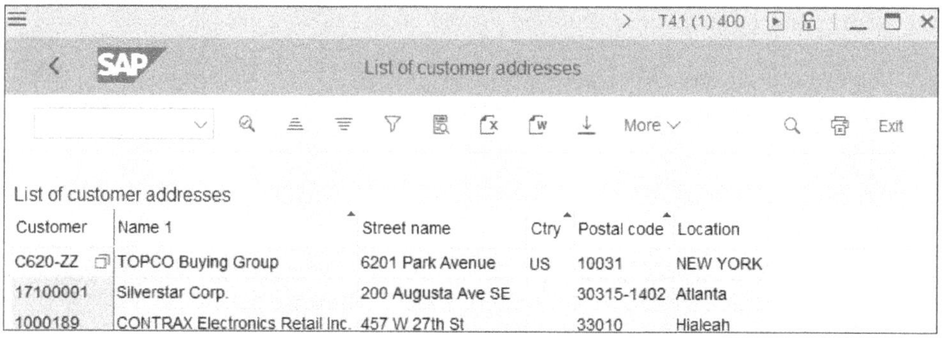

Figure 4.3 Address List

4.1.2 Save the Report Variant

In our example, we assume that you'll need this selection more often in the future. After you've made your selection specifications in the selection screen, you therefore now save them as a report variant. Follow these steps:

1. In the selection screen in the header line, click the **Save As Variant** button. The **Variant Attributes** screen appears, as shown in Figure 4.4. Don't be confused by the multitude of settings; concentrate on the upper window area. Make entries in the **Variant Name** and **Description** fields for the short description of the report variant addition.

2. Select the **Protect Variant** checkbox. This and all other checkboxes are optional.

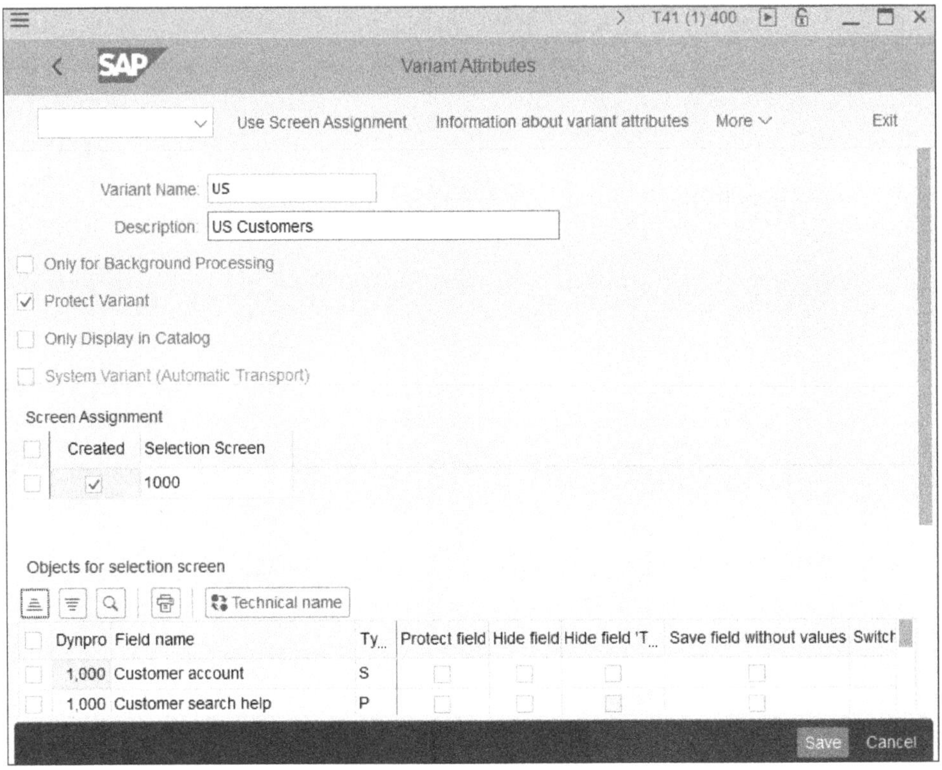

Figure 4.4 Variant Attributes

[+] **Protect Your Important Variants**

Report variants apply across users, which means that other users not only can access and use your report variants but also change and delete unprotected variants. So, to protect a variant from being edited, activate the **Protect Variant** checkbox. If this is checked, only you can change and delete the protected report variant.

---

1. Click the **Save** button in the footer line. After saving, you're back in the selection screen. The report variant is saved, and a corresponding message appears in the footer line.

2. Click ◄ (**Back**) to return to the **SAP Easy Access** window.

In these instructions, you've only filled in the fields **Variant Name** and **Description**. However, we don't want to withhold the following setting options from you at this point:

- **Only for Background Processing**
  Selecting this checkbox doesn't make sense as you wouldn't be able to use the variant after calling the report. This is because it would then only be available for report execution in background processing.

- **Objects for Selection Screen**
  You can also make settings for individual fields of the selection screen in this the lower section of the screen. The fields are listed here line by line. For example, individual fields of the selection screen can be protected against changes or can be hidden.

[«]

Can I Create Multiple Report Variants?

You can create any number of variants for a report. In our case, you can create several report variants for the address list, for example, for different postal code areas or countries.

4.2   Call Up a Report Variant

After you've saved the report variant, you can use it in the selection screen. You always start a report variant from the selection screen of the report:

1. Start Transaction S_ALR_87012180 (Address List) again for which you've just saved the variant, as shown in Figure 4.5.

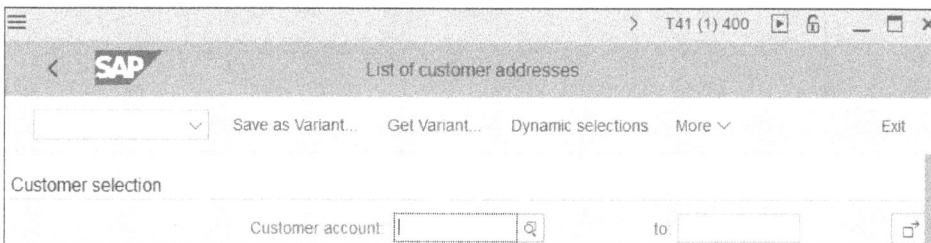

Figure 4.5 Transaction S_ALR_87012180: Initial Screen

2. In the header line of the selection screen, now click the **Get Variant** button. The **ABAP: Variant Directory** window appears.

[+]

Missing the Get Variant Function?

The ▢ (**Get Variant**) function is only displayed if there is at least one variant for this report. If you don't have this icon, you must first save a variant for this report as described in Section 4.1.

3. In the **ABAP: Variant Directory** window, double-click the report variant you created. Here, you'll basically see all the variants for the current report, including the variants created by other users (see Figure 4.6).

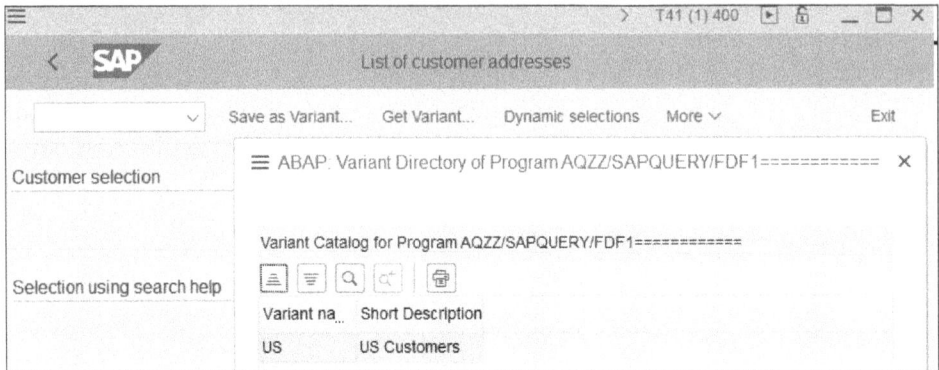

Figure 4.6 ABAP: Variant Catalog

4. After selecting the report variant, you'll see the corresponding selection specifications as default values in the selection screen. In our case, the output form **SAP List Viewer** is selected, and in the header line, you see the entry **1 Active**, as shown in Figure 4.7. This is your default for the **Country** field.

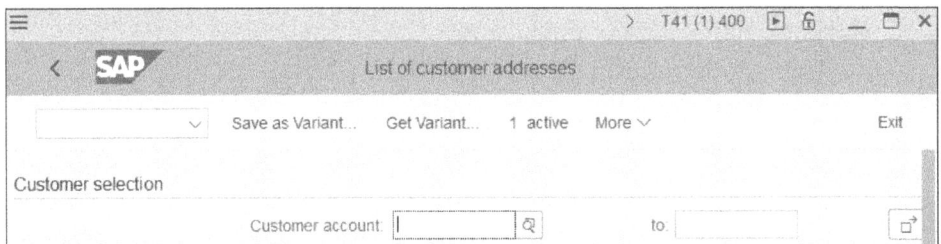

Figure 4.7 Report Variant Screen

5. Thanks to your variant, you save yourself from having to enter these selection criteria again manually. You can now click the **Execute** button in the footer line and get the corresponding list on the screen. In the list, click ⟨ (**Back**) because the selection screen will continue in a moment.

[»]   **Does a Search Help Appear Instead of the Variant Catalog?**

If there are many stored variants, a search help appears first instead of a variant directory. Your username is already entered in this search form. If you confirm the search form with the Enter key, the SAP system displays all the variants stored with your username.

[«]

Does Neither a Variant Catalog nor a Search Help Appear?

If you don't see the catalog or the search help, look at your selection screen because the saved selection values were automatically applied there. How so? You've created only one variant. And because there is only one variant, the SAP system cleverly copies the values from this variant to the selection screen if there is a corresponding default setting for this.

4

4.3   Maintain Report Variants

Nothing is as constant as change. Therefore, in this section, you'll learn how to subsequently change the selection in already saved report variants, adjust the variant name, and delete report variants that are no longer required. We'll walk through the steps to change and delete variants in the following sections.

4.3.1   Change a Report Variant

To do this, use the report variant created in Section 4.1. The procedure is similar to calling a text file in Microsoft Word, changing it, and saving it again under the same file name. The following prerequisite exists for changing a report variant: you've already called up the variant to be changed according to the instructions in Section 4.2 by using the **Get Variant** button. Your starting point is the selection screen of the corresponding report. Follow these steps:

1. In the selection screen, as shown in Figure 4.8, make the desired changes.

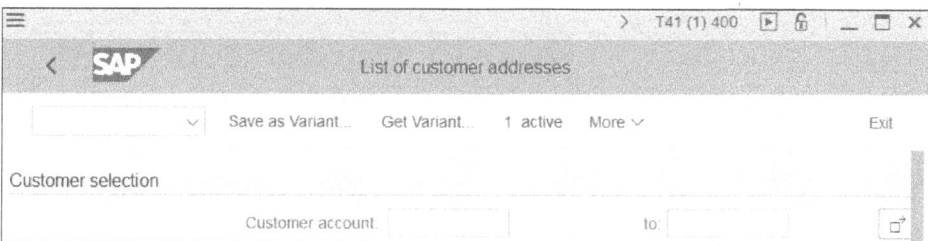

> T41 (1) 400

SAP — List of customer addresses

Save as Variant...   Get Variant...   1 active   More ∨   Exit

Customer selection

Customer account: to:

Figure 4.8 Making Changes in the Selection Screen

2. In the header line, click **Save as Variant**. The **Variant Attributes** screen appears, as shown in Figure 4.9.
3. Click the **Save** button in the footer row.

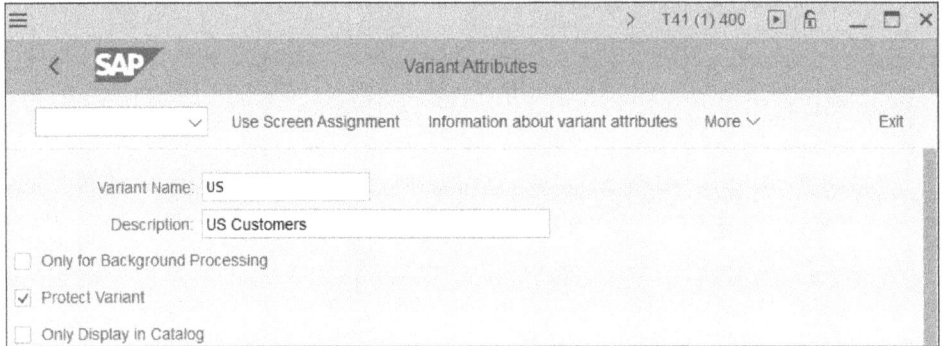

Figure 4.9 Variant Attributes

4. You'll now be asked in a dialog box whether the existing variant should be overwritten, as shown in Figure 4.10. If you want to do so, click the **Yes** button; otherwise, create the report variant under a changed name.

Figure 4.10 Overwriting the Existing Variant

This saves the changes. The next time the report variant is called, the changed values will be used.

4.3.2   Change the Variant Name

There is no special function for changing variant names, so you just save the report variant under the new name and then delete the report variant with the old name.

4.3.3   Delete a Report Variant

Delete report variants that only you use when you no longer need them. This keeps the variant directory clear and prevents it from growing out of control. To delete report variants, proceed as follows:

1. Start the corresponding report. You don't need to call the report variant beforehand—unlike if you want to change it.

2. In the selection screen, choose **More • Goto • Variants • Delete**, as shown in Figure 4.11.

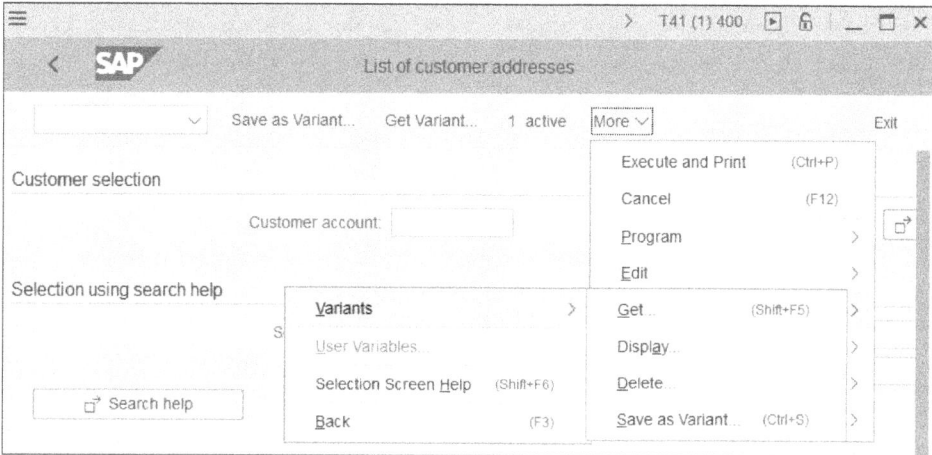

Figure 4.11 Opening the Delete Variants Window

3. The **Delete Variants** window appears, as shown in Figure 4.12. Double-click the report variant you want to delete.

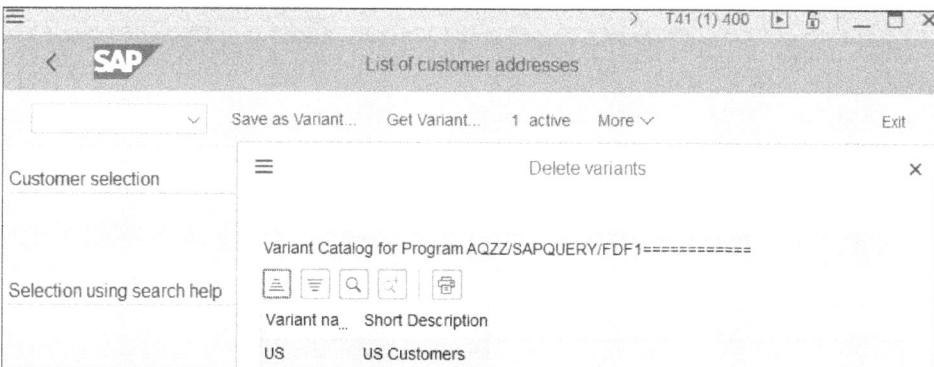

Figure 4.12 Delete Variants Window

4. A dialog box appears asking whether the deletion is to be carried out across all clients. Select the **In Current Client Only** radio button, and click the **Continue** button.

5. A confirmation prompt appears. Click the **Yes** button to confirm that you really want to delete the report variant. The report variant is now deleted and is no longer displayed in the selection.

[»] **Report Variants**

Unfortunately, in some companies, you as a user aren't authorized to create report variants; only key users and SAP administrators have that authority. If this is the case for you, the preceding three sections weren't in vain because you now know the possibilities. On this basis, you can contact your key users or your SAP administrators so that they can create appropriate report variants for you if necessary.

4.4   Get Creative: Design the Layout

The prefabricated standard reports don't fit, there are superfluous columns in the list, or important information is missing? Then get to work—you can now create your own lists using the SAP List Viewer functions.

SAP describes SAP List Viewer as "a flexible tool for list display," and this is spot on because it offers you many options to compile the information exactly as you need it with just a few clicks.

Use the icons when changing the layout, or it won't be any fun. You show the icons, as described in detail in Chapter 1, Section 1.2, by choosing **More • SAP GUI Settings and Actions • Options**. Open the **Visual Design** folder in the upper-left corner, click the **Theme Settings** entry, and select **Show Toolbar Buttons with Icons**. Open with **Str + +** a new SAP GUI window to see the icons.

[+] **Use Predefined Layouts**

Before you reinvent the wheel and create your own layouts, check with a click on [icon] (**Select Layout**) to see if there are any ready-made layouts that meet your requirements. This procedure will be described later in Section 4.6.

In many reports, there are already ready-made layouts that someone has created for you—either SAP itself or a nice person from your company, such as a consultant or key user. In the standard system, however, there are no ready-made layouts for our address list, so you may continue right here.

For the corresponding settings, call a list, and use the icons from the header line as well as the context menu for column headers. As an example, use the address list for customers again. You can see the initial state in Figure 4.13.

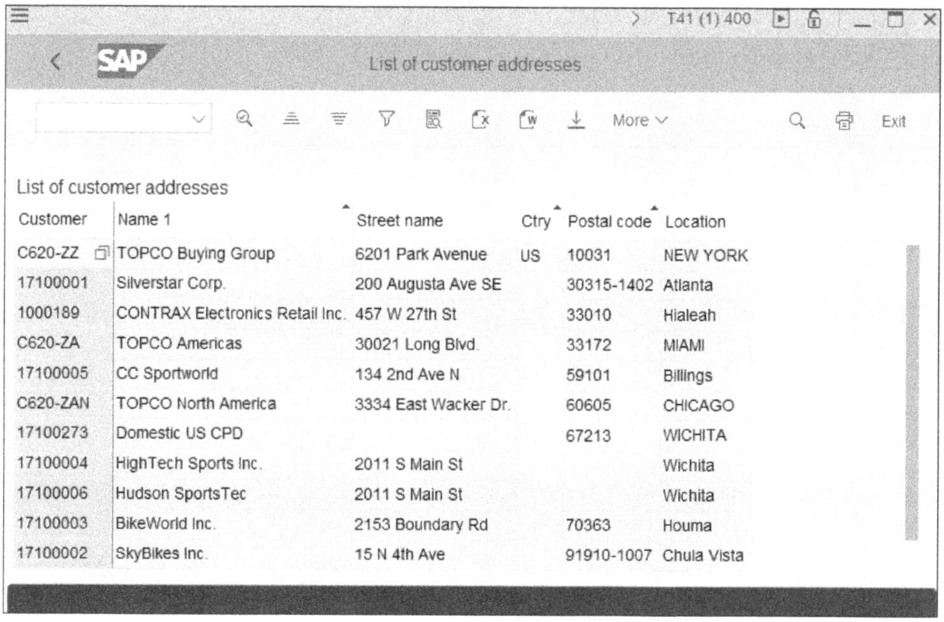

Figure 4.13 Address List for Customers before the Changes

You'll now make the following adjustments in this sample address list:

1. Sort the list by the **Name 1** column.

2. Move the **Customer** column with the customer numbers to the far right.

3. Hide the **Ctry** and **Street Name** columns. The **Ctry** column is superfluous because it contains the same entry "US" everywhere.

4. Display the **Street Name** column again afterwards.

The sum of these and other settings is also called a *layout* or *display version*. You'll learn how to save them in Section 4.5, so that you can retrieve a finished layout—similar to a report variant—at any time with just a few clicks.

The instructions described in the following sections work in the modern output format **ALV Grid Control**. In the older and less frequently used format, **ABAP List** (*ALV classic*), there are partly other procedures, which aren't described in this basic course.

4.4.1   Design the Layout

Open the address list that you've used for the report variants (see the instructions in Section 4.1.1).

Follow these steps:

1. Double-click the transaction entry **S_ALR_87012180 – Address List**. (You don't have the transaction in your **Favorites** list? You can find it in the **SAP Menu** by choosing

Accounting • Financial Accounting • Accounts Receivable • Information System • Reports for Accounts Receivable Accounting • Master Data).

The selection screen appears, as shown in Figure 4.14.

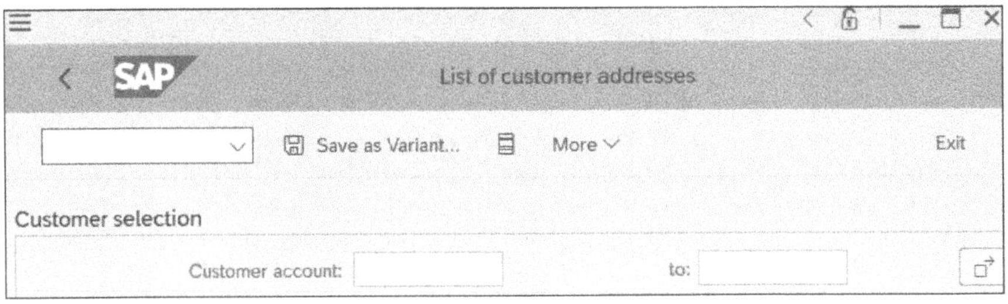

Figure 4.14 Selection Screen

2. Here you enter your selection. You need a list of all customers from Germany. So first click on ▤ (**Dynamic Selections**).

3. In the upper part of the window, enter "US" in the **Country** field on the right, as shown in Figure 4.15.

   It's important that you select the **SAP List Viewer** output form further down the screen because different rules apply to the layout design in the standard **ABAP List** setting.

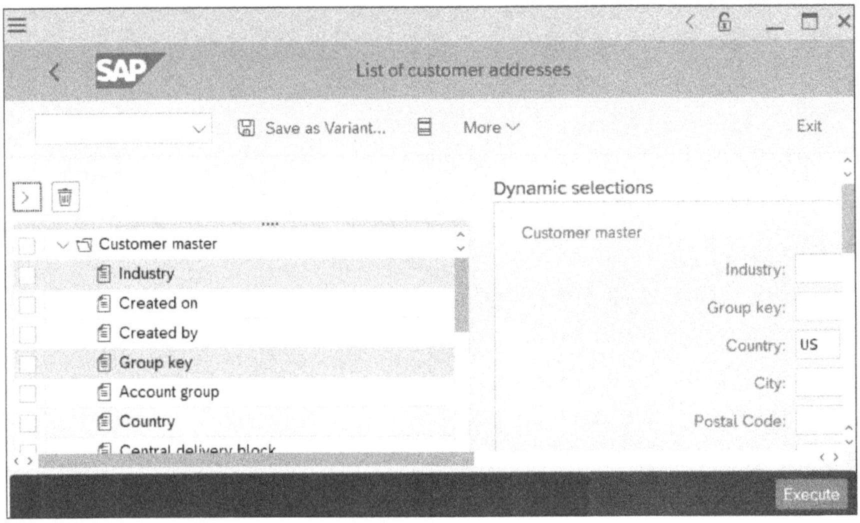

Figure 4.15 Dynamic Selections

4. Click the **Execute** button in the footer line. If a prompt for the number of records to read appears, confirm it with ⌨ Enter .

The result is your address list, as shown in Figure 4.16. In this list, you'll now sort, move columns, change column widths, and show and hide columns according to the instructions in the next sections.

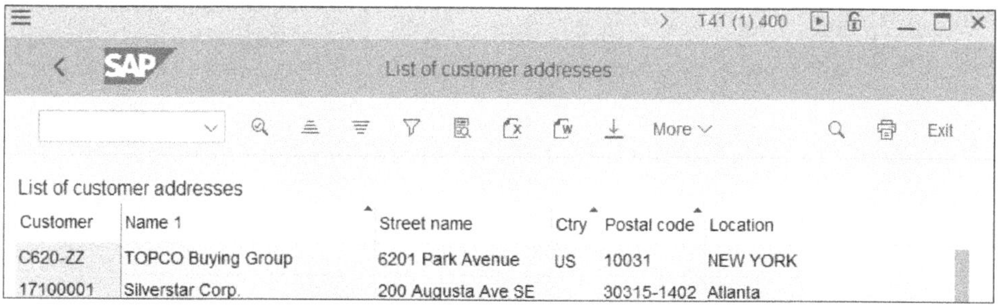

Figure 4.16 List of Customer Addresses

4.4.2   Sort by Columns

You can sort by a column by first selecting the sort column. Follow these steps:

1. In our example, click on the **Name 1** column heading to select this column.
2. Then to use an ascending sort, click ☰ (**Sort Ascending**), and the list is sorted. The sorting column is marked with a small red triangle right above the column header.

You'll get a descending sort after marking the corresponding column heading and clicking ☰ (**Sort Descending**).

4.4.3   Move Columns

The **Customer** column with the customer numbers should be on the far right. To move the column, use the drag and drop technique. Click on its column header—in our case, **Customer**—and drag it to the desired position while holding down the left mouse button. Once there, release the mouse button.

Figure 4.17 shows the current appearance of our list.

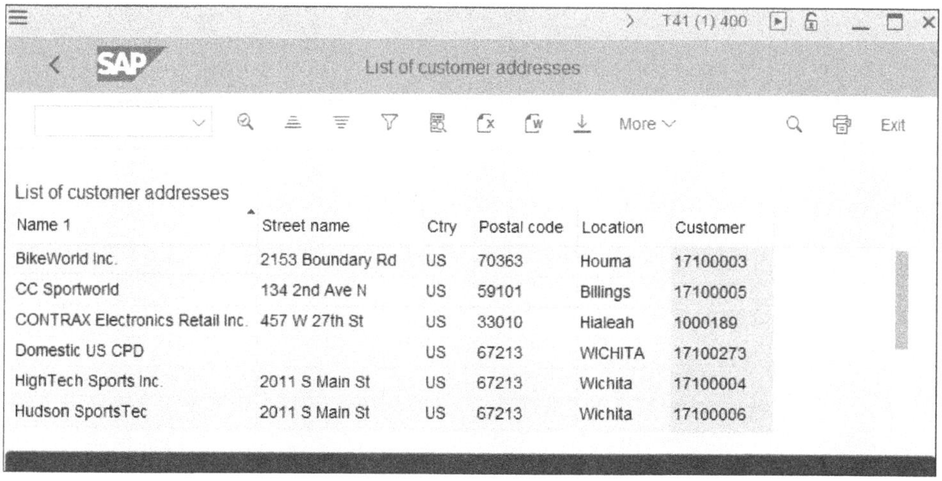

Figure 4.17 Customer Address List after Sorting and Moving a Column

4.4.4 Change the Column Width

Columns are often unnecessarily wide. Drag the separator between two column headers to the left to reduce the column width or to the right to widen the column.

Here's another little trick: double-click on the hyphen to get the optimal column width corresponding to the longest entry, just like in Microsoft Excel.

4.4.5 Hide Columns

To hide a column, right-click on its column header, in our case, **Ctry**. The context menu will open, as shown in Figure 4.18.

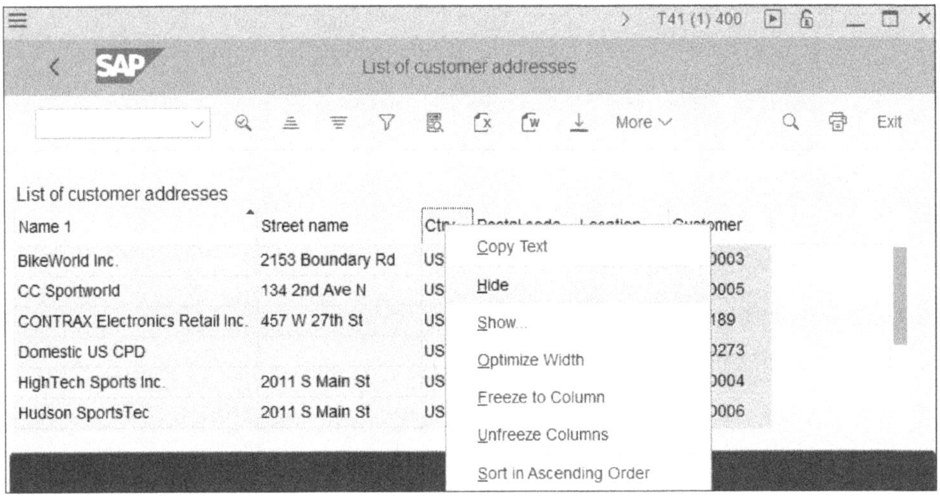

Figure 4.18 Context Menu for a Column Heading

In the context menu, select the **Hide** command. Now hide the **Street Name** column in the same way.

4.4.6 Show Columns

You can use this function to show a column again that you've previously hidden. In addition, in many reports, but not in our address list, there are many additional columns that are hidden by default and thus not visible. These often contain valuable additional information that is just waiting to be used by you. To show the hidden columns, follow these steps:

1. Right-click on any column header.
2. Select the **Show** command in the context menu, as shown in Figure 4.19. The new **Change Layout** window appears.

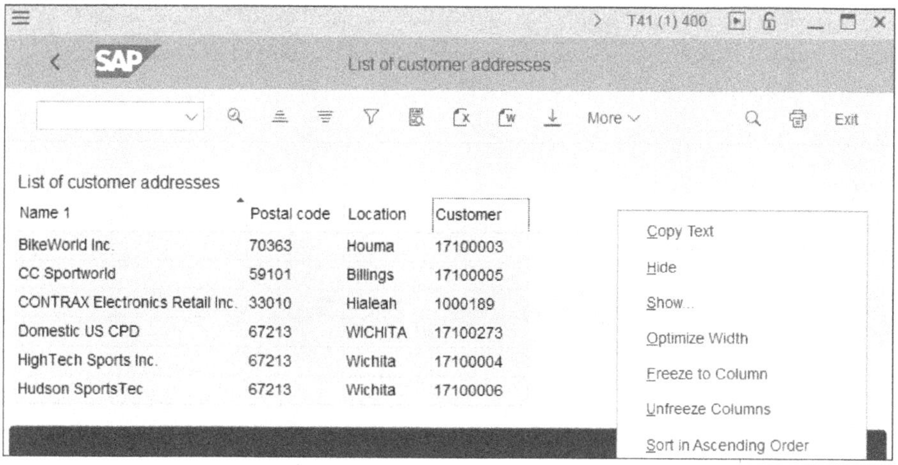

Figure 4.19 Show Command in the Context Menu

3. In the **Change Layout** window, you'll find five different tabs, as shown in Figure 4.20. To show and hide the columns, you only need the first tab, **Column Selection**, which has the following features:

   – **Displayed Columns**
   You can see the columns that are currently displayed in this list in the left part of the window.

   – **Column Set**
   In this right part of the window, you can see the hidden columns.

4. To show a column, select its name in the **Column Set** area by clicking on it, and then click ◁ in the center. After that, you can find the column name in the **Displayed Columns** area on the left side.

5. Use this method to also show the **Street Name** (or in this case, **Street and House Number**) column again.

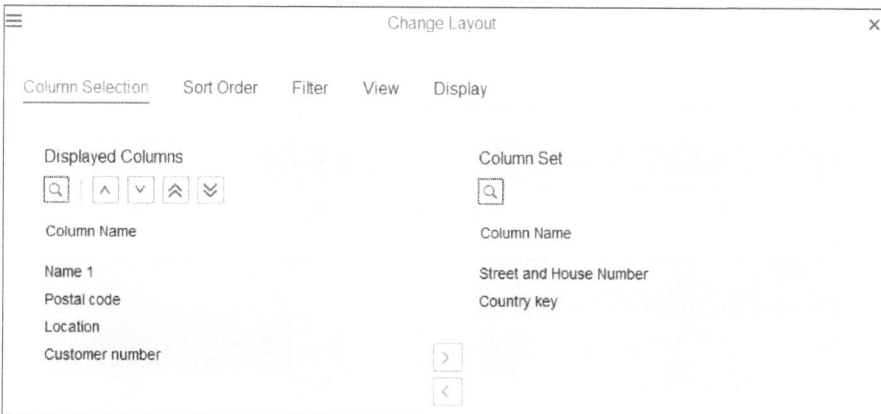

Figure 4.20 Change Layout with Columns

6. Activate your changes by clicking ✓ in the bottom-right corner. The **Change Layout** window will disappear, and you'll immediately see the changes in the list.

7. In the list, drag the **Street name** column before the **Customer** column.

[+]

Change the Column Order

While you're here, you can also change the order for all columns in the **Change Layout** window. To do this, click on the column name in the **Displayed Columns** area, and move it by clicking on ⋀ or ⋁ according to the desired order. Moving the column name also often works by dragging and dropping.

4.5   Save a Layout

As soon as you've changed layout settings and leave the list, the SAP system unfortunately forgets them. But, of course, you don't want to start from scratch every time you call up the list. That's why you now save your layout settings so you can quickly recall them when you need them. We'll discuss how in the following sections.

4.5.1   Save the Layout

To save your layout for future use, follow these steps:

1. Click in the list on 🔲 (**Save Layout**). In some other lists, this icon doesn't exist, so you need to **More • Save Layout** on the header line. Now the **Save As** window appears, as shown in Figure 4.21.

2. In the **Save As** window, layouts that have already been saved are displayed in the middle part of the screen, if there are any. In the lower part of the screen, you make the following settings:
   - Enter a short name in the **Layout** field; this must start with a letter for a user-specific layout.
   - Specify a long name in the **Name** field.
   - Check the **User-Specific** box because the layout is used only by you.
   - Check the **Default** box if you want your report to be displayed in this layout immediately after execution. We don't set a checkmark here for this example.

3. In the **Save As** window, press the ⌨Enter key. This completes the backup and saves your settings.

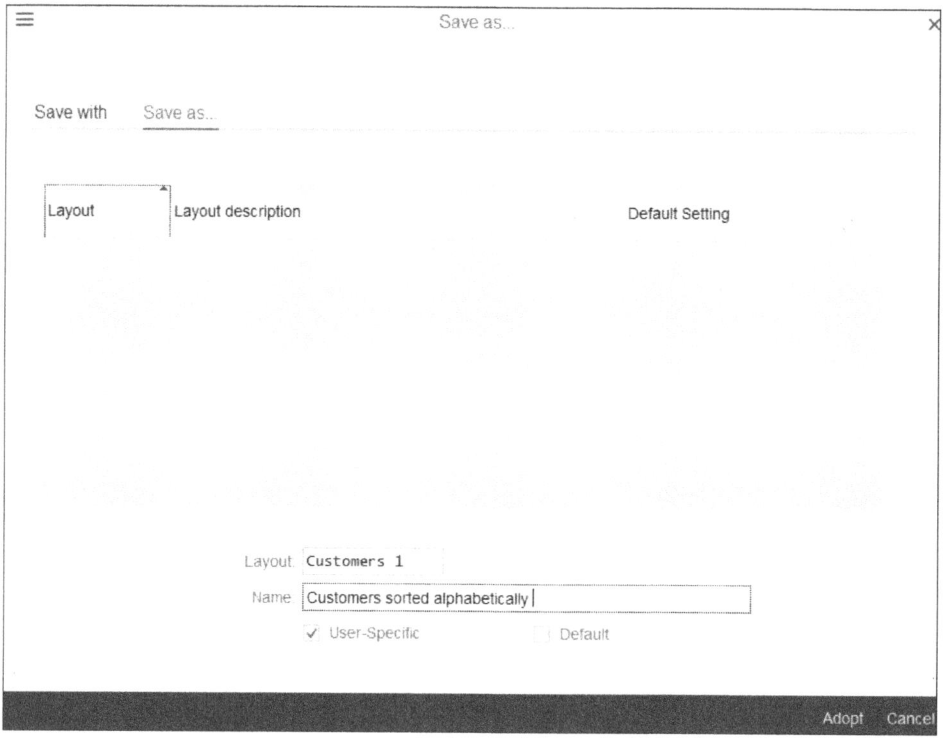

Figure 4.21 Saving the Layout

4.5.2   Create a Second Layout

For the instructions in Section 4.6, you need two different layouts, so you'll now change the first layout a little and then save it again. You can do this, for example, with ≦ (**Sort Ascending**) icon to sort by the **Postal Code** column and drag and drop it to the beginning of the list, as shown in Figure 4.22.

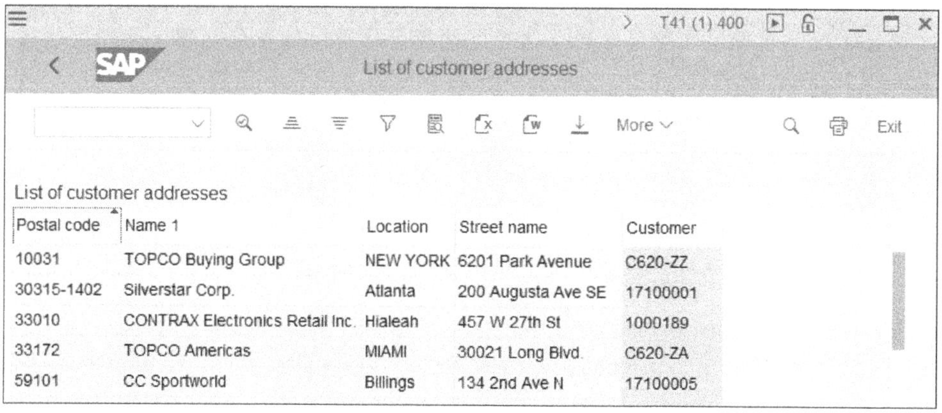

Figure 4.22 Our Address List after the Changes

After that, run the previous instructions again for user-specific saving, as shown in Figure 4.23. The only differences are as follows: You save the layout this time under the short name "CUSTOMERS 2" and with a different name. In addition, you activate the **Default** indicator in this round.

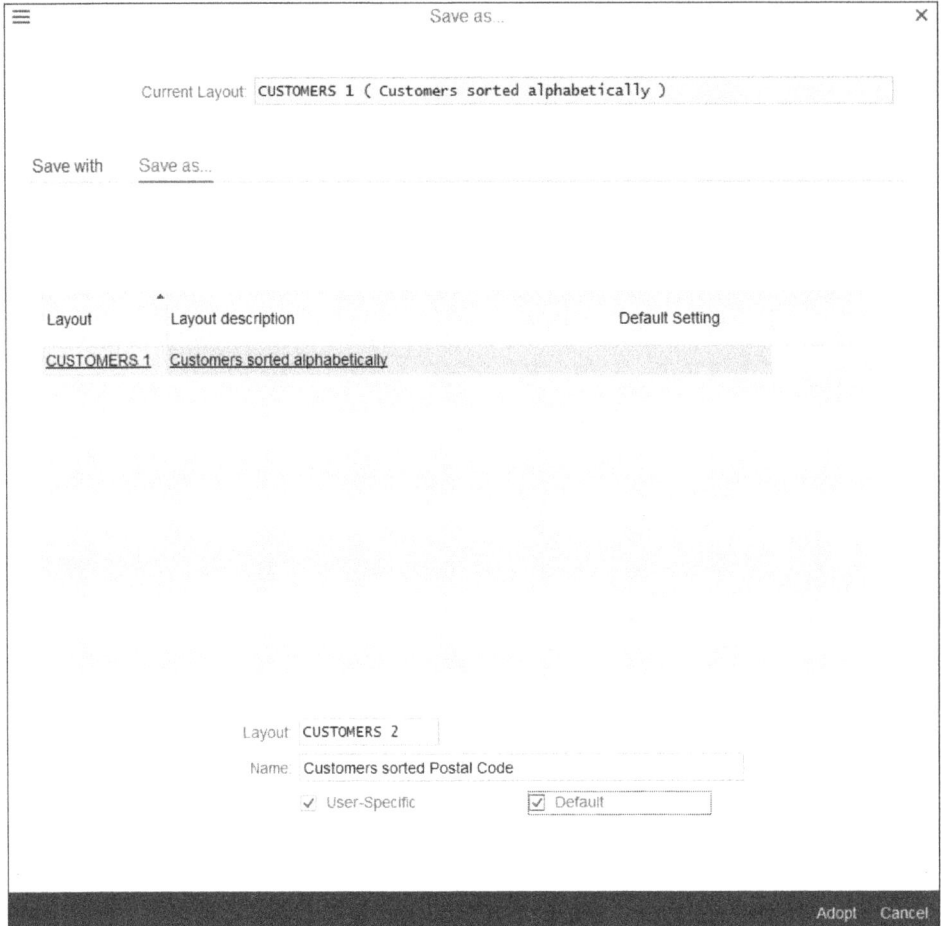

Figure 4.23 Settings When Saving the Second Layout Customer Zip Code

After you've pressed `Enter` or clicked ✅ **Adopt** at the bottom right, two layouts are now saved for your user.

4.5.3   Save a Global Layout

Here's another case from practice: You've created a fantastic layout, and a colleague comes over and wants to use your layout as well. However, this only works if you save the layout as a *global layout*. In Section 4.5.1, however, we still asked you to activate the **User-Specific** indicator.

In practice, users often don't have permission to save global layouts (another term: *standard layouts*). In this case, there is only one official way for your colleague to use your layout: a key user or application administrator must give you permission to save global layouts.

But just try to save the current layout additionally as a global layout. If you're in a training system, there is a good chance that you have permission to do this (if not, you'll get an error message). Follow these steps:

1. Click in the list on ▦ (**Save Layout**).

2. In the **Layout** field, as shown in Figure 4.24, give the short name, but this time put a slash in front of it (the character for the global layout): "/Standard".

3. Don't check the **User-Specific** and **Default** checkboxes.

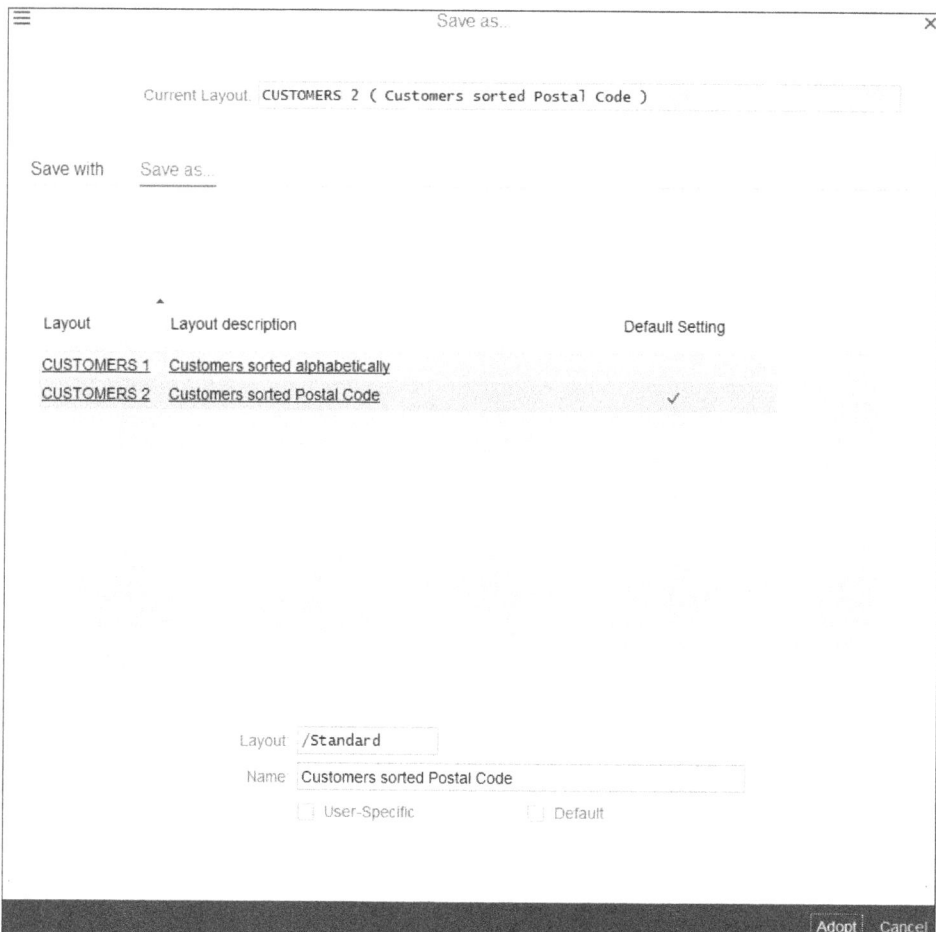

Figure 4.24 Settings When Saving the Third Layout /Standard

Press ⌈Enter⌉ or click on the ✅ **Adopt** button to save the global layout. Your colleague can then call it up in the usual way.

4.6   Call a Layout

Have you participated in our instructions so far? Then you have at least two saved layouts in the address list after the previous section. In this section, we'll show you how to change the layout.

Follow these steps:

1. In the address list, click ⊞ (**Select Layout**) in the header line. (In some reports, the ⊞ icon is missing. In this case, choose **More • Select Layout** from the header line, as shown in Figure 4.25, or try the key combination ⌈Ctrl⌉+⌈F9⌉).

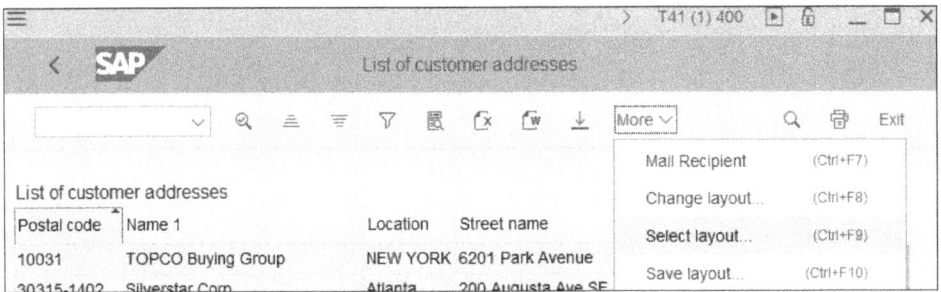

Figure 4.25 Choosing the Menu Path from the Address List

2. No matter which way you go, you'll get a list of layouts to choose from, as shown in Figure 4.26. In the **Choose Layout** window, you should first make sure that the layouts of all layout types are displayed. If **All** isn't entered in the topmost **Layout Setting** field, click ▼ to open the dropdown list, and then select the **All** entry. (We'll explain the layout types in more detail next.)

3. Click further down on the desired layout, for example, **/STANDARD**. This closes the **Choose Layout** window, and the display of the list is adjusted immediately.

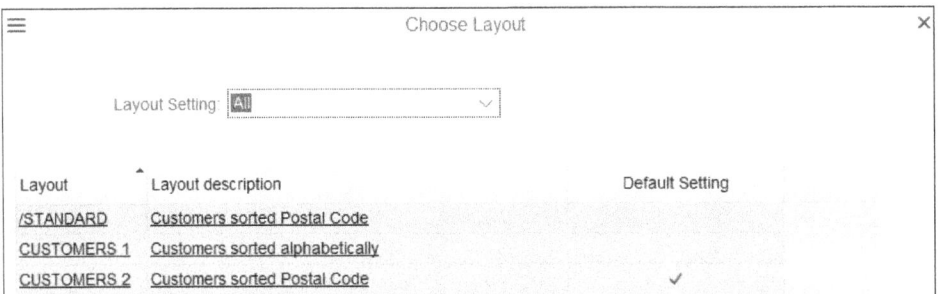

Figure 4.26 Choose Layout Screen

Now let's look at the layout types that are available for selection in the **Layout Setting** field. As mentioned, there are different layout types:

- **User-specific layouts**
  User-specific layouts (also called user layouts) are those you've created yourself and that can only be accessed with your user. Other users don't see these layouts. These layouts always start with a letter.

- **Global layouts**
  Global layouts are those that have been created in your company, for example, by a key user, and can be used by all employees. Every user sees these layouts. They always start with the character "/", for example, "/Standard".

- **SAP layouts**
  A special type of global layouts are *SAP layouts* (also called *SAP default display variants*). These are layouts created and supplied by SAP itself. There are more than 1,000 of them, but not in every report. The names of the SAP layouts always start with a number, for example, "1SAP".

4.7   Maintain Layouts

You've created different layouts? Let's learn how to change and delete them and how to change the preset.

4.7.1   Change the Default Layout

In Section 4.5, you learned that you can activate the **Default** indicator when saving a layout (see Figure 4.27). This causes the layout in question to automatically apply to your list, and it will appear in the desired layout immediately after execution. You should activate the **Default** checkbox for the layout you need most often.

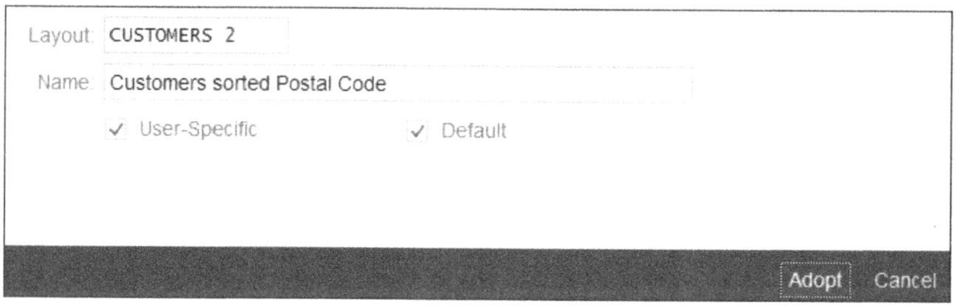

Figure 4.27 Default Checkbox in the Save As Window

In the following instructions, we'll show you the official procedure to change this default setting.

Follow these steps:

1. In the address list, from the header line, choose **More • Settings • Layout • Administration**, as shown in Figure 4.28. (In other lists use a different but similar path, for example, **More • Settings • Layout • Layout Management**.)

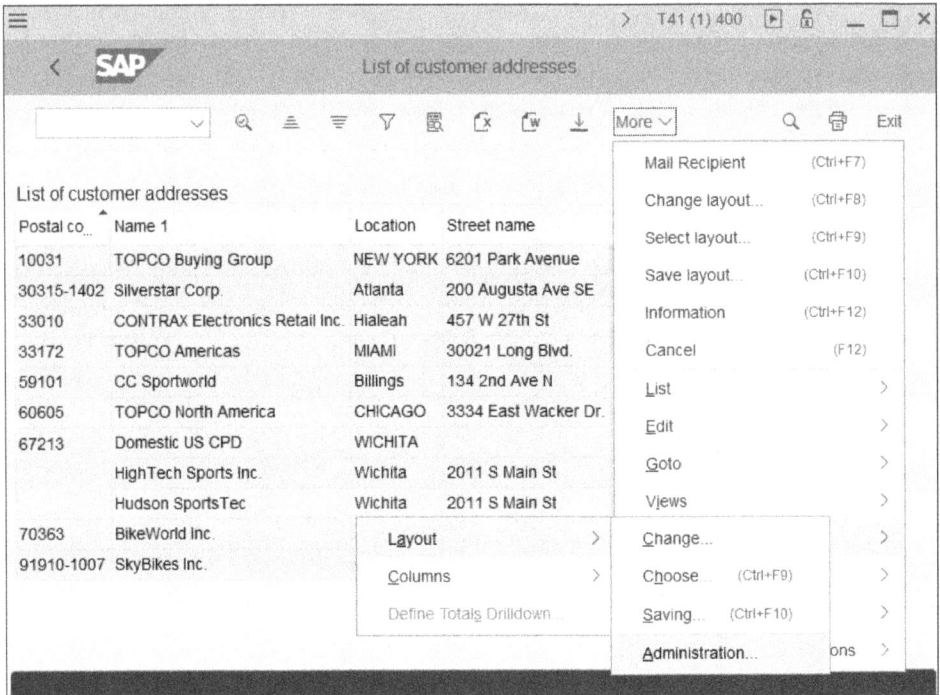

Figure 4.28 Navigating to Layout Administration

2. The **Layout: Management** window appears in all cases, as shown in Figure 4.29. Here, either only global layouts *or* user layouts are shown. If you see the **User Layout** button in the header row, click it.

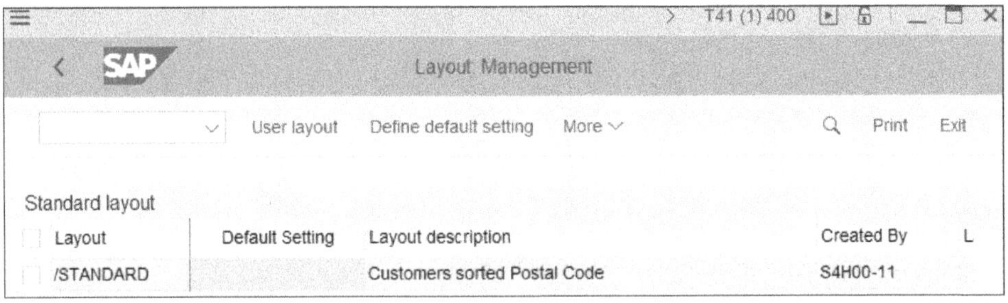

Figure 4.29 Layout: Management

This click displays your user layouts, and the button is now called **Standard Layout**. It's used to switch between standard and user layouts.

3. In the row with the desired layout, click in the green-colored field in the **Default Set-ting** column. A green checkmark will appear there, as shown in Figure 4.30.

4. Save your preset in the footer bar by clicking the **Save** button.

5. That's it, so exit the **Layout: Management** window by clicking (Back).

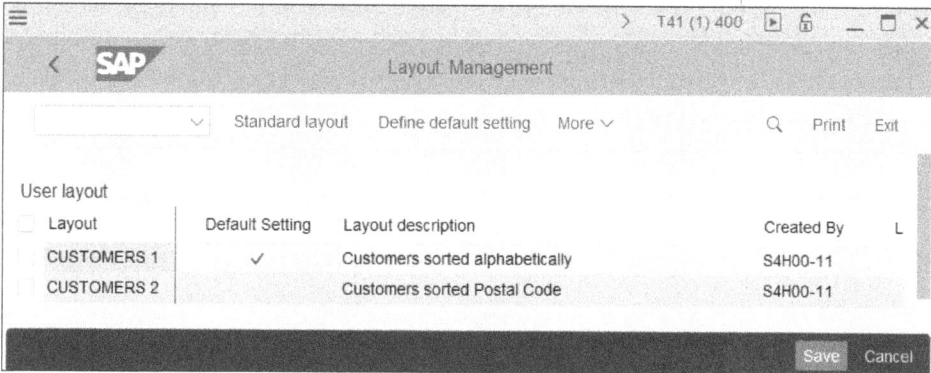

Figure 4.30 Selecting Default Setting

6. Now test your new preset by leaving the list with a click on (Back), and restart it in the selection screen by clicking the **Execute** button.

User Layout Trumps Standard Layout

Have you preset a default layout, and the SAP system ignores your setting? This may be because you can preset both in the **User Layout** view and in the **Standard Layout** view. So, you have two presets, but which one is valid? In such a case, the SAP system decides to use the preset user layout and ignores the preset standard layout. Eliminate the user layout preset if you want to have a standard layout preset.

4.7.2   Change a Layout

Do you need a different sort order or a different column order, for example? The proce-dure for changing a layout is the same as for changing a report variant. Follow these steps:

1. Call the layout.

2. Change the settings, for example, the sorting or the column order.

3. Save the changes under the same layout name. When doing so, confirm that the existing layout should be overwritten.

4.7.3   Change the Layout Name

There is no special function for changing layout names. But you're flexible and use the following method, which you already know from the names for report variants: You save the layout under the new name and then delete the layout with the old name.

4.7.4   Delete a Layout

Delete layouts that you no longer need so that the listing of layouts remains clear. You benefit from a clear selection window, especially when you quickly switch to another layout.

When deleting layouts, it's important that you don't accidentally delete a global layout (intentional deletion of these layouts is reserved for key users). But it can easily happen because there is a little trap in the process: there is no security question before the deletion is finally saved. Follow these steps:

1. In the ALV list, from the header line, select **More • Settings • Layout • Administration** or **More • Settings • Layout • Layout Management**. In both cases, the **Layout: Management** window appears, as shown in Figure 4.31.

2. In the **Layout: Management** window, only standard layouts may be listed at first. In this case, click the **User Layout** button in the application bar.

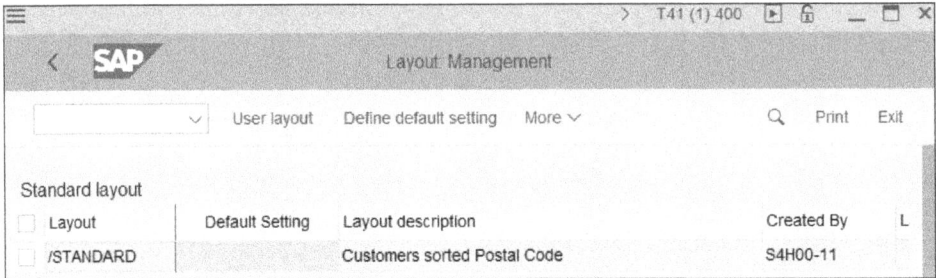

Figure 4.31 Layout: Management

3. Only your user-specific layouts will then be listed, as shown in Figure 4.32. Mark the layout you want to delete by clicking on its checkbox.

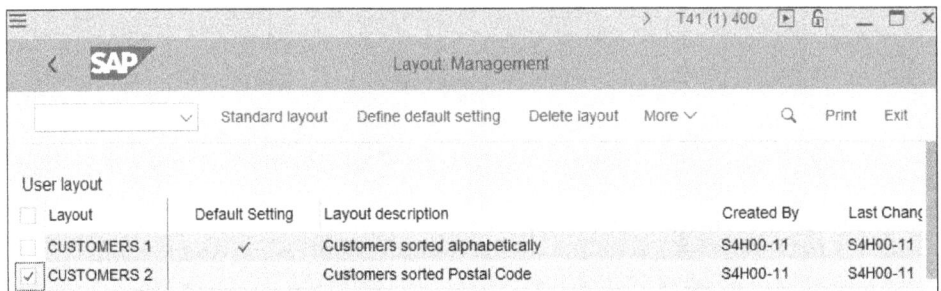

Figure 4.32 User-Specific Layouts

4. Kick the layout into the garbage can with a click on 🗑 (**Delete Layout**). The corresponding row will then be deleted from the table without further prompting, as shown in Figure 4.33.

Figure 4.33 User-Specific Layouts after Deletion

5. Save your changes by clicking the **Save** button in the footer line.

6. Exit the **Layout: Management** window by clicking ◁ (**Back**).

4.7.5   Overview of the SAP List Viewer Functions

There are even more functions for designing and using lists. You start most functions with the icons from the header line, as shown in Figure 4.34.

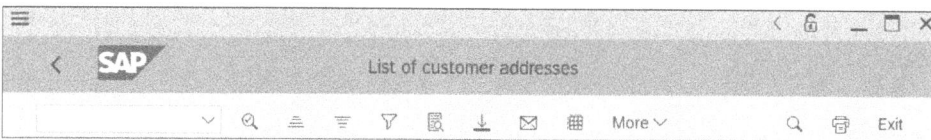

Figure 4.34 Header Line of the Address List with Functions for SAP List Viewer

In Table 4.1, you'll find an overview of the icons covered so far as well as other functions.

Icon	Meaning
◵	The **Details** icon shows details about the current line in the form of an index card. This function is only exciting for very wide lists, where not all information about a line fits on the screen.
≜≡	These icons sort in ascending or descending order according to a previously selected column.
▽	The **Filter** icon leads to a separate window where you can make additional selections without returning to the selection screen.
▦	The **Print Preview** icon displays the table as a print preview.
⬇	The **Local File** icon transfers the records from the PC to a local file, for example, a text file or the clipboard.

Table 4.1 Standard Functions in SAP List Viewer

Icon	Meaning
	The **Spreadsheet** icon transfers the records to Microsoft Excel.
	The **Mail Recipient** icon is used to send the displayed table as an email.
	The **Change Layout** icon leads to a separate window where you can show columns and perform various other layout changes such as multiple sorting.
	The **Select Layout** icon activates saved representations for the layout.
	The **Save Layout** icon saves settings such as sorting and column selection in a layout that can be activated later.
	This icon informs about the functions of SAP List Viewer.
	The **Refresh** icon updates the displayed data (see the upcoming text box).
	Of course, the icons **Sum** for grand total or **Subtotal** work only for numeric columns.
	To calculate a grand total, first select the numeric column, and then click on the $\sum$ icon (**Sum**).
	To calculate subtotals, first calculate a grand total. Then select the column for which the subtotals are to be calculated, and click (**Subtotal**).

Table 4.1 Standard Functions in SAP List Viewer (Cont.)

Don't be surprised if you call different lists: not every list has the full range of functions according to the preceding table, and vice versa; sometimes there are additional list-specific functions. The icons offered in the header line differ from report to report.

[Ex]

Why Refresh the Display?

Here is a small example from practice: Colleague Smith has opened the list of orders and is missing an order for an evaluation. She asks her colleague Brown to enter this order as quickly as possible, which he does. After half an hour, Smith looks at the list again and still doesn't see a new order. But why? Data isn't updated automatically, but only after clicking (**Refresh**).

However, in some lists, this icon or a corresponding menu function is missing. Then there is only one way to update a list: exit the list by clicking (**Back**), and call it up again with the **Execute** button.

Also note that you can sometimes find additional functions in the context menu and the **More** menu for your lists.

4.8 Summary

You've seen that your reports can be adapted very flexibly using report variants and layouts without the help of key users or programmers.

In the selection screen, you save the default values for reports in report variants.

In the list, the prefabricated standard reports often don't fit, so you sort by columns, move and hide columns, change the columns widths, or show additional columns. Then you save these settings as layout variants so you can quickly recall them when you need them.

Now, let's complete our discussion of the SAP GUI user interface with personalization options in the next chapter.

Chapter 5
Personalizing and Optimizing SAP GUI

So, you know the basics of the user interface and are ready to work with it if everything is set up optimally. If everything *is* set up optimally for you, that's great! Then you don't really need to read this part of the book. However, we've never seen everything perfectly set in practice because each user needs different settings to be able to work optimally.

This implementation of settings is also called *personalization*. An optimal personalization results from the sum of all personal *presets*. These are presets that are specific and exclusive to your user. It's like adjusting the seats and mirrors in your new car before you drive it for the first time, selecting the radio stations for the car radio, and adjusting the air pressure of the tires.

You made the following default settings in the first chapter on SAP GUI (see Chapter 1):

- Selected a theme and your favorite color
- Created favorites
- Set the display of the transaction codes in the **SAP Menu**
- Entered user data, such as output devices

In this chapter, you'll make other important presets that will help you work smoother, faster, and safer in practice.

What You'll Learn

- How to optimally set up the **Favorites** menu in the **SAP Easy Access** window
- How to set cross-transactional defaults as user parameters and save yourself lots and lots of typing
- How to reduce larger selection lists to personal value lists that contain exactly what you need in practice

---

In this chapter you hide the icons, as described in detail in Chapter 1, Section 1.2, by choosing **More • SAP GUI Settings and Actions • Options**. Open the **Visual Design** folder in the upper-left corner, click the **Theme Settings** entry, and deselect **Show Toolbar Buttons with Icons**. Open with **Str + + a** new SAP GUI window to see the result.

5.1   Optimize the Favorites Menu

In practice, many SAP users enter every transaction they need in their daily work as a favorite in the **SAP Easy Access** window the first time they use it. This is exactly what we recommend and why we've already shown you how to add some transactions to your **Favorites** menu in the first SAP GUI chapter. This will save you the time of searching for transactions in the **SAP Menu** or in any documentation later on.

However, over time, the transactions start to pile up, and then this menu quickly becomes confusing. Some people use 50 or more transactions! Therefore, in this section, you'll learn how to maintain order in the **Favorites** menu.

5.1.1   Create Folder

The first step to better order in the SAP **Favorites** menu is to create folders.

Follow these steps:

1. Switch to the **SAP Easy Access** window.

2. Right-click the top **Favorites** folder, and choose **Insert Folder** from the context menu, as shown in Figure 5.1.

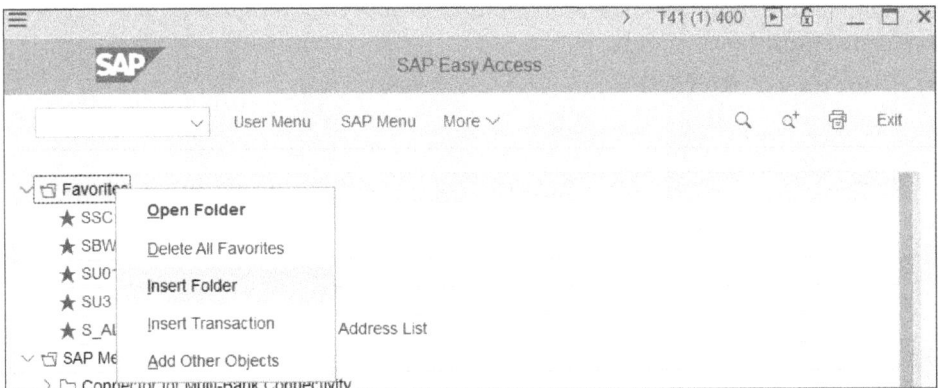

Figure 5.1 Choosing Insert Folder from the Context Menu

3. When the SAP system asks for the name of the folder, enter it in the **Folder Name** field, as shown in Figure 5.2, and press the `Enter` key. The new folder ☐ then appears immediately in the **Favorites** list.

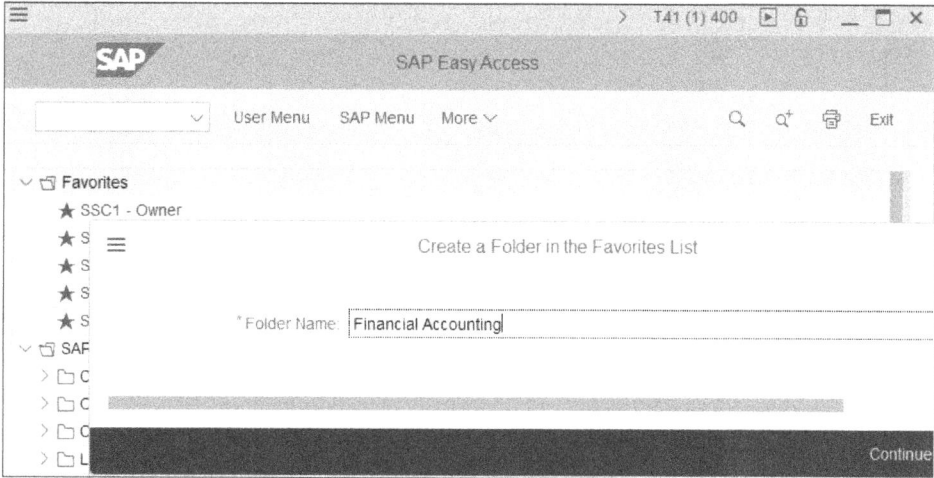

Figure 5.2 Entering the Name of the Folder

5.1.2   Move Favorites and Folders

Drag the entry to the new location via drag and drop with the left mouse button pressed. Dragging a favorite into a folder or dragging it out of a folder only works with drag and drop.

Don't Hide Favorites in the Folders

You make your own **Favorites** menu so you don't have to open and close folders. Avoid the temptation to drag all your favorites into folders because it's no fun and costs unnecessary time to open a folder every time you want to start a transaction. It's better to drag and drop the favorites themselves into a meaningful order.

5.1.3   Rename Favorites and Folders

In Chapter 3, Section 3.1.2, you added the transaction entry **S_ALR_87012180 – Address List** to your **Favorites** using a command from the context menu. However, you're not quite satisfied with the long text containing the path from the **SAP Menu**, so now you want to change it to "Customers: Address List". Follow these steps:

1. Right-click on the entry you want to rename.

2. Select the **Change Favorites** command from the context menu. The SAP system shows you the previous text in a window.

3. In the **Text** field, change the name of the favorite, as shown in Figure 5.3, and press ⎡Enter⎤ to confirm.

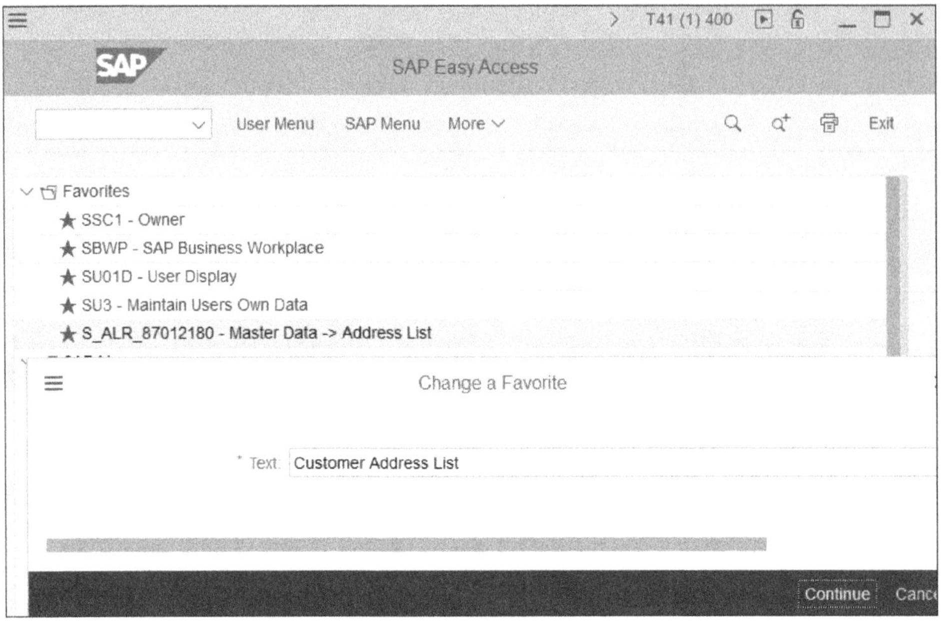

Figure 5.3 Renaming a Favorite

5.1.4   Create Web Addresses and Files as Favorites

You can also add web addresses and even links to individual files to the **Favorites** list. Adding a web address to the **Favorites** list goes like this:

1. Right-click to open the context menu for the **Favorites** folder.

2. Select the **Add Other Objects** option from the context menu, as shown in Figure 5.4.

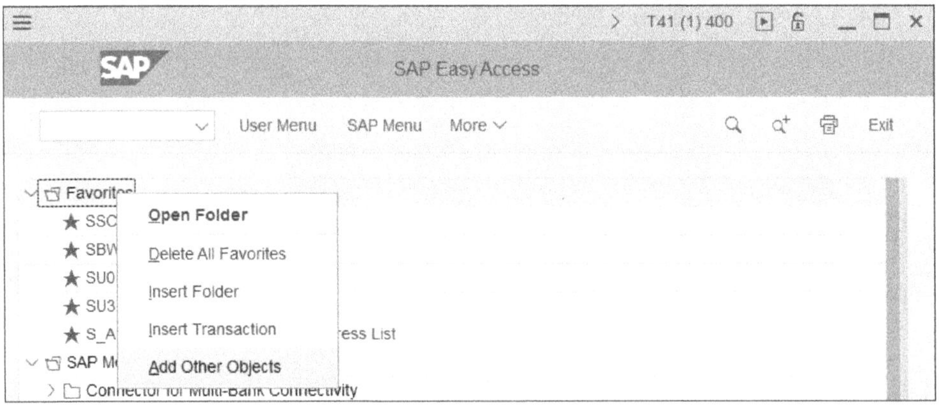

Figure 5.4 Add Other Object from the Context Menu

3. The SAP system shows the **Add Additional Objects** window (see Figure 5.5). Here, the topmost selection field, **Web Address or File**, is already selected. Confirm this with `Enter` or double-click on this field.

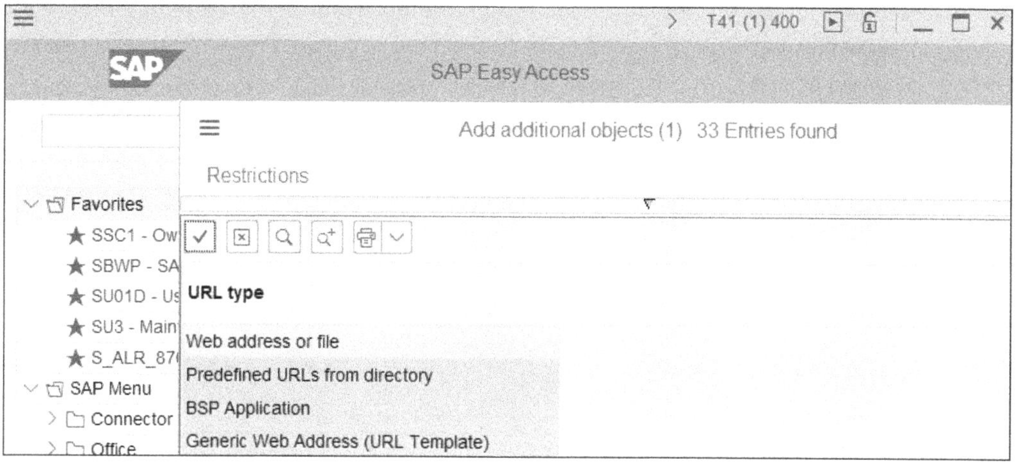

Figure 5.5 Inserting Additional Objects

4. Now the **Add a Web Address or File Path** window appears, as shown in Figure 5.6. Enter a name for the new favorite in the **Text** field. In the second field, enter the web address (URL), for example, "www.sap.de", or simply copy it from the internet browser via the clipboard.

5. Press `Enter`, and the new entry `URL - SAP Homepage` appears in the **Favorites** list.

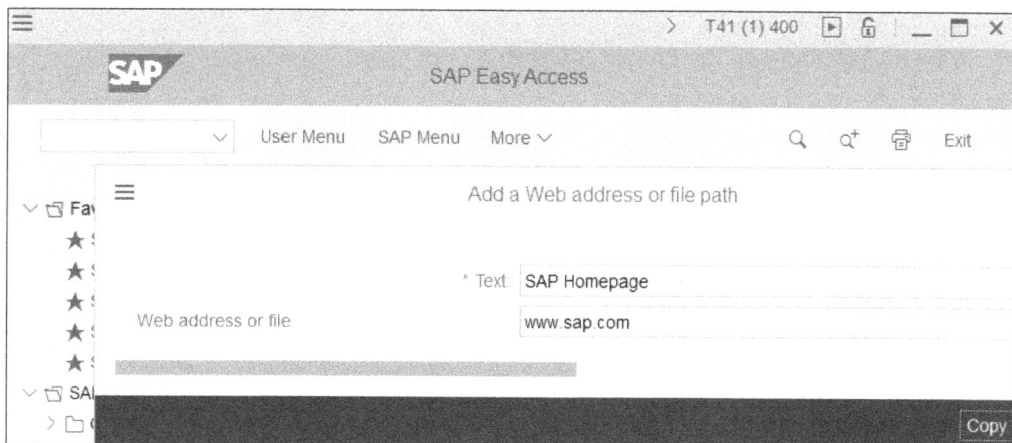

Figure 5.6 Adding a Web Address or File Path

If you've created a web address as a favorite, after double-clicking on it and confirming if necessary, your default web browser is started (e.g., Microsoft Edge). It immediately shows you the corresponding website in a new window.

[»]

How Do I Link a File?

In step 4, you can also enter a path to a file, for example, "C:\Documents and Settings\ User\Addresses.doc". But typing is no fun, is it? It's better to use 🔲 (**Value Help**) to the right of the field. Now you can double-click on the file in Windows Explorer. The rest of the previous instructions work identically for the files. A double-click on this favorite starts the corresponding program and opens the associated document. For example, a favorite containing a Microsoft Word file with the extension *.docx* will be opened with Microsoft Word.

As soon as you rename or delete the file or its Windows folder, or move the file to another Windows folder, the favorite no longer works. It only saves the name of the file and the respective path.

5.1.5   Delete Favorites and Folders

Now, let's quickly cover how to delete favorites and folders. Follow these steps:

1. Select the favorite entry to be deleted by clicking it; in our case, click **SSC1 - Own**.
2. In the header line, press the [Del] key.

A favorite or an empty folder is deleted immediately. A confirmation prompt appears only when a folder contains favorites.

5.1.6   Functions of the Favorites List

In Table 5.1, you'll again find the functions with which you maintain your **Favorites** list.

Function	How to Use
Create favorites	• Right-click on the **Favorites** folder, and then choose **Insert Transaction**. • Drag and drop from the **SAP Menu** into the **Favorites** list. • Right-click on the transaction in the **SAP Menu**, and then select the **Add to Favorites** menu command.
Create folder	Right-click on the **Favorites** folder, and then choose **Insert Folder**.
Rename entries	Right-click on the favorite entry, and then choose **Change Favorites**.
Delete entries	Right-click on the favorite entry, and then choose **Delete Favorites**.

Table 5.1 Functions for Maintaining the Favorites List

Function	How to Use
Move entries	<ul><li>Click on either ☑ or ⋀ in the header line.</li><li>Drag and drop the entry into the **Favorites** list.</li></ul>
Create link to web address or file	Right-click on the **Favorites** folder, choose **Add Other Objects**, and then double-click on the **Web address or file** entry.

Table 5.1 Functions for Maintaining the Favorites List (Cont.)

The example of creating the **Favorites** menu shows that you can set up SAP GUI like your living room. Take advantage of these possibilities; it makes working with the SAP system more fun—just like living in a nicely furnished living room.

5.2   Save Typing with Default Values (User Parameters)

Do you enter the same numbers in different transactions every day over and over again? Wouldn't it be nice if the SAP system filled in these fields for you? Especially with organizational units such as company code, purchasing organization, plant, or sales organization, the entries are repetitive.

This section shows how you can decisively reduce your typing effort with *default values*. It's thus the best prevention against carpal tunnel syndrome and tendonitis. Take advantage of the option to define central default values. They are called *parameters* in the SAP system and are unfortunately used far too rarely in practice.

But you can do it better: you now specify your organizational units and other frequently needed field entries once as central default values, so you don't have to enter them in many transactions. The default values apply permanently and specifically to your user. The nice thing is that they *basically* apply across all transactions with the corresponding field. For this purpose, here are some representative examples from different SAP modules:

- In materials management, you often need the same plant and storage location when displaying material stocks (Transaction MMBE) and in many other transactions.
- In sales and distribution, on the initial screen for entering inquiries (Transaction VA11), quotations (Transaction VA21), orders (Transaction VA01), and many other transactions, you always enter the same sales organization, distribution channel, and division.
- In accounting, many users work with only one company code. You specify this company code in the item displays (Transactions FAGLL03, FBL1N, FBL3N, FBL5N), in the balance display (Transactions FAGLB03, FD10N, FK10N, FS10N), and in many other transactions.

5.2.1 Define Default Values

No matter which SAP module you work with, the method for defining central default values is always the same. We'll use the already known Transaction SU3 (Maintain User Own Data).

Follow these steps:

1. To call the transaction, double-click the **SU3 - Maintain User Own Data** entry in your **Favorites**, or enter the code "/nSU3" in the command field, as shown in Figure 5.7, and press `Enter`.

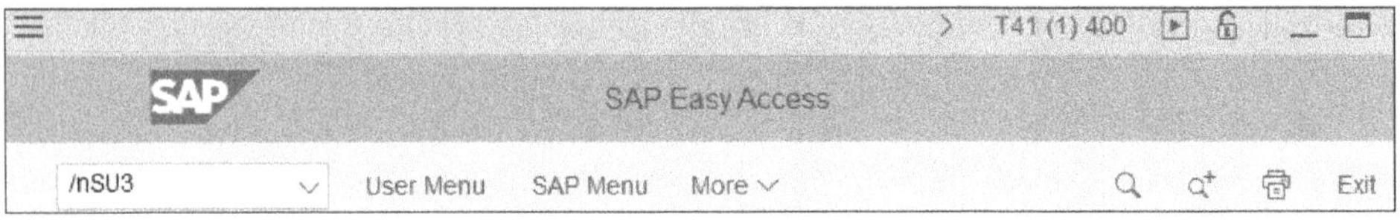

Figure 5.7 Calling Transaction SU3

2. The **Maintain User Profile** window will appear. Click the right tab labeled **Parameters**.

3. On the **Parameters** tab, click in the first free field in the **SET/GET Parameter ID** column on the left. Enter the parameter ID "BUK" here, as shown in Figure 5.8.

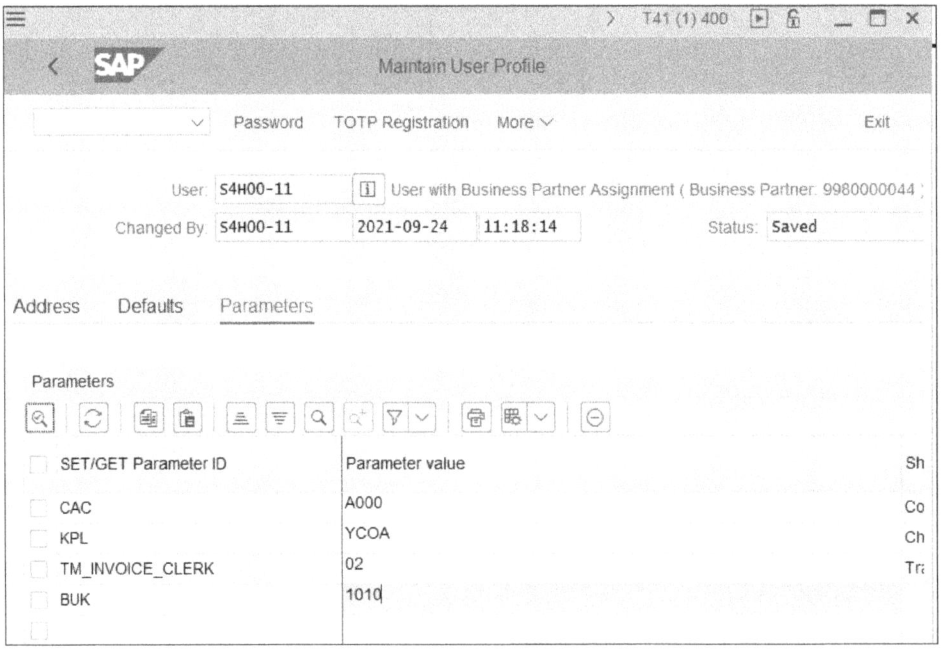

Figure 5.8 Entering the Parameter ID

4. Enter the desired default value in the **Parameter Value** column to the right (in our case, "1010").

5. Press the ⌜Enter⌝ key. Now the SAP system displays the meaning of the parameter ID in the **Short Description** column.

6. Repeat steps 3 and 4 until you've defined the desired default values. Use the parameter IDs from Figure 5.9. (In Section 5.2.3, we'll show you how to find the parameter IDs yourself.)

7. Save your entries by clicking on the **Save** button in the footer line at the bottom right, as shown in Figure 5.9. You'll then find yourself in the **SAP Easy Access** window.

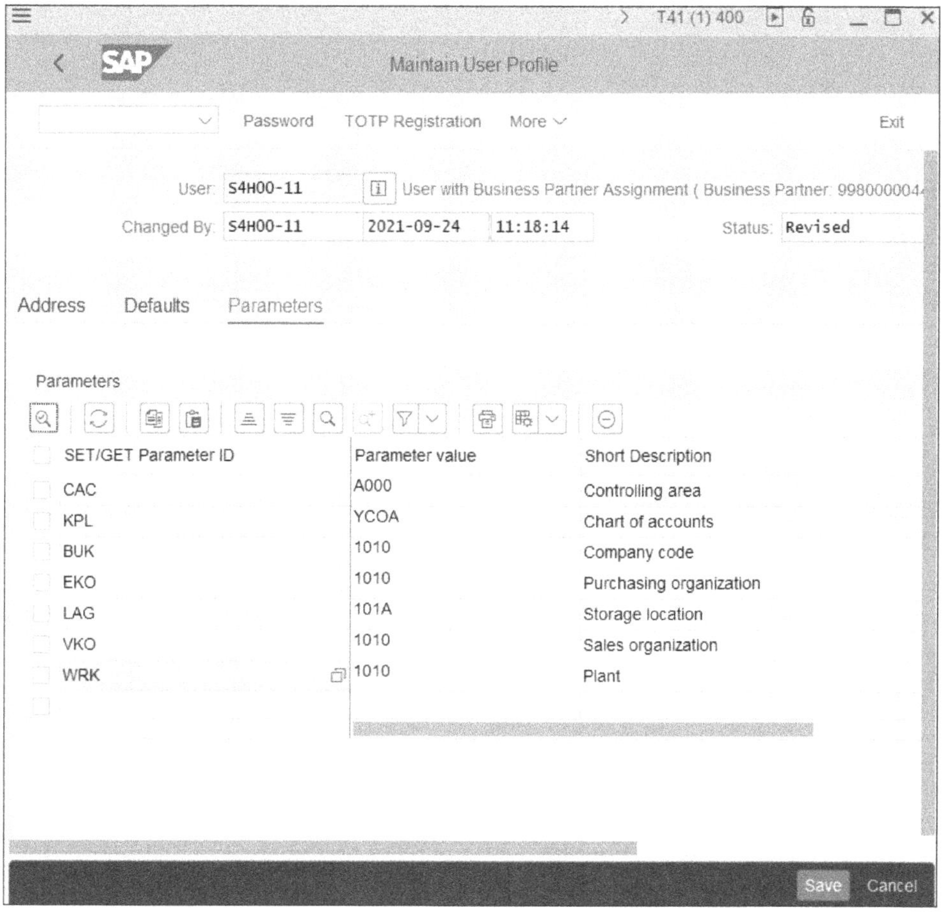

Figure 5.9 Parameters

5.2.2   Test Default Values

Seeing is believing, so test the new settings now. To do this, stay in the same SAP window. The test doesn't work in previously opened windows because they don't yet know your new parameter settings. Follow these steps:

1. Start Transaction MMBE (Stock Overview) by entering "/nMMBE" in the command field. In the initial screen of this transaction, as shown in Figure 5.10, you'll now find the fields **Plant** and **Storage Location** already filled.

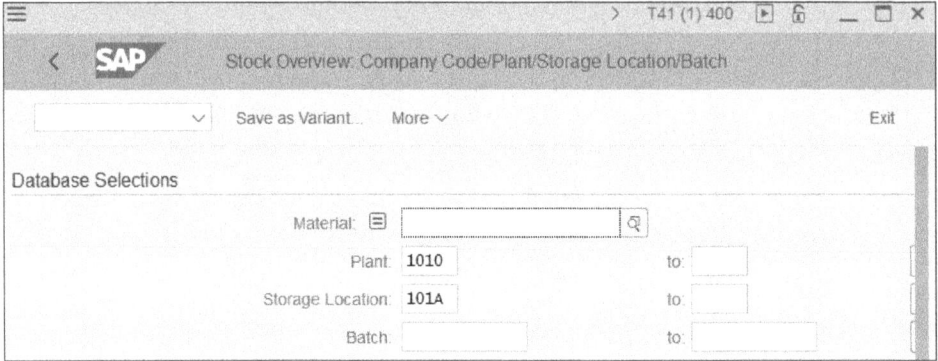

Figure 5.10 Default Values for Stock Overview

2. If you enter a **Material** number above it and click the **Execute** button at the bottom right, you'll see the stocks of the material. To search for the material number, you can use F4 or Q (**Value Help**) after clicking in the **Material** field.

[!]
The Exceptions to the Rule

We have good news and bad news for you. First the good news: the default values also apply in the "fiorized" transactions in SAP Fiori. Now, the bad news: At the beginning of the section, we had written that the default values *basically* apply across the board to all transactions with the corresponding field. However, *basically* means that there are exceptions, and the top SAP GUI rule "no rule without exceptions" applies.

Don't be disappointed if the user parameters don't work everywhere. Especially in the selection screens of many reports, you'll often not find any default values. For these selection screens, however, there is an even more powerful function for defining default values: *report variants*. We covered this topic in Chapter 4.

5.2.3 Find Parameter IDs with the Technical Info

There are more than 10,000 different fields in the SAP system for which you can set default values to save yourself unnecessary typing effort. You're probably wondering how to find the required parameter IDs, for example, BUK, WRK, or LAG. You might think the obvious answer is with F4 or 🗇 (**Value Help**). Unfortunately, this is wrong. If you search for a parameter ID with F4 or by clicking on 🗇, you often find yourself at a loss in front of a confusing list. As an example, we've searched for a parameter ID with the search specification "Plant" and get 15 entries, as shown in Figure 5.11. That is too many results because you don't know which is the "normal" parameter ID for the **Plant** field.

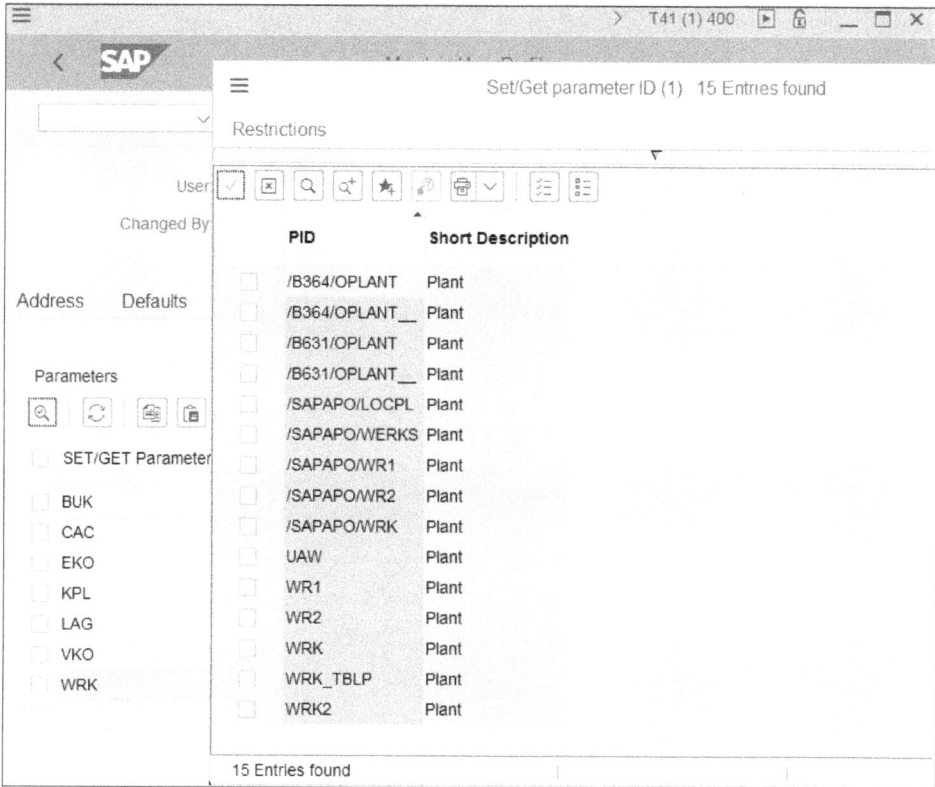

Figure 5.11 List of Values of the Parameter IDs for the Plant Short Description

In this case, there is a better method. The royal road to finding a parameter ID isn't via the input help, but via the technical info, which is called using the *field help* (F1 help). Although this way seems to be a little bit more cumbersome, it almost always leads to success:

1. Start the transaction that contains the desired field, for example, Transaction MMBE (Stock Overview) with the **Plant** field.

2. On the initial screen, place the cursor in the field whose parameter ID you want to determine. In our example, click in the **Plant** field, as shown in Figure 5.12.

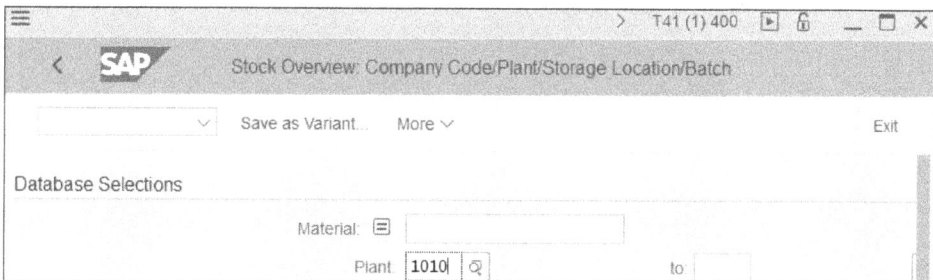

Figure 5.12 Transaction MMBE (Stock Overview) with the Plant Field

3. Press ☐F1☐. In the **Performance Assistant** window, as shown in Figure 5.13, the field help appears with an explanation of the field. Click 🔍 (**Technical Info**). The **Technical Information** window appears.

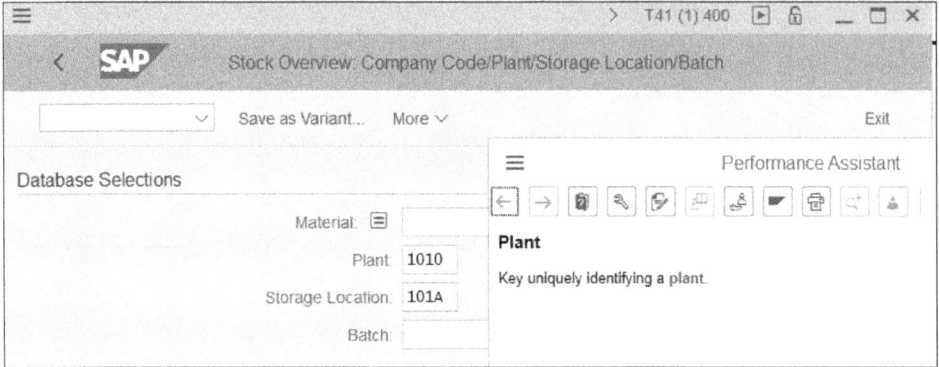

Figure 5.13 Performance Assistant

4. In the **Technical Information** window, as shown in Figure 5.14, you can see (at the very bottom) that our field has the parameter ID **WRK**.

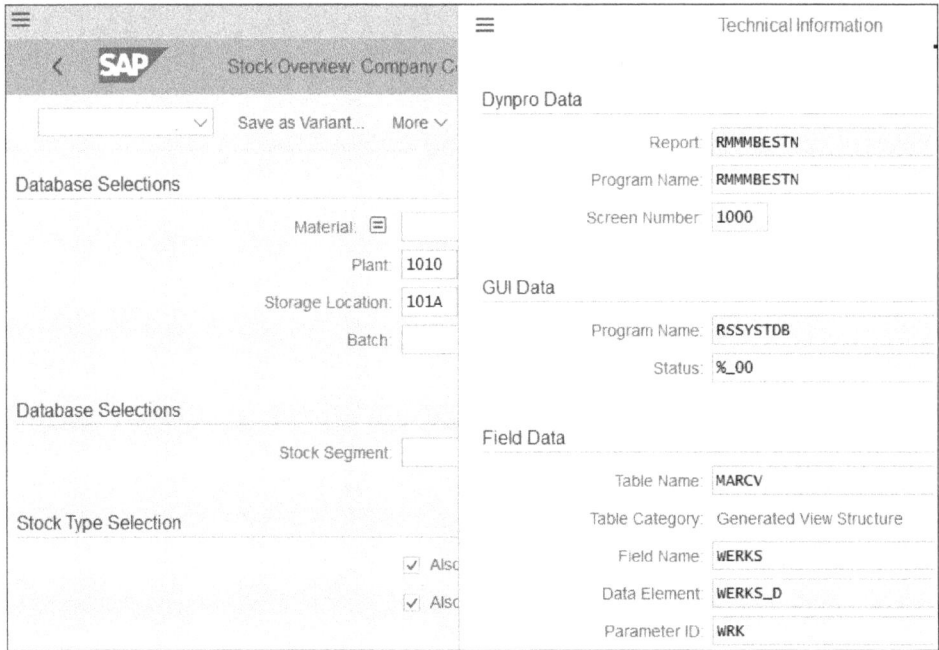

Figure 5.14 Technical Information

There are also fields for which there is no parameter ID. In these cases, the corresponding **Parameter ID** field is missing in the **Technical Information** window.

5. Close the windows with the technical information and the field help by clicking ☒ (**Close**).

Now you know the ultimate method to find the parameter ID to define the default value in Transaction SU3 (Maintain User Own Data), as described in Section 5.2.1.

[+]   5

Default Values for Individual Transactions

In addition to the parameter IDs, it's possible to set *transaction-specific default values* in some transactions. These include frequently used transactions such as Transaction ME21N (Create Purchase Order), Transaction MIGO (Goods Movement), or posting transactions in financial accounting such as Transaction FB50 (G/L Account Posting Single Screen Transaction), Transaction FB60 (Enter Incoming Invoices), or Transaction FB70 (Enter Outgoing Invoices). Sometimes, there is a menu command called **Settings** or **Personal Settings** in the **More** menu for this, or a button called **Editing Options** or **Personal Settings**.

5.3   Personal Value Lists

Whether it's material groups for purchase orders, order types in sales orders, or general ledger accounts in accounting, from large lists of values with several hundred entries, you often only need one or two dozen entries. So, let's get rid of the unnecessary ballast! You can replace a list that appears by default with your own list that contains only the values you need. This is then your *personal value list*. We'll discuss how to compile and display a personal value list in this section.

5.3.1   Compile a List of Values

In this example, you want to display only the countries of the Caribbean islands in the list of values of the countries, instead of the more than 200 countries that are created in the SAP system.

Follow these steps to compile your list of values:

1. Start any transaction that contains the **Country** field. For this example, start Transaction SU3 (Maintain User Own Data) by entering "/nSU3" in the command field and then confirming with ⌨ Enter .

2. In the **Address** tab shown in Figure 5.15, in the **Person** field group, click ⊞ (**More Fields**) on the far right of the screen.

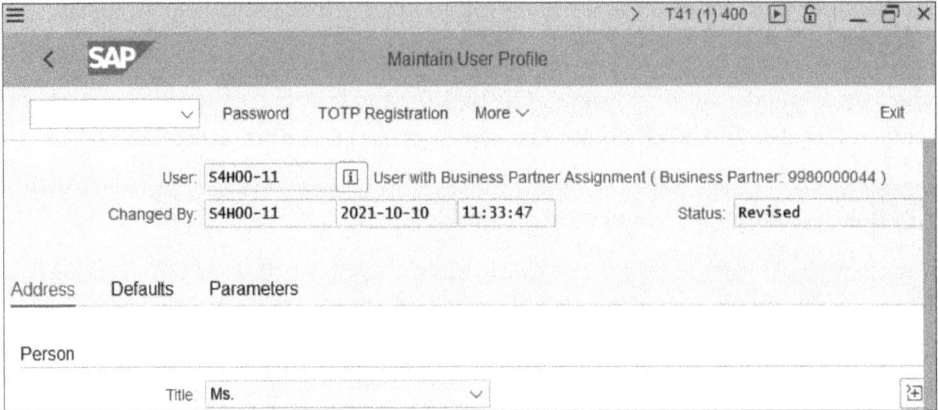

Figure 5.15 Address Tab

3. Click in the **Format Country** field as shown in Figure 5.16, and press `F4`, or click 🗗 (**Value Help**). The SAP system shows all countries in the value list (see Figure 5.17).

4. Click the first country you want to add to your personal list of values to select it, and then click ⭐ (**Insert into Personal List**).

5. Repeat the previous step until you've transferred all the desired countries to the personal value list. This creates this value list, but it's not yet displayed.

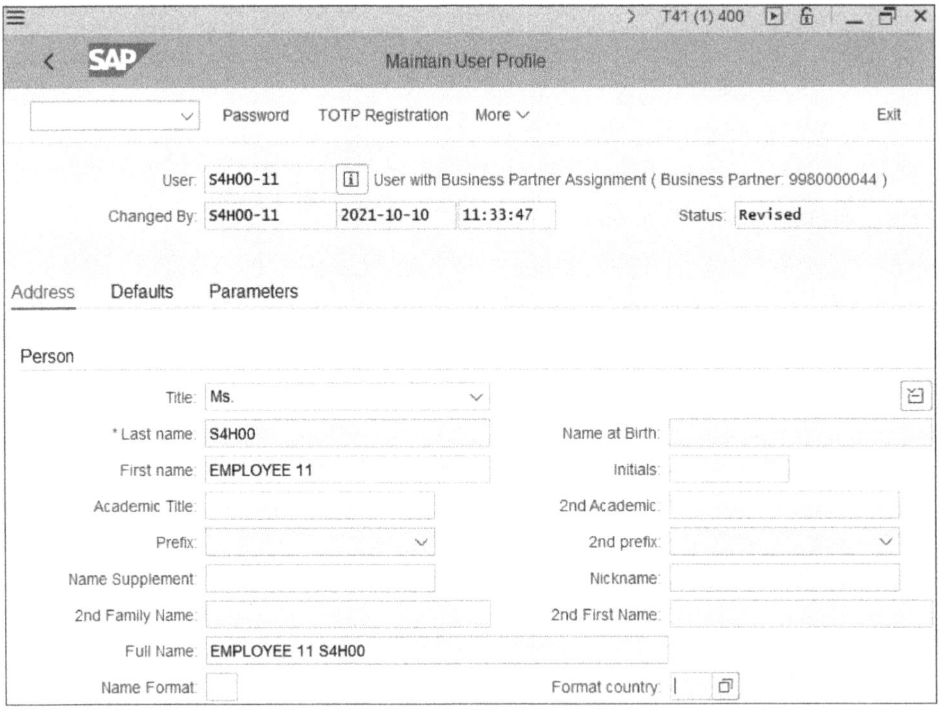

Figure 5.16 More Options for Person

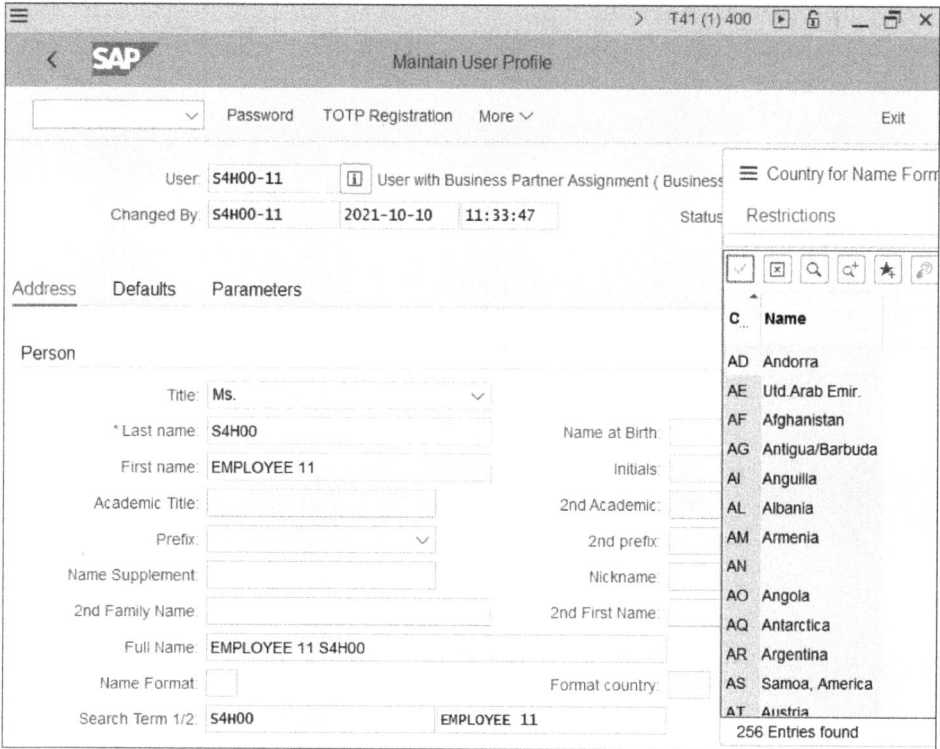

Figure 5.17 Value List for Countries

6. Click [?] (**Personal Value List**) to display the personal list of values, as shown in Figure 5.18.

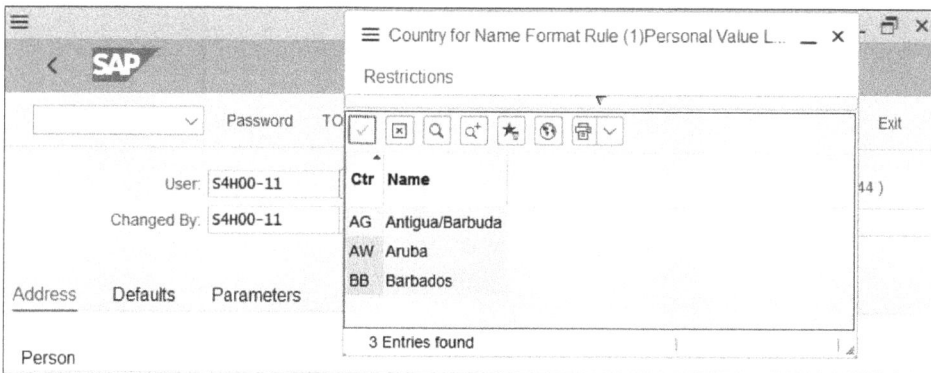

Figure 5.18 Personal List of Values

7. Close the value list window, and exit the transaction by clicking the **Exit** button.

Another bit of good news is that in the personal value lists, you have all the options that are also available to you in normal value lists.

And we have even *more* good news: the personal value lists apply across transactions, so it doesn't matter which transaction you're in. If you open the input help in the same field in another transaction, the personal value list you created will always appear.

5.3.2   Display the Personal Value List

Now, we'll try displaying a personal value list in Transaction BP (Business Partner), which you use to manage customers and suppliers. Follow these steps:

1. Start Transaction BP (Business Partner) by entering "/nBP" in the command field and pressing Enter to confirm. You'll arrive at the screen shown in Figure 5.19.

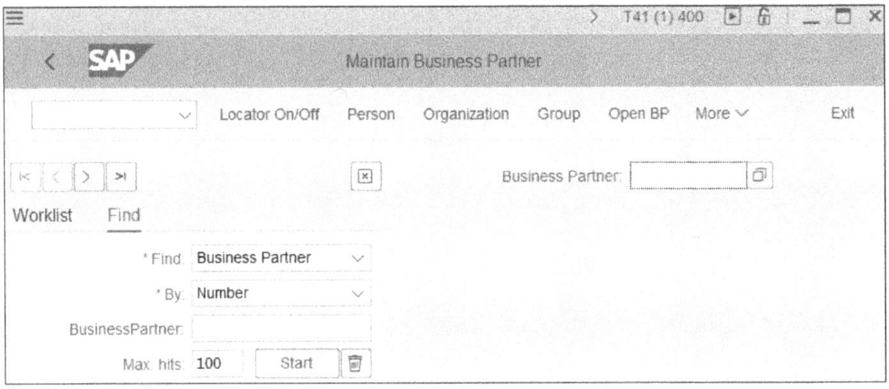

Figure 5.19 Maintain Business Partner Initial Screen

2. Click on the **Start** button. Now the business partners are listed in a table, as shown in Figure 5.20.

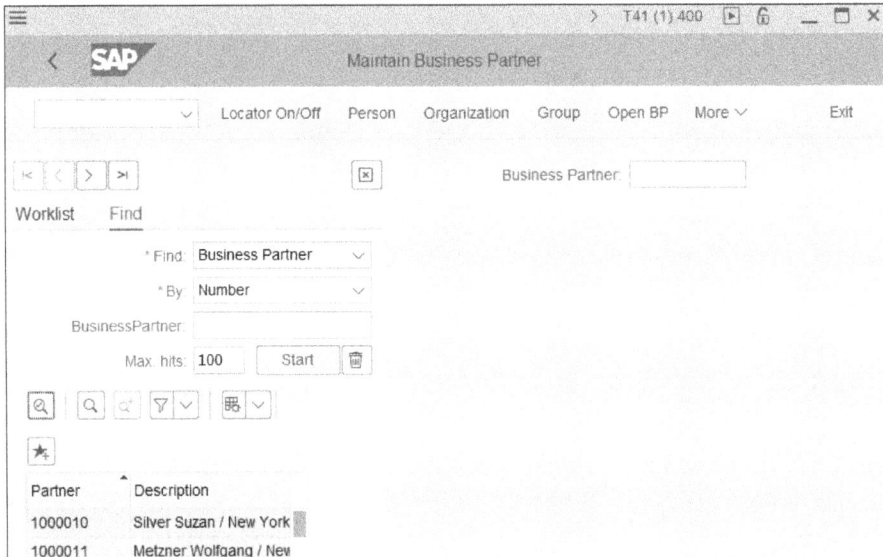

Figure 5.20 Table for Business Partners

3. Double-click on a business partner in the table. This will display the address data of the business partner, as shown in Figure 5.21.

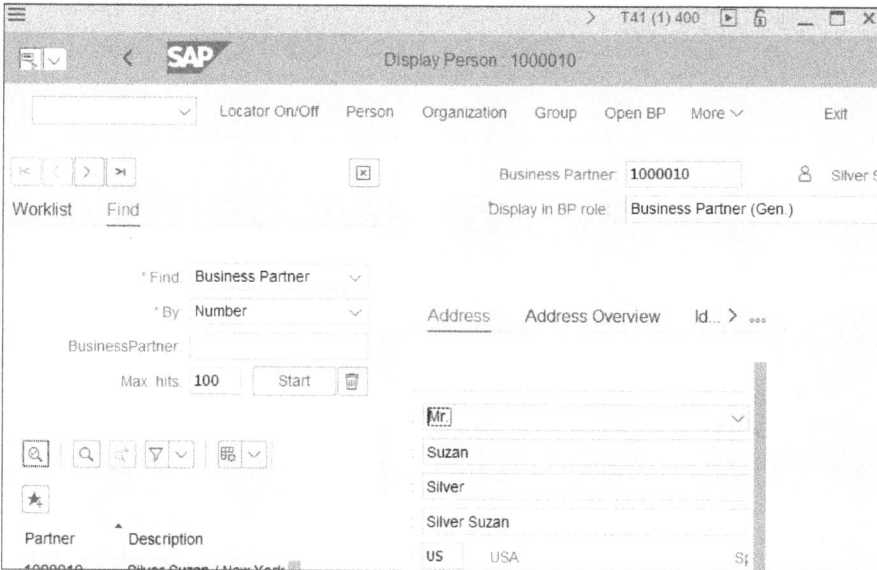

Figure 5.21 Displaying Business Partner Data

4. Further down in the **Street Address** field group, click in the **Country** field. Press the F4 key here, or click ⬚ (**Value Help**). The SAP system immediately displays the personal value list if there is a default setting, as shown in Figure 5.22.

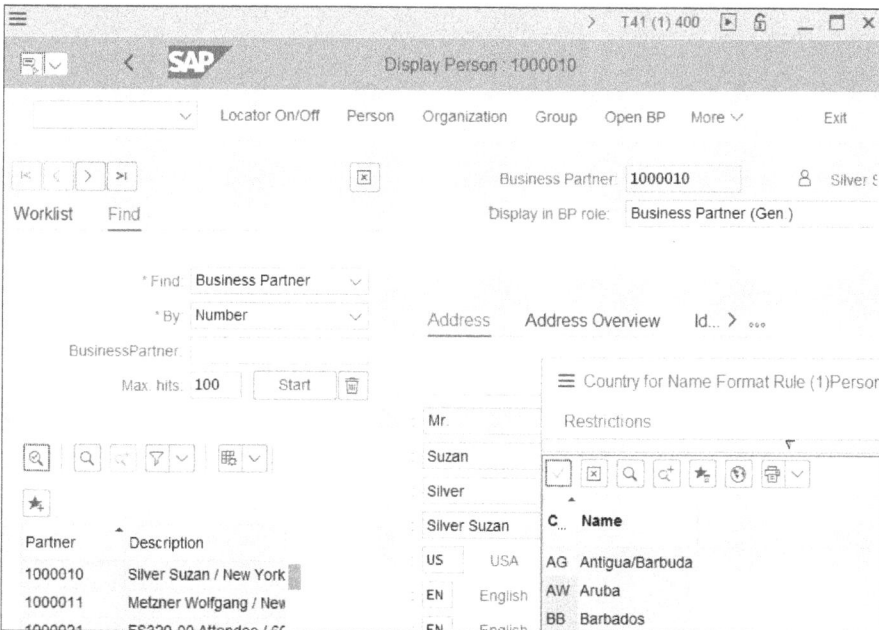

Figure 5.22 Displaying the Personal Values List

5. Do you need the complete list again? No problem! Just click 🌐 (**Show All Values**).

6. After that, close the value list window by pressing the ⌷Esc⌷ key, and exit the transaction by clicking the **Exit** button.

[»]

Delete Superfluous Values

As we've said before, nothing is as constant as change. Even personal value lists sometimes need to be updated. If you no longer need a value, first click on this value, and then click ★ (**Delete from Personal List**).

[+]

When Is a Personal Value List Worthwhile?

Of course, setting up such personal value lists requires some work. However, this effort is worthwhile for input fields that you use often and for which the value lists don't have to be constantly maintained. These are usually fields with master data or organizational units that aren't changed so often. Examples are cost centers, general ledger accounts, units of measure, production lines, storage locations, or personnel subareas. Using a personal value list in these instances saves you from searching and scrolling in long lists.

If the number of values is small, there is a simple alternative to the personal value list: the *input history*, which you get by pressing the ⌷Space bar⌷ or the ⌷Backspace⌷ key in an empty field.

5.4   Summary

You've now learned some good methods for saving time: starting transactions faster using an optimized **Favorites** menu using default values and personal value lists.

You may only save a few seconds on each action, but seconds become minutes, minutes become hours, and hours become days. Just do the math: If you save just 3 minutes per day using these methods, that's 600 minutes over the course of a year, which is significantly more than an entire workday!

This concludes our discussion of the classic user interface, SAP GUI. Now, let's move on to SAP's new user interface, SAP Fiori.

PART II

Using SAP Fiori

Chapter 6
First Steps with SAP Fiori

We welcome you to the user interface of SAP in the 21st century. Be curious because a lot has changed compared to SAP GUI! Perhaps you've already been able to catch a glimpse of it through door-to-door conversations with colleagues or through technical articles that include screenshots. While SAP newbies probably think: "Cool design—I like that," experienced SAP users will be surprised at their first encounter with the new design; after all, everything looks different at first glance! But don't worry, the core remains the same. Starting in this chapter, we'll explore the new SAP system with you in a structured, step-by-step manner. Have fun!

What You'll Learn
- How to log in and log out
- How to optimize the appearance of SAP Fiori on your screen
- How an SAP Fiori screen is structured
- How to find and launch your apps
- How to add and remove apps

We assume that you've read through Chapter 1 and have already performed your initial logon in SAP GUI, because we'll now build on that.

6.1 Log On and Off with SAP Fiori

"All roads lead to Rome"—yes, and two lead to SAP Fiori: the first from SAP GUI and the second from your Windows desktop.

Are you ready? In this section, you'll learn everything chronologically from login to logout. Here we go!

6.1.1 Log On from SAP GUI

If the SAP system is configured accordingly, you can jump from SAP GUI directly to the SAP Fiori launchpad using Transaction /UI2/FLP (SAP Fiori Launchpad).

Let's walk through the steps:

1. In Chapter 1, you learned how to log on to the system using ![icon] (**SAP Logon**) on your Windows desktop. Repeat the procedure described in Chapter 1, Section 1.1.

2. In the **SAP Easy Access** screen, you'll see the *command field* at the top left. This is familiar to you from Chapter 1 as a launching pad to all transactions within the SAP system. Enter "/n/UI2/FLP" in the command field, as shown in Figure 6.1, and confirm your entry with the ⌈Enter⌋ key. After a short moment of waiting, your browser will open and display the SAP Fiori launchpad, as shown in Figure 6.2 (for details on SAP Fiori launchpad, see Section 6.3).

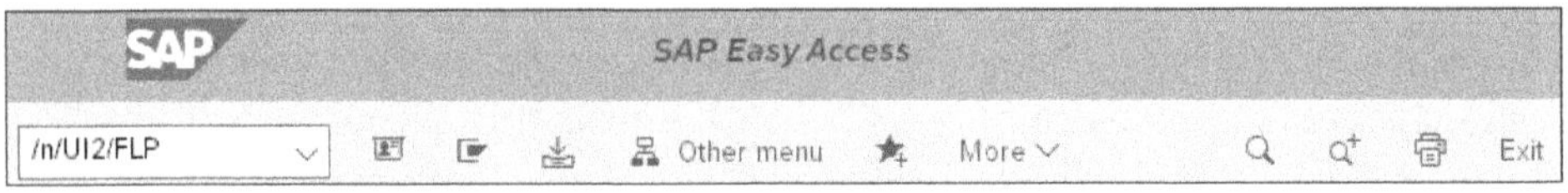

Figure 6.1 Entering the SAP Fiori Launchpad Transaction

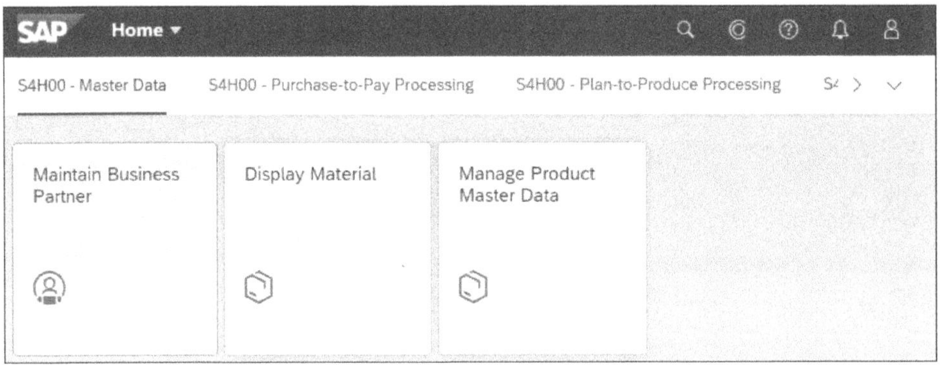

Figure 6.2 SAP Fiori Launchpad

6.1.2 Log On from Microsoft Windows

When you log in with the desktop shortcut, unlike the SAP GUI logon, you're not prompted to specify the SAP system you're using because it's already defined in the link. Follow these steps:

1. Double-click the **SAP Fiori Launchpad** tile on your Windows **Start** menu or the SAP Fiori shortcut ⌈e⌋ on your desktop. If everything worked, you're now in the SAP Fiori login screen, as shown in Figure 6.3.

2. You'll see several fields here. Enter the following:
   - **User**: Enter your username.
   - **Password**: Enter your password in correct uppercase and lowercase letters. The input is masked by asterisks.
   - **Language**: Select **EN – English** from the list.

– **Client**: In the last field, enter the number of the client. This field does not exist in every SAP Fiori logon screen because the client can be defined in the link.

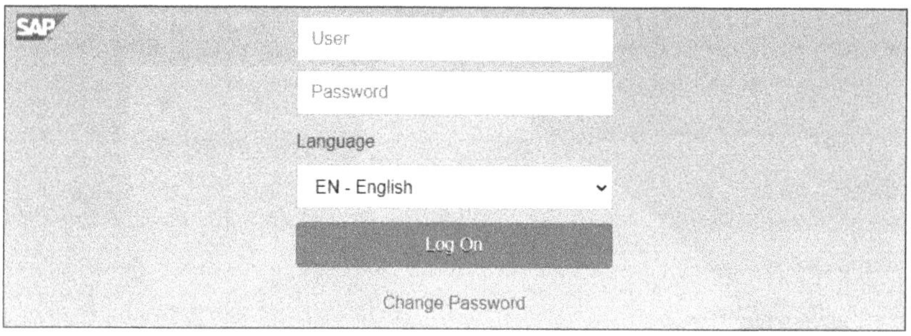

Figure 6.3 SAP Fiori Login Screen

3. Done? Then either press the ⌈Enter⌋ key or click the **Log On** button. Now SAP Fiori will check your data, and if everything is correct, it will grant you access to the SAP Fiori launchpad.

Which Browser Can I Use?

You can use all popular browsers, such as Mozilla Firefox, Google Chrome, Windows Edge, or Apple Safari (on macOS). On mobile devices, Apple Safari on iOS or Google Chrome on Android work. Note that enterprises often mandate the use of a specific browser.

6.1.3   Change Your Password

If you want to change your existing password, this is also very easy! In the completed SAP Fiori logon screen (see step 2 of the preceding section), this time don't click the **Log On** button, but click **Change Password** below it, and the window for changing the password will appear, as shown in Figure 6.4.

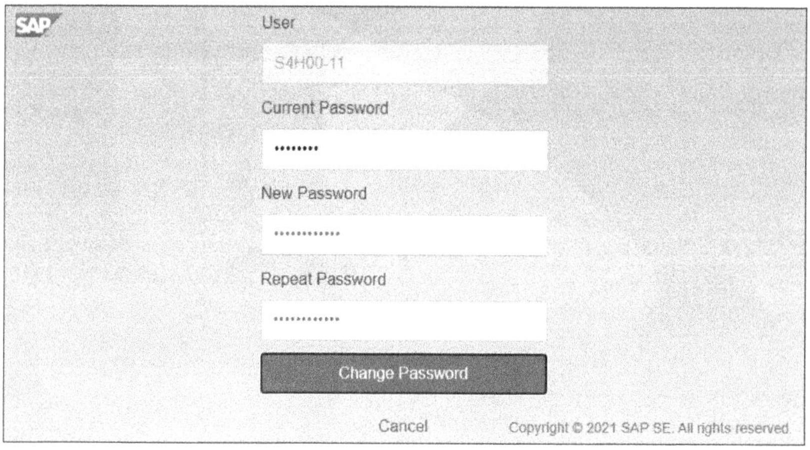

Figure 6.4 Password Change Window

You'll now be prompted to define a new password. You can click 👁, and SAP Fiori will switch from masked display to plain text display.

Confirm your entry when you're done by clicking the **Change Password** button. This way, you'll immediately land in the SAP Fiori launchpad. The password change is valid for your user and also for the logon in SAP GUI.

[»]

Multiple Logins

In contrast to SAP GUI, SAP Fiori allows logging in on multiple PCs. On the same PC, another login via the browser causes an additional tab to be opened in the browser.

6.1.4   Log Off

Logging out is a function that you'll need no later than your lunch break. Follow these steps:

1. Click 👤 in the upper-right corner to open the *user menu*, as shown in Figure 6.5.

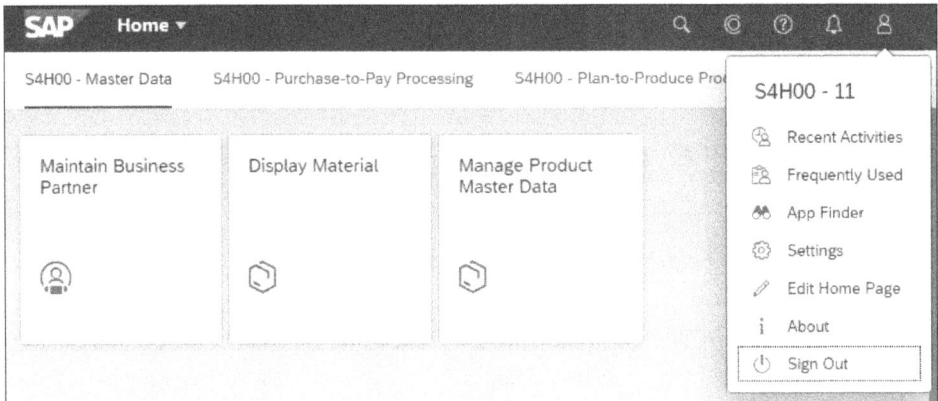

Figure 6.5 User Menu

2. Click on the last menu item, ⏻ **Sign Out**. A window will appear asking, "Are you sure you want to sign out?", as shown in Figure 6.6

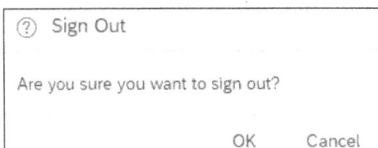

Figure 6.6 Logout Confirmation

3. If you're sure that you won't lose any data that hasn't yet been saved by logging off from the system, you can confirm without worrying by clicking **OK**. You'll then be logged off from the SAP system. If you're not sure, click **Cancel**.

When Logging Out, Avoid Closing the Window with the Icon   [+]

Logging out by simply closing the browser window with ☒ (**Close Window**) is a quick and convenient way to log out. But do you just let the front door of your home fall into the lock, or do you prefer to lock it properly? Locking properly is safer, and that also applies to the internet! Therefore, always use the menu item ⏻ (**Sign Out**), as this is the only way to reliably cut the data connection.

6.2   Customize the Appearance

Appearance is defined as an external image of someone or something that affects the viewer. So, it's about what you see—and what you see can be customized in SAP Fiori through various settings. We'll explore your options in this section.

6.2.1   Select Themes

In SAP Fiori 3, you have several *themes* to choose from. SAP Quartz Light is set as the default, which is what we used for the screenshots in this book.

You can change the theme by following these steps:

1. Start SAP Fiori using the browser you trust.

2. Click ⚇ (**User Menu**) in the upper-right corner to open the user menu, as shown in Figure 6.7.

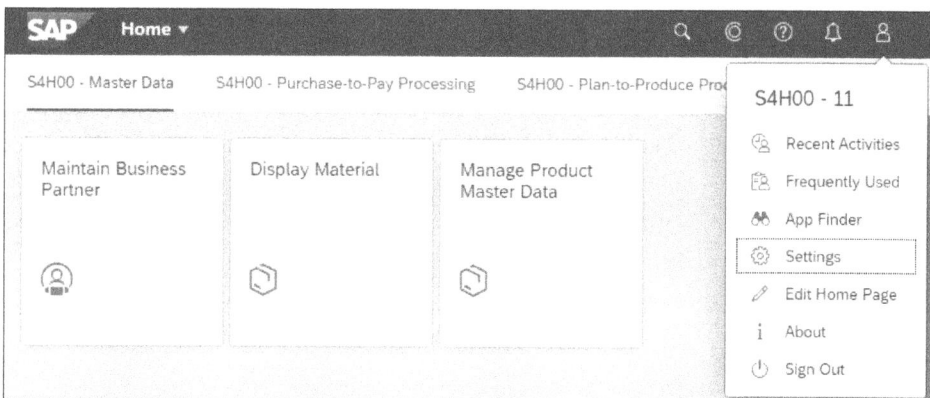

Figure 6.7 Opening the User Menu

3. Click the ⚙ (**Settings**) menu item. The **Settings** window opens.

4. Click 🎨 (**Appearance**) in this window, as shown in Figure 6.8.

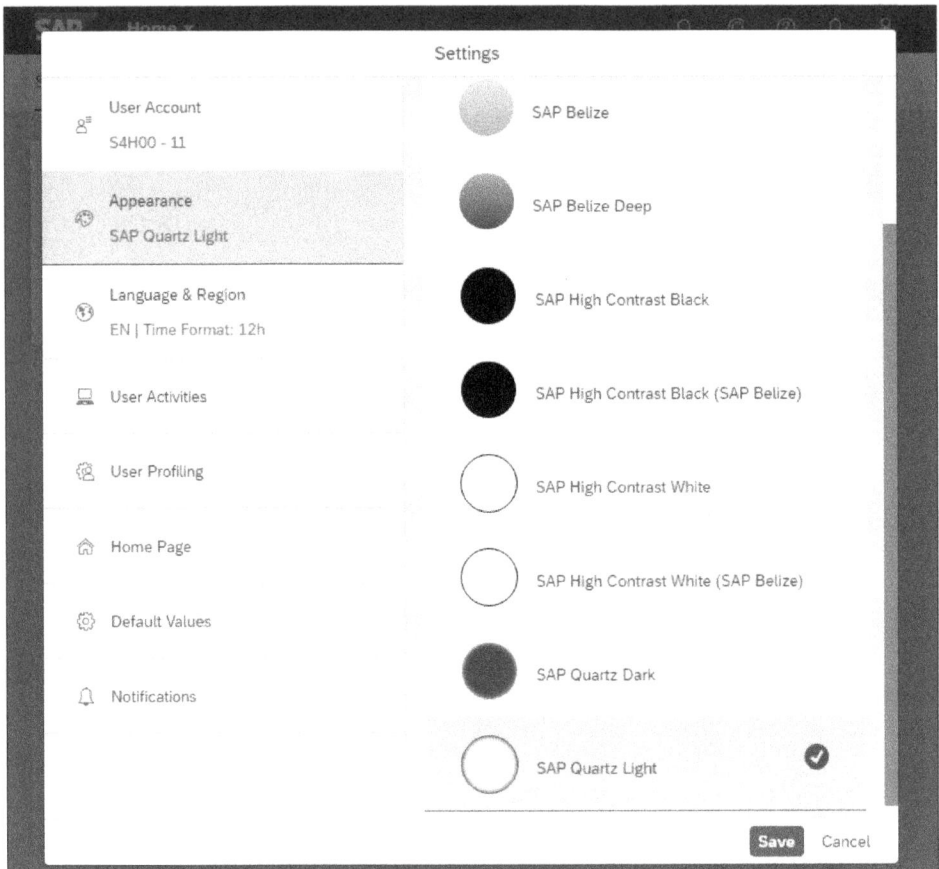

Figure 6.8 Appearance Settings

5. The appearance settings are divided into two tabs in the right part of the window: **Theme** and **Display Settings**.

   Currently, the **Theme** tab is open. All available themes are listed here. You can identify the currently selected theme by the fact that it's highlighted and has a ✓ on the far right.

6. Click on the **SAP Quartz Light** theme to select it. Don't click **Save** or **Cancel** yet, because the next section continues with the **Display Settings** tab.

[+]
Suitable Themes for Every Occasion

The SAP High Contrast Black and SAP High Contrast White themes can be a good choice in case of light sensitivity or extreme lighting conditions. The dark themes are recommended in dark environments such as poorly lit factory floors.

[«]

Oops, I See More Themes

It's possible that you have more themes to choose from besides the default ones. Companies can create their own company-specific themes in SAP Fiori. However, to make it easier for you to work with this book, we recommend the SAP Quartz Light theme, at least for the beginning.

6

6.2.2   Display Settings

You can make more settings for the appearance. You can optimize SAP Fiori for *touch input* by following these steps:

1. You're still in the appearance settings? Great, then now switch to the **Display Settings** tab with one click, as shown in Figure 6.9.

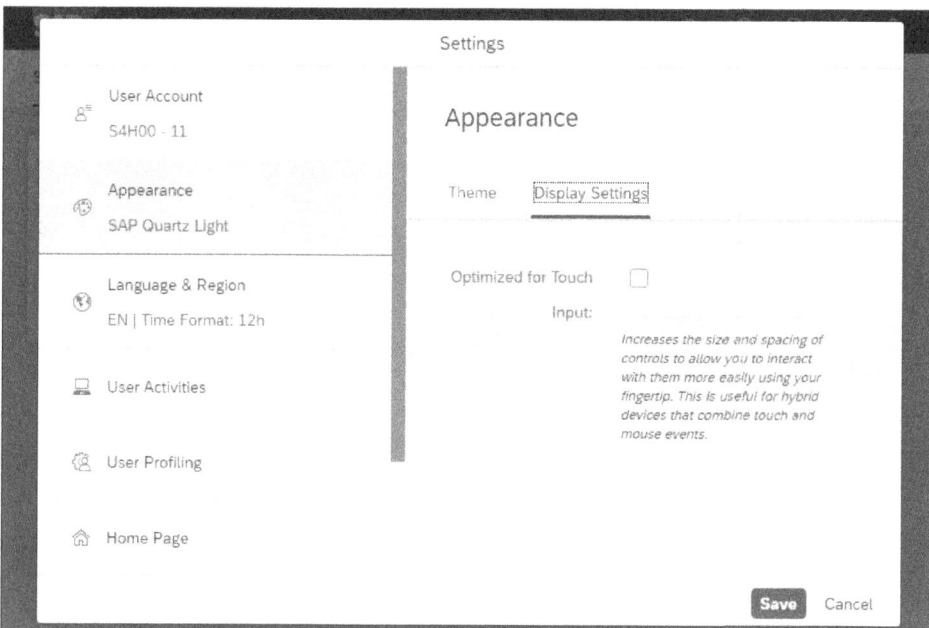

Figure 6.9 Display Settings

2. If you navigate in SAP Fiori with an end device such as a smartphone or a tablet that is operated by touch input, activate the **Optimized for Touch Input** setting. SAP Fiori then automatically increases the distances between the individual elements. This allows you to navigate to individual elements more easily with your fingertip.

3. You may now save your changes by clicking **Save** at the bottom right. After the click, the window will close, and SAP Fiori will signal the successful saving of changes with its own message.

6.3   Your New Home: The SAP Fiori Launchpad

What you could see after the successful login is called the *SAP Fiori launchpad*. Like any system user interface, the SAP Fiori launchpad consists of several components that you should know and understand to use it. Therefore, in this section, we'll give you a brief theoretical introduction and overview. Later, you'll explore the individual functions in detail.

SAP describes the SAP Fiori launchpad and apps as follows:

- **Role-based**
  This means that only those applications are available to you for which you also have the necessary user roles and privileges.

- **Personalizable**
  With a few tricks and a few clicks, you can customize SAP Fiori launchpad to your needs and wishes.

- **Responsive**
  That means you can access the SAP Fiori launchpad and real SAP Fiori apps regardless of device or browser, even with mobile devices such as smartphones or tablets.

- **Coherent**
  A coherent app structure means that the apps are built according to the same rules.

6.3.1   SAP Fiori Launchpad Start Page

The *start page* of the SAP Fiori launchpad (another term: *homepage*) is what you see immediately after successfully logging in, as shown in Figure 6.10. You could also call this page the heart of the SAP Fiori launchpad, comparable to the **SAP Easy Access** window in the SAP GUI.

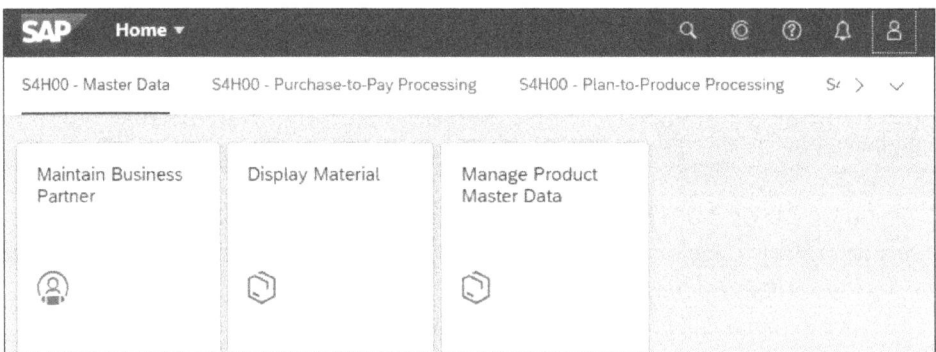

Figure 6.10 SAP Fiori Launchpad Homepage

The launchpad homepage consists of the following elements:

- **Shell bar**
  Bar at the top of the screen showing different icons.

- **Group selection bar**
  Bar just below the shell bar for faster navigation to the desired tile, but is only displayed if there are multiple groups.

- **Tile groups**
  Grouping of tiles and links with a title (e.g., **S4H00 – Master Data**, as shown in Figure 6.10) that you use to launch your apps.

We'll present these screen elements in more detail in the following sections.

6.3.2   Tiles

We'll start, contrary to the arrangement on the screen, with the tiles, because they are the most exciting image elements for you. Square, practical, and simple, each tile usually represents an app. In addition, some tiles contain additional information in the form of a small graphic (another term: *micro chart*) or a key figure (another term: *key performance indicator* [KPI]) that provides you with interesting information right on the SAP Fiori launchpad homepage. For example, the **Manage Sales Orders** tile (see Figure 6.11) shows the number of orders as a KPI.

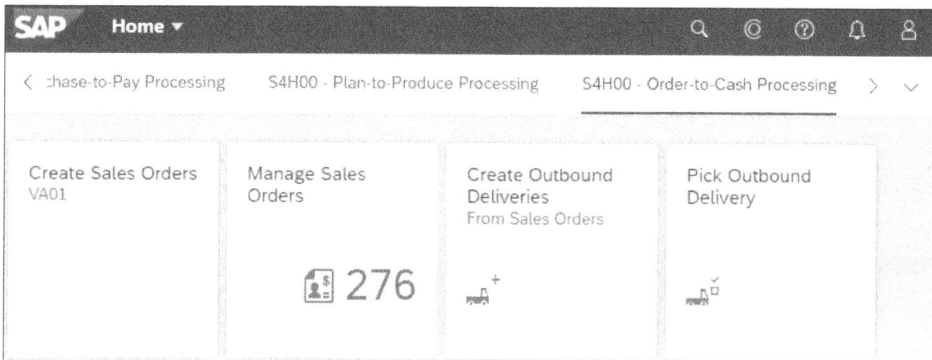

Figure 6.11 Key Figure in the Manage Sales Orders Tile

From a user perspective, a tile in SAP Fiori corresponds to an entry in the **Favorites** menu of SAP GUI, and an SAP Fiori app corresponds to a transaction.

Give it a try by clicking on any tile on the SAP Fiori launchpad homepage. A different screen will appear, and you're immediately in the corresponding app. In the shell bar, you'll now see the app's name, as shown in Figure 6.12.

Figure 6.12 Shell Bar after Launching the Manage Purchase Orders App

Click ◄ (**Back**) in the upper-left corner of the screen to go back to the homepage. You'll only see this icon if you're not on the SAP Fiori launchpad homepage.

[+]

Show More Tiles

You want to see more tiles in the screen for a better overview? In contrast to the SAP GUI under Windows, this works very elegantly and quickly in SAP Fiori because here you're in a browser and can therefore zoom. Move the mouse wheel toward yourself while holding down the Ctrl key: with the reduced zoom, you'll immediately see more tiles. You can also change the zoom level with the shortcuts Ctrl + + or Ctrl + - or with a setting in the browser.

6.3.3   Tile Groups and Group Selection Bar

In practice, users often have dozens of tiles on the SAP Fiori launchpad homepage. For a better overview, tiles are bundled into tile groups. The names of the groups are displayed individually horizontally in the *group selection bar*. This is located directly below the shell bar, as shown in Figure 6.13. If you click on one of the entries in the bar, you can easily access the tiles of the respective group. Alternatively, you can also scroll down in the screen display if there is a corresponding preset.

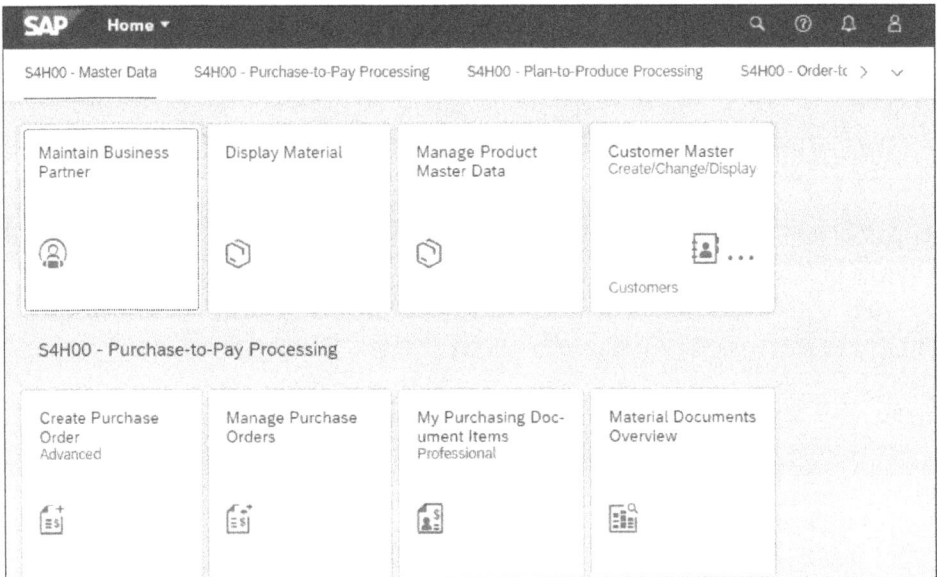

Figure 6.13 Group Selection Bar below the Shell Bar, below Some Related Tiles and a Heading

You can clearly distinguish the beginning and the end of a group by a heading line above the tiles that belong together (see Figure 6.14).

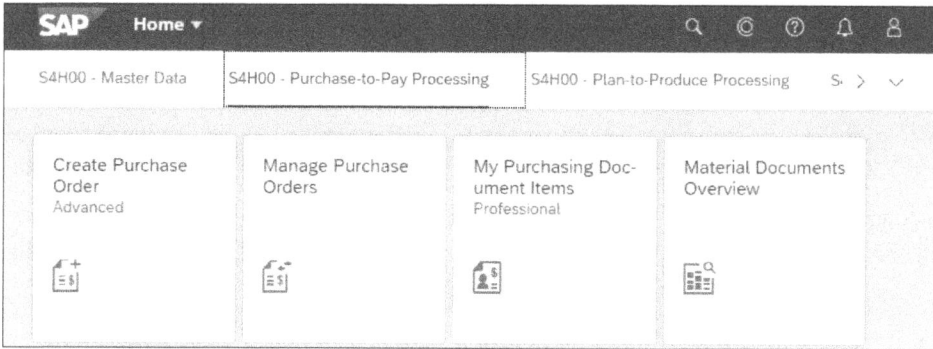

Figure 6.14 Tile Group S4H00 – Purchase-to-Pay Processing, below Some Related Tiles without a Heading

---

Can't See Header Lines for Tile Groups?

If you don't see headings for the tile groups as shown in Figure 6.14, that's not a bug! On the **Home Page** tab in your settings, you only have the **Show One Group at a Time** option enabled. This ensures that only one tile group is displayed to you at a time. To switch to another group, click on the respective group in the group selection bar with this setting.

6.3.4   Links

As on a smartphone, there are apps in SAP Fiori that you need much more rarely than others. In SAP Fiori, the tiles of these apps can be turned into space-saving links, and thus more fits on the screen, as shown in Figure 6.15. Each group includes its own optional area for links below its tiles. For now, there are no links in the SAP Live Access training system. In Chapter 10, Section 10.2.2, you'll learn how to create links.

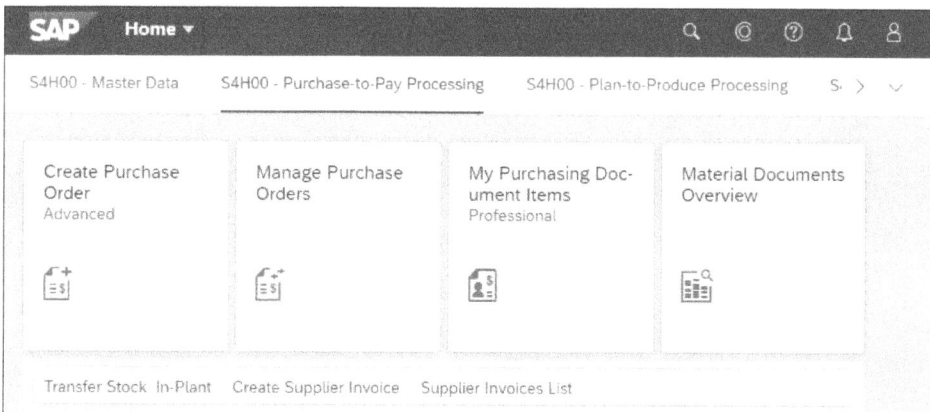

Figure 6.15 Links Appearing below the Tiles

6.3.5   Shell Bar

Now we're moving on to the top bar—the shell bar—of the beautiful SAP Fiori interface! It's your most loyal companion while working with SAP Fiori because no matter which app you're in and how far down you scroll, the shell bar always remains fixed at the top of your browser.

In Figure 6.16, you can see the shell bar almost fully configured. You probably won't see all the icons as this depends on the configuration of your system. We'll start on the right:

❶ 🗓 (**User Menu**): User menu with various commands, for example, for logging off.

❷ 🔔 (**Notifications**): Notifications, for example, about an order that is waiting for your approval.

❸ ⑦ (**Open Help**): To display the definition of fields, for example.

❹ 📢 (**Submit Feedback**): To send written feedback.

❺ 🔍 (**Search**): To search for data or apps.

❻ **Navigation Menu**: To display the navigation history with the last opened windows since the start of the last app. When you're in an app, the stations of your journey are displayed.

❼ SAP (**Navigate to Home Page**): To return to the homepage when you're outside of the homepage. If your admin hasn't placed another image, you can admire the SAP logo here. With one click, you travel—no matter where you're—directly to the SAP Fiori launchpad homepage.

Figure 6.16 Shell Bar Icons

After this overview of the shell bar, you'll now get an overview of the user menu. We'll discuss the important functions of the shell bar in more detail later.

6.3.6   User Menu

By clicking on 🗓 (**User Menu**), a menu will open, as shown in Figure 6.17. It contains a colorful mixture of commands that you'll need more often.

> **[»]**
>
> **Can't See All the Menu Items**
>
> You'll only see the **Edit Home Page** menu item if you're in the SAP Fiori launchpad homepage. In addition, the **Contact Support** and **Give Feedback** menu items are only listed if they have been enabled by your administration.

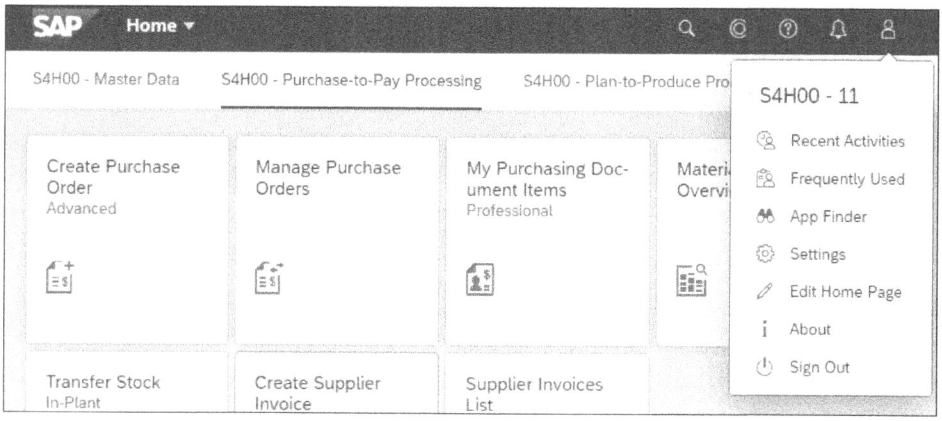

Figure 6.17 User Menu

The following is a brief overview of the functions of each menu item. Don't worry, more detailed information on this will also follow in the course of the book.

- 🔍 (**Recent Activities**): Opens a window with a listing of up to 30 of your most recently accessed apps in reverse chronological order.

- 📖 (**Frequently Used**): Opens a window with a list of the apps you've used most frequently in the past 30 working days.

- 🔭 (**App Finder**): Opens the App Finder. This lists all the apps that have been enabled for you.

- ⚙️ (**Settings**): Opens a window where you can make settings regarding your user, the start page, or the appearance.

- ✏️ (**Edit Home Page**): Switches to the edit mode of the SAP Fiori launchpad homepage. This mode allows you to personalize the homepage. For example, you can use it to create your own tile groups or change the tile descriptions.

- ✉️ (**Contact Support**): Opens a window that allows you to contact your support directly. An input field offers the possibility to describe the problem.

- 📣 (**Give Feedback**): Opens a window that allows you to evaluate your personal SAP experience and send it as written feedback.

- ℹ️ (**About**): Opens a window that contains all sorts of technical information, such as the version number of SAP S/4HANA and the technical name of the app you're in.

- ⏻ (**Sign Out**): Logs you out of the system.

6.3.7   Footer Toolbar

You don't see a *footer toolbar* on the homepage in the last line of the window? Neither do we. You'll only see it when you need it! To do this, launch any app. Most apps have a footer toolbar; if this is the case, it will automatically appear at the bottom of the screen.

In our example, we've started the Manage Business Partner Master Data app. In Figure 6.18, you can immediately see the typical components of a footer toolbar:

- All actions such as **Continue**, **Save**, or **Cancel** are always located on the right.
- On the left, you'll find messages. Whenever you see a message in the footer, be sure to pay attention—SAP Fiori wants to tell you something! The number right next to the icon shows you the number of messages. You can easily view the messages by clicking on the icon.

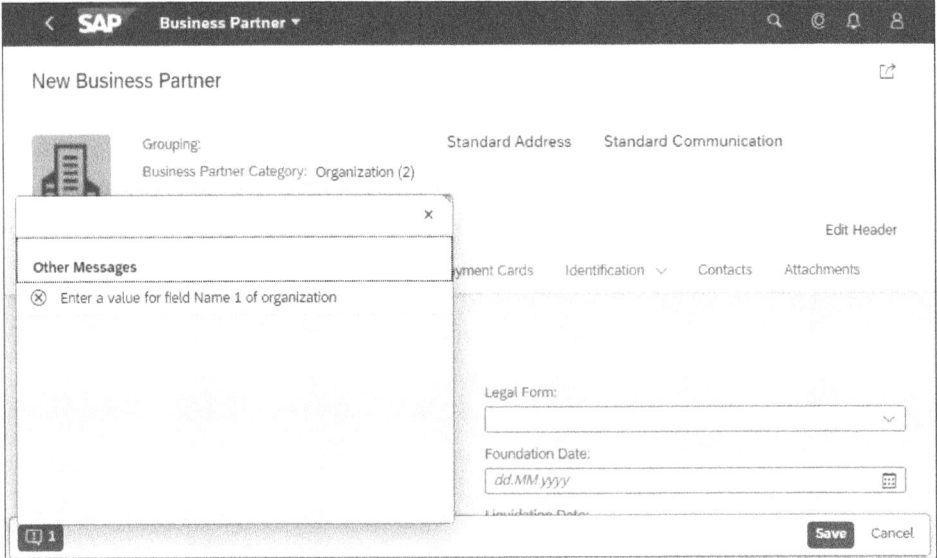

Figure 6.18 Footer Toolbar with Message Icon at the Bottom Left

6.4   Find and Run SAP Fiori Apps

After your first login to SAP Fiori, you'll probably see some preselected apps in the SAP Fiori launchpad. Usually, these also cover the needs in the company that you need for your daily work. However, sometimes some are missing! In this section, you'll learn how to find more apps.

[+]

Frequently Used SAP Fiori Apps

This section simply tells you how to find and run an app. If an app is in frequent use, it deserves a place on your homepage. To learn how to add an app to your home page, see Section 6.5.

In the following instructions, you'll search for the Customer Master app via the *App Finder*. It works much like the Apple App Store or Google Play, in that you'll find all the apps listed here that you have permission to use and that you can launch in SAP Fiori.

(If you don't need the Customer Master app, which we're going to look for here as an example, just use another app—this procedure applies to all apps.)

Follow these steps:

1. Start SAP Fiori in the browser, and click ⌷ (**User Menu**). The user menu you're already familiar with will open, as shown in Figure 6.19.

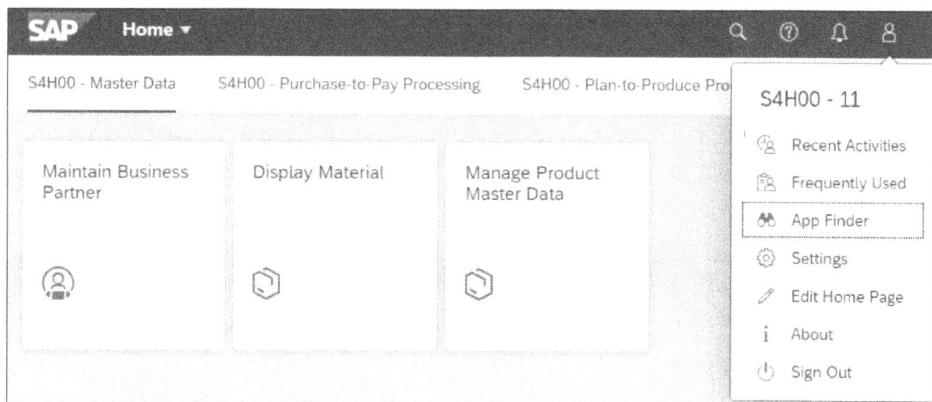

Figure 6.19 Opening the User Menu

2. Click the 🔍 (**App Finder**) menu option. This will take you out of the SAP Fiori launchpad and allow you to take a first look at the **App Finder** screen, as shown in Figure 6.20.

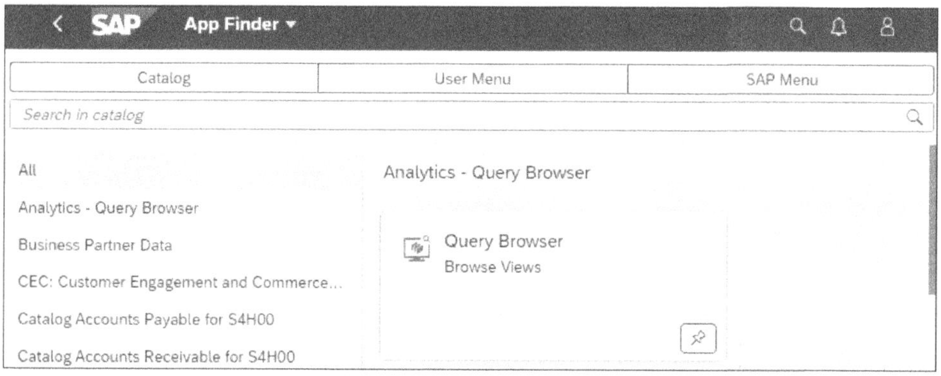

Figure 6.20 App Finder

3. This is a divided window. In the left part of the window, you see a list of all catalogs. As soon as you click on one of these catalogs, the right part of the window is filtered and will only show you the apps of the respective selected catalog.

   A *catalog* is a group of apps. Grouping apps into catalogs allows you to search for apps by topic and subject in a bundle.

> **[»]**
> ## Why Can't I See the Left Part of the Window?
> If you don't see the left part of the window with the list of all catalogs, it's because either the browser window is too narrow or you're accessing SAP Fiori with a tablet/smartphone. To show the left part of the window, simply click ≡ (**Show Menu**) directly below the shell bar on the left.

4. You now want to search *all* catalogs for apps that allow you to view and maintain customer data. Therefore, click the **All** entry in the upper-left corner.

5. Directly below the shell bar on the far right is the Q (**Search in Catalog**) field, as shown in Figure 6.21. Enter the name or part of the name of an app in the field, in our case, "Customer".

< **SAP** App Finder ▼			Q ⏰ 8
Catalog	User Menu	SAP Menu	Customer × Q

Figure 6.21 Search in Catalog Field

6. Click Q (**Search**) or alternatively press the ⌜Enter⌝ key. The **App Finder** screen will then display the search results, as shown in Figure 6.22.

> **[»]**
> ## The App Finder Isn't Finding Anything?
> If you don't have permission for an app, it won't be displayed for you. The app then isn't included in any of the catalogs available to you.
>
> Another possible reason is that only the catalog selected in the left part of the screen is searched. If you want to search all catalogs, scroll all the way up on the left, and click the **All** entry at the top, as shown in Figure 6.22.

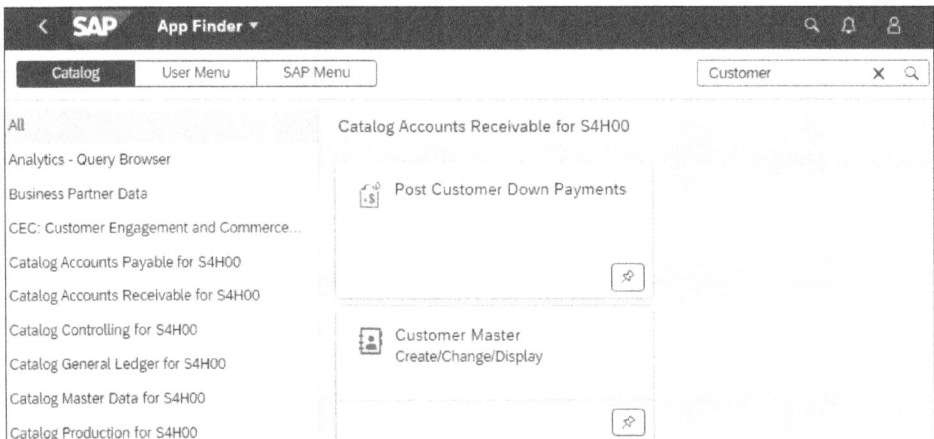

Figure 6.22 Search Results in the App Finder

7. To launch the app, simply click on it (in our example, it's the Customer Master app, as shown in Figure 6.23). Voilà! You've now completed your first search for an app and successfully launched it.

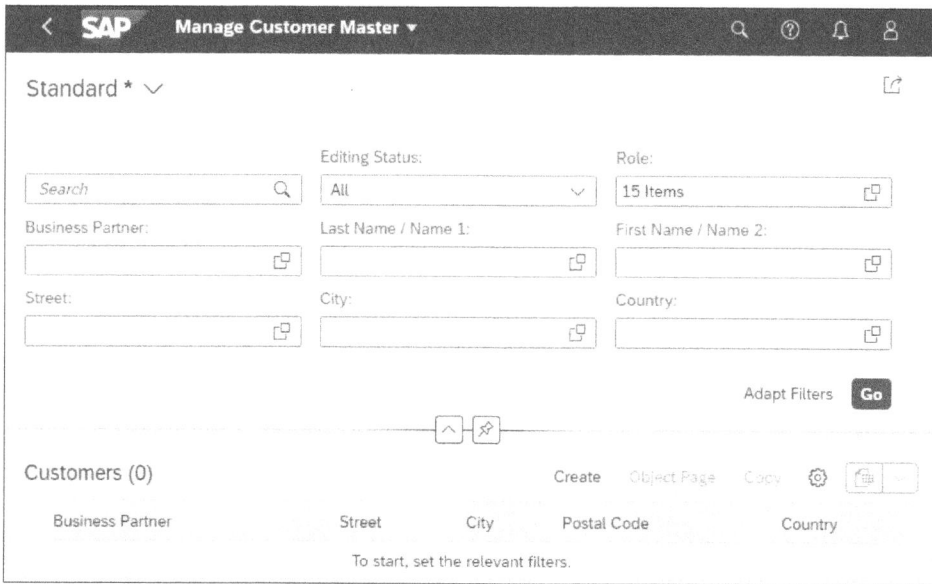

Figure 6.23 Launching the Customer Master App

8. Click ❮ (**Back**) in the upper-left corner of the screen to return from the app to the launchpad homepage.

How Can I Find the SAP Fiori Apps for My SAP GUI Transactions?

Depending on the launchpad homepage implementation, the **App Finder** screen is divided into three sections: **Catalog**, **User Menu**, and **SAP Menu**. With the **User Menu** and **SAP Menu** buttons, you can find your favorite transactions as apps in the corresponding menu trees of SAP GUI. These buttons can also be disabled depending on the configuration of your SAP system.

SAP has already turned many, but not all, SAP GUI transactions into apps. However, you can't use smartphones or tablets to access the apps created directly from SAP GUI transactions via the SAP Fiori launchpad homepage. If you launch the launchpad homepage on a mobile device such as a phone, these apps aren't displayed.

Find More Apps in the SAP Fiori Apps Reference Library

With the App Finder, you only see the apps for which you're authorized. In contrast to SAP GUI with its complete **SAP Menu**, there is no way within SAP Fiori to see what other exciting apps are available for your area of expertise. However, there is a freely accessible

SAP Fiori apps reference library at https://fioriappslibrary.hana.ondemand.com. Here, you can choose to track down the available apps by component, role, or department.

You can use the App Finder to find and run apps. But you can also use it to add apps as tiles to your personal SAP Fiori launchpad homepage, as we'll discuss next in Section 6.5.

6.5   Add and Remove SAP Fiori Apps

In the previous section, you learned how to find and run apps using the App Finder. However, this procedure is only practical for apps that you rarely need! Some apps deserve a permanent place on your SAP Fiori launchpad homepage because of their frequent use. We'll discuss how to add SAP Fiori apps, move them, and remove them in this section.

6.5.1   Add an SAP Fiori App from the App Finder

In the following steps, we'll use the Customer Master app to show you how to permanently add an app to your SAP Fiori launchpad homepage.

Follow these steps:

1. Click [8] (**User Menu**) to open the user menu, and select the command [👓] (**App Finder**) in it.

2. As described in Section 6.4, you'll now search for the app you want to add using [Q] (**Search in Catalog**), as shown in Figure 6.24. In our example, we'll search for the Customer Master app. Of course, you're welcome to search for another app from your field of activity to follow these instructions.

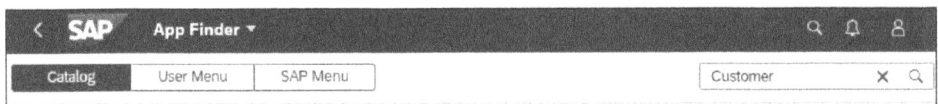

Figure 6.24 App Finder: Search in Catalog

3. After you've found the appropriate app, click [✗] (**Add Tile**) to add the app to your SAP Fiori launchpad homepage.

4. A menu will open where you can select a group or groups to which you want to add the app as a tile, as shown in Figure 6.25.

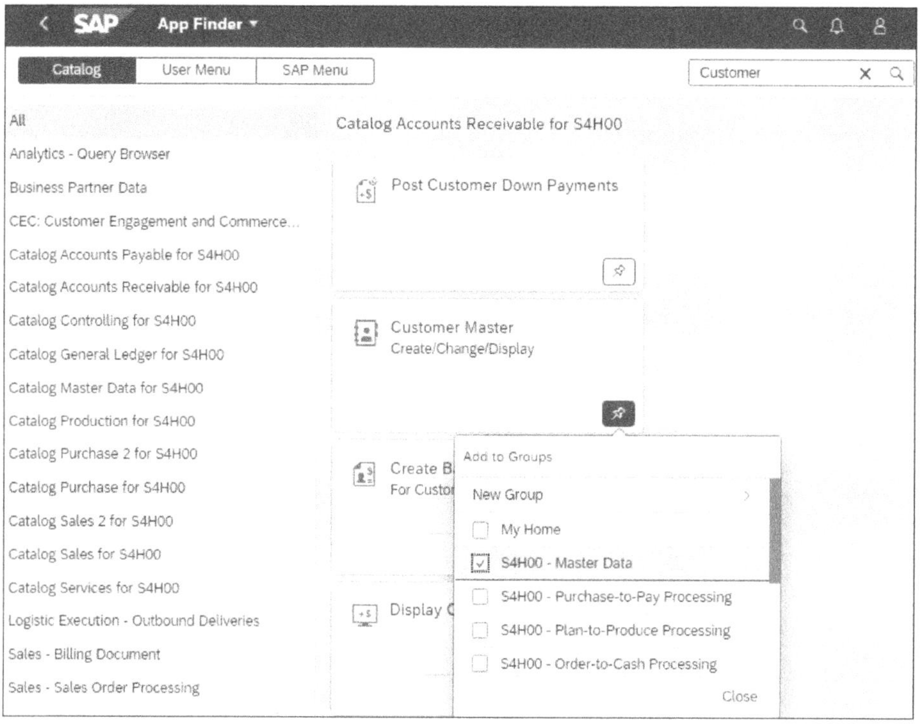

Figure 6.25 Adding the App to a Group

5. Select one or more tile groups as desired. To do this, simply click on the names of the respective groups. As soon as you've selected a group as a target, you'll see a check-mark ✓ in front of the respective name of the group.

How Can I Realize if an SAP Fiori App Is Already Assigned to a Group?

You can realize that an app is already assigned to a group by the fact that the box next to the group's name already has a checkmark ✓. In addition, the pin gets a colored background 📌.

---

6. To successfully complete your task and save the change, click the **Close** button. SAP Fiori will signal the successful change to you as usual with its own message, for example, "Customer master was added to 2 groups."

What Does the Entry "New Group" Mean?

If you click on the **New Group** entry, you can create a new tile group directly from the App Finder and also assign the app to it directly.

---

7. Click on the logo icon from the shell bar, and you'll be back on the SAP Fiori launch-pad homepage.

6.5.2   Move Tiles

Don't like the order of the apps? Drag a tile to the desired place with the left mouse button pressed and—once there—release the mouse button again. You may also move tiles from one group to another.

6.5.3   Remove Tiles with the App Finder

Now you know how to add apps in the form of tiles to your SAP Fiori launchpad homepage. Next, we'll show you how to remove unnecessary tiles from it.

Follow these steps:

1. Click ⌂ (**User Menu**), and select 👓 (**App Finder**) to open the catalog, as shown in Figure 6.26.

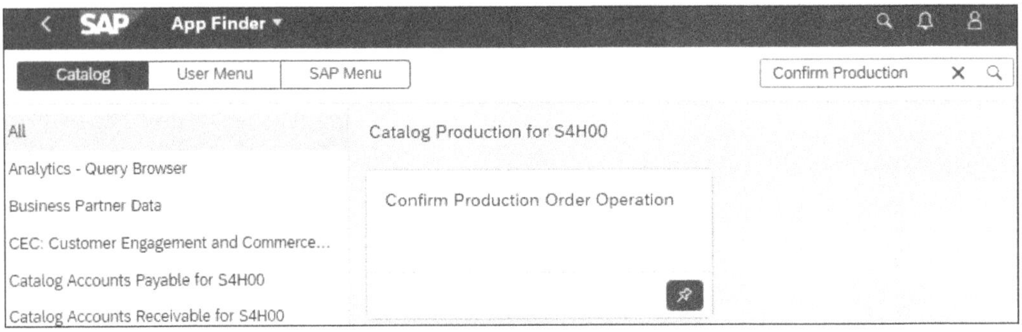

Figure 6.26 Opening the Catalog

2. Select an app that you want to remove from the SAP Fiori launchpad homepage; for example, in the SAP Live Access training system, select the Confirm Production Order Operation app with the search term "Confirm Production".

3. You can tell if an app already exists as a tile on your SAP Fiori launchpad homepage by the fact that the 📌 icon has a colored background: 📌. Now simply click on this pin. The **Add to Groups** menu will open, as shown in Figure 6.27.

4. In this window you can not only add apps but also move them to another group or remove them from a group:

   – To move a tile to another group, uncheck ✔ the old group and check the new group instead.

   – To remove a tile from a group and thus from the launchpad homepage, simply remove all checkmarks ✔.

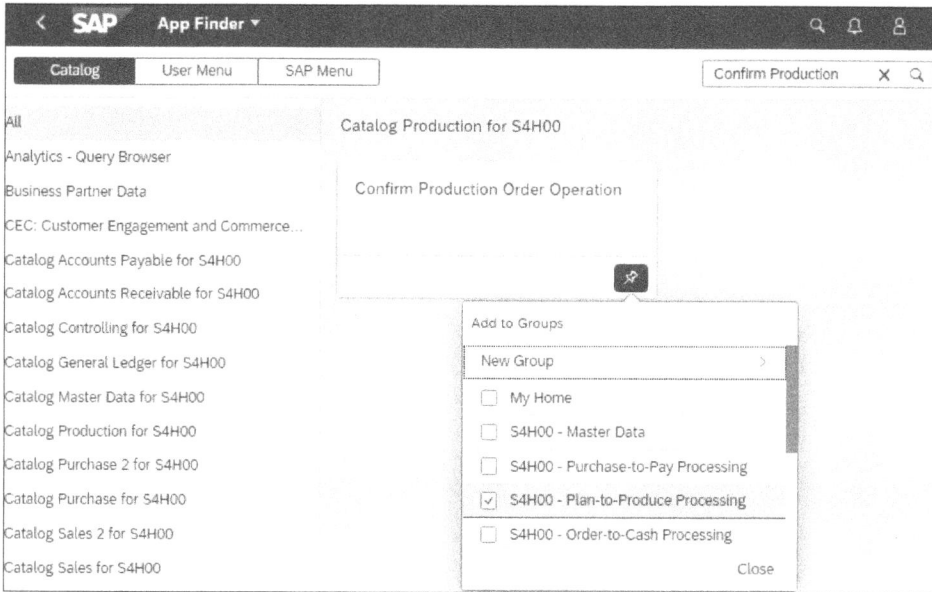

Figure 6.27 View Groups

When you're done, click the **Close** button. You've now successfully moved the tile or removed it completely from your SAP Fiori launchpad homepage!

How to Make Your SAP Fiori Launchpad Homepage Faster

Are you counting the seconds until the SAP Fiori launchpad homepage on your screen finally shows all the tiles? This may be because you're using a lot of tiles and especially a lot of tiles with KPIs, that is, with additional up-to-date information. When loading the launchpad homepage, the KPIs are recalculated each time. Even the fastest database needs time for this if the KPIs are calculated from many data sets.

The SAP Fiori launchpad homepage will be faster and more organized if you remove all the tiles you don't need.

Still too slow? Open the user menu by clicking 🔲 in the upper-right corner, and click on ⚙ (**Settings**). Select the **Home Page** option on the left, and then click the **Show One Group at a Time** on the right. This will allow you to speed up the loading process of the start page once again.

Click on the logo from the shell bar, and you'll be back on the SAP Fiori launchpad homepage.

6.6   Summary

You've now gotten started with SAP Fiori and seen what SAP's new user interface has to offer. After reading the next chapter, you'll be well prepared to start your daily practice and use SAP Fiori apps to display and maintain data.

Chapter 7

Displaying and Maintaining Data with SAP Fiori Apps

So far, you've been floating relaxed on the surface of SAP Fiori. Now you can dive a little deeper and explore the term *app*, which is already frequently used in this book. But don't worry, this is a book for users—you don't need a diving suit! For now, we'll use very clear apps.

What You'll Learn

- What types of apps are available
- How to launch and navigate within apps more quickly
- How to use apps to view, create, and change data
- How to use filters and enterprise search to find your documents and master data
- Where to get help resources for apps and fields
- How to make the best use of your browser for SAP Fiori

Do you want to click along but you can't find an app on the SAP Fiori launchpad home-page? Use the App Finder from the user menu, already described in Chapter 6, Section 6.4, to launch the app in question.

7.1   Legacy Apps and Real SAP Fiori Apps

Certain SAP Fiori apps are called "real" because they have been completely rebuilt in line with the new SAP operating and design concept using the SAPUI5 framework for SAP Fiori and largely function according to clear and uniform rules. SAPUI5 stands for *SAP User Interface 5* and is based on the standard HTML5 and JavaScript web technolo-gies. You can look forward to these new, "real" apps in particular.

Are There Any Fake Apps?

Well, the term *fake app* isn't used, but there are many apps that aren't built on SAPUI5. These apps are even in the majority, and they are based on classic SAP GUI transac-tions. An SAP Fiori wrapper has been "put over" the SAP GUI transactions, so to speak, and that is why they are also called *fiorized* transactions or *legacy apps*.

SAP GUI lives on in the legacy apps. Because these "fiorized" transactions aren't based on the new SAPUI5 concept, the colorful world of GUI transaction rules still applies here, for example, the F4 help, the **More** button from the SAP GUI header, and the input history.

Before we get to the real SAP Fiori apps, let's explore the legacy apps and see how to check the type of an app.

7.1.1 Navigate in a Legacy App

We'll first show you a legacy app, namely the Display Material app, which corresponds to Transaction MM03 (Display Material) in SAP GUI. There are hardly any differences in the operation of this app compared to the corresponding SAP GUI transaction.

Follow these steps:

1. Start the Display Material app.

2. As shown in Figure 7.1, on the **Display Material (Initial Screen)**, search for a material number using the value help, as in SAP GUI, and confirm your entry with the Enter key. The **Select View(s)** screen will appear.

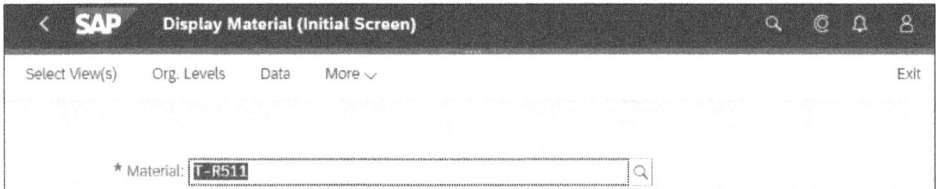

Figure 7.1 Display Material Initial Screen

How to Find a Material Number

Click in the **Material** field and then on Q (**Value Help**) to the right of this field. In the **Material Number** window that follows, click on the **Find** button at the bottom right without entering any data. Now you can take over any material by double-clicking on the material in the list that appears.

3. In the **Select View(s)** window, as shown in Figure 7.2, select only the **Basic Data 1** view, and then press Enter. Now the detail screen will open.

4. In the detail screen shown in Figure 7.3, you can see some material data, and everything is the same as in SAP GUI. Leave this view open because this is where we'll continue in the next section.

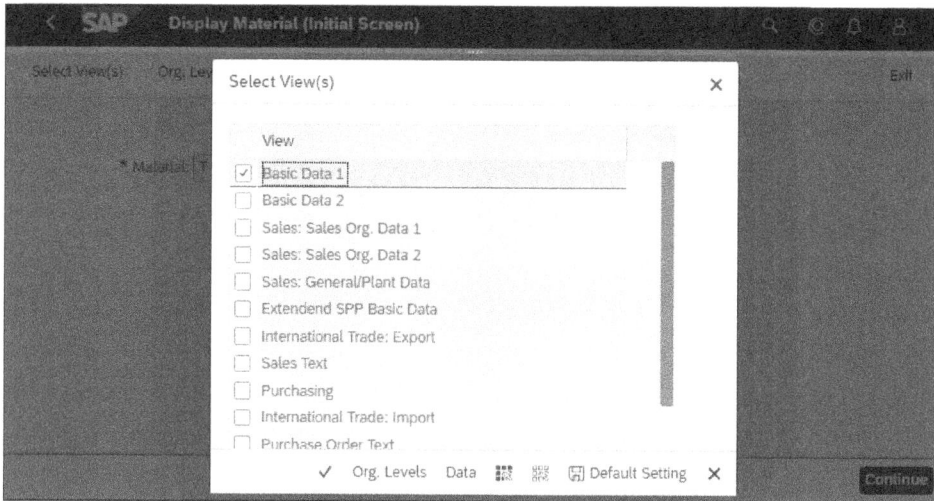

Figure 7.2 View Selection

Figure 7.3 Display Material: Detail Screen

How Can I Identify a Legacy App?

This is a valid question because you want to know what rules legacy apps or "fiorized" transactions follow. Legacy apps basically have the following characteristics:

- The **More** button is provided.
- On the right you *won't* see ⬀ (**Share**). (We'll discuss this exciting icon later in Chapter 9, Section 9.4.)

You can also use the **About** command from the user menu to check what kind of app it is, which you'll learn about next.

7.1.2   Use the About Command to Check the Type of App

We'll now show you how to check if the Display Material app is a "fiorized" transaction or a legacy app. As described in the previous section, you can currently see the details screen of this app. Then, follow these steps:

1. Open the user menu via ⌸ (**User Menu**) in the shell bar, and select the ⓘ (**About**) menu item from the list. This will open a new window, as shown in Figure 7.4.

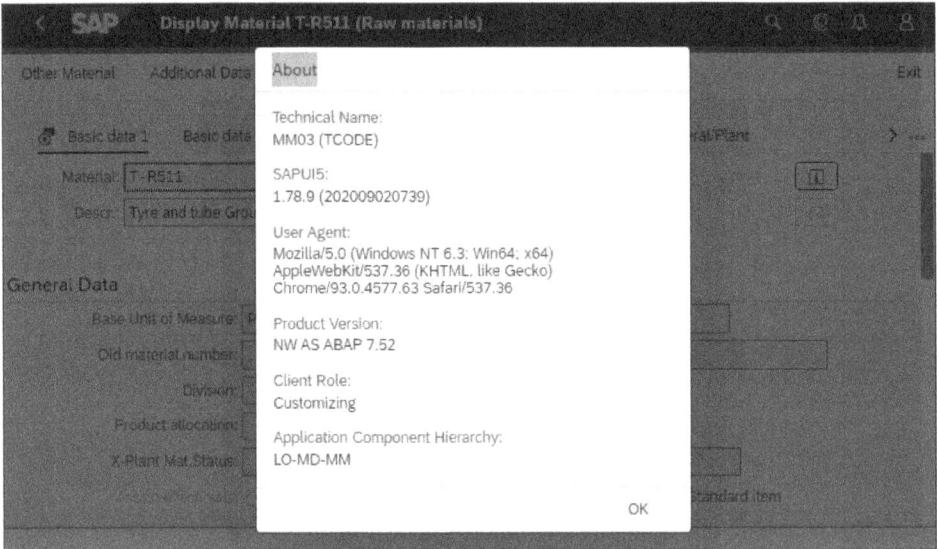

Figure 7.4 About Window in the Display Material App

The **About** window shows you all sorts of technical information. The most relevant information is at the top. Here, you see MM03, the transaction code of the SAP GUI Display Material transaction, and behind it, the specification (**TCODE**). So, because there is a transaction code, it's as clear as day that the app is a "fiorized" SAP GUI transaction, that is, a legacy app.

You can also recognize a real SAP Fiori app by the fact that an app ID, such as F1574, is specified in the **About** window.

2. Close the **About** window by clicking the **OK** button and close the app by clicking the **Exit** button in the upper-right corner of the window or on the logo icon from the shell bar.

Much more exciting for you than the old legacy apps are the real SAP Fiori apps that don't exist in SAP GUI. So, from now on, in this chapter, we'll only show you such apps.

[»]

Is There a Real SAP Fiori App for Every SAP GUI Transaction?

No, not every SAP GUI transaction has a real SAP Fiori app counterpart. At the time of writing, there were about 12,000 real SAP Fiori apps, while the number of SAP GUI transactions is more than 100,000. However, there are individual real SAP Fiori apps that each replaces several SAP GUI transactions.

7.2   Launch and Navigate an SAP Fiori App Faster

Do you remember the components of the SAP Fiori launchpad we talked about in Chapter 6, Section 6.3? If not, scroll back briefly and get an overview again.

Back again? Great, because on this basis, you'll now get a deeper insight into navigation.

7.2.1   Recently Used SAP Fiori Apps

You can use the function described here not only when your boss asks you what you've been up to all day or when you ask yourself what you've been up to all day, but also when your SAP Fiori launchpad is overflowing with apps or you're looking for a rarely used app and have recently launched it. The ⌨ (**Recent Activities**) option (in previous versions, called **Recently Used**) in the user menu is the quickest way to get there.

Follow these steps:

1. From anywhere in the SAP Fiori launchpad, click 🔒 (**User Menu**) to open the user menu and then select the ⌨ (**Recent Activities**) menu item.

2. If you were successful, you'll then see the **Quick Access** window with two different tabs: **Recent Activities** (or **Last Used**) and **Frequently Used**, as shown in Figure 7.5.

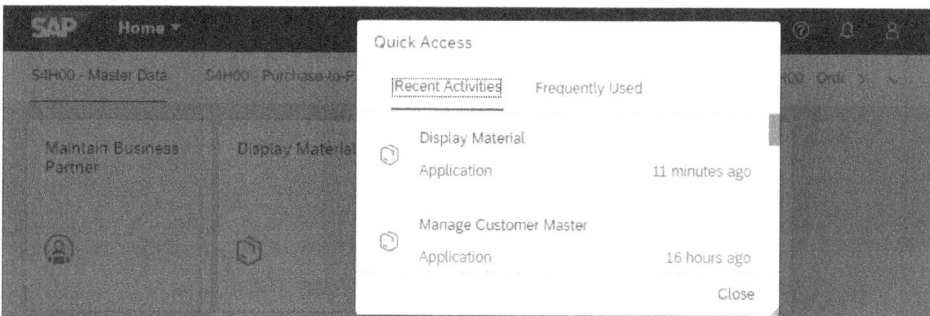

Figure 7.5 Quick Access Window

Because you've selected ⌨ (**Recent Activities**) in the user menu, the **Recent Activities** tab is opened immediately accordingly.

Here you can see your most recently accessed apps sorted by time (as well as searches performed with the enterprise search function, which we'll discuss further in Section 7.8).

3. Click one of the listed apps. This launches the app, and the **Quick Access** window will disappear.

[»]

What Happens When I Click on the Frequently Used Tab?

If you click on the **Frequently Used** tab, you'll remain in the same window and switch to this tab. You'll then see directly what you would see if you had selected the 🔖 (**Frequently Used**) menu item in the user menu.

Each list item in the **Quick Access** window consists of three pieces of information, as shown in Figure 7.6:

- A title (e.g., the name of the app and an icon)
- An object type such as **Application**, which tells you whether the list item is an app or a search
- A timestamp that tells you the amount of time that has passed since you last performed the activity (only in the **Recent Activities** tab)

After you've called a list item, the timestamp updates and moves the list item to the top of the list.

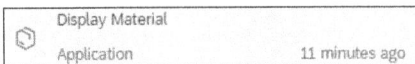

Figure 7.6 List Item in the Quick Access Window

7.2.2   Frequently Used SAP Fiori Apps

To save you the trouble of searching for an app, the SAP system also makes it easier for you to call up apps in other ways. In addition to the **Recent Activities** list, there is also the **Frequently Used** list, which collects the apps that you work with regularly. You call up this list by choosing 🔖 (**Frequently Used**) in the 👤 user menu.

The 🔖 (**Recent Activities**) and 🔖 (**Frequently Used**) menu items both open the same window, just each with a different tab. Thus, 🔖 (**Frequently Used**) opens the window with a hit list of the apps you've used most frequently in the past 30 working days, as shown in Figure 7.7. This list is sorted in descending order, so the most frequently used app is at the top.

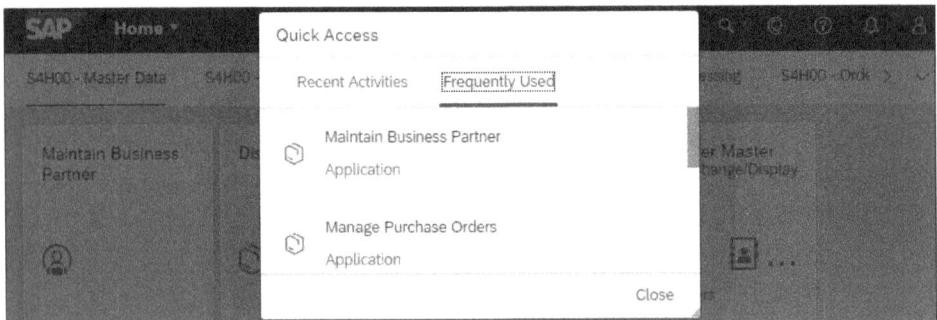

Figure 7.7 Quick Access Window with the Frequently Used Tab

7.3   Navigate in a Transactional SAP Fiori App

There are three different *app categories* (another term: *app type*), which you'll encounter more frequently throughout this book:

- **Transactional**
  These apps are used to perform data maintenance, that is, the entry and modification of data records. At the same time, you can also just use it to display the data.
- **Analytical**
  These apps are used for reporting and creating graphical evaluations.
- **Info sheets**
  These apps (another term: *fact sheets*) give you an overview of all the important information about an object. An object is either a document (e.g., sales order) or a master record (e.g., supplier).

Is a Transactional SAP Fiori App a "Fiorized" Transaction?

Don't get confused by these terms. A transactional app isn't a "fiorized" transaction, but belongs to the group of real SAP Fiori apps.

7.3.1   The Stations of Your Journey

In this section, we'll explain the process, structure, and navigation in a real transactional SAP Fiori app using a clear example. You're displaying customer master data.

The following sample instructions also apply to other transactional real SAP Fiori apps, so we'll give you an overview first for orientation:

1. Start the app to get to an initial screen.
2. In the *initial screen*, click on the **Go** button, and you'll immediately see all the data records—in our case, the customers—in a list. You can also select one or more data records for this list in the initial screen before clicking the **Go** button.
3. In the *list*, you can see the most important information in the form of a table. If you want to see more information about a customer, click on the corresponding customer in the list. Then all the information will be displayed in a form.
4. In the *form*, you have the following options for navigation on the far left of the shell bar:
   - The ◁ (**Back**) icon from the shell bar takes you back to the previous screen; in our case, this is the list.
   - Use the SAP icon from the shell bar to exit the current app and immediately land on the SAP Fiori launchpad homepage.

This is the basic flow. Too abstract? Ok, then let's just start with actual practice to experience the flow just described. You'll also get to know some additional navigation options.

Follow these steps:

1. Start the Customer Master app with the tile titled **Customer Master**. You're now in the initial screen, as shown in Figure 7.8. (Each screen has a title, and the title is displayed to you on the left side of the shell bar.) At the top in the header, you can see the filters you can use to select customers for the list.

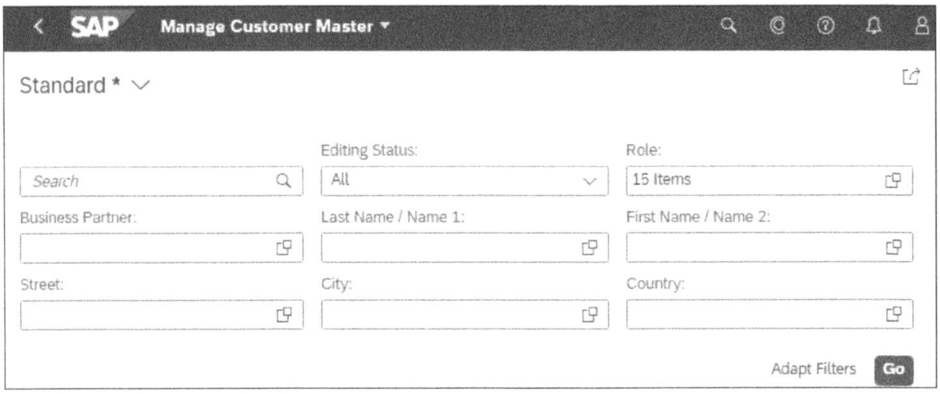

Figure 7.8 App Customer Master with Filters in the Header

[+]

If You Can't See the Filters

The filters can be hidden. To show them again, simply click ⌄ (**Expand Header**). After you click it, the icon then looks like ⌃ (**Collapse Header**) and causes the filter fields to collapse.

---

2. On the initial **Manage Customer Master** screen, press the [Enter] key immediately or click the **Go** button on the far right. This means that you don't make a selection, and all customers are displayed in the list on the lower half of the screen, as shown in Figure 7.9.

   If you want to display only one customer, you can easily filter it out using the fields in the upper part of the window. For example, you can enter the customer number in the **Business Partner** field before clicking **Go**. (We'll discuss filtering functions in more detail in Section 7.7.)

3. From the list of all customers, clicking on a customer line will take you to the **Customer** form with all customer master data, as shown in Figure 7.10.

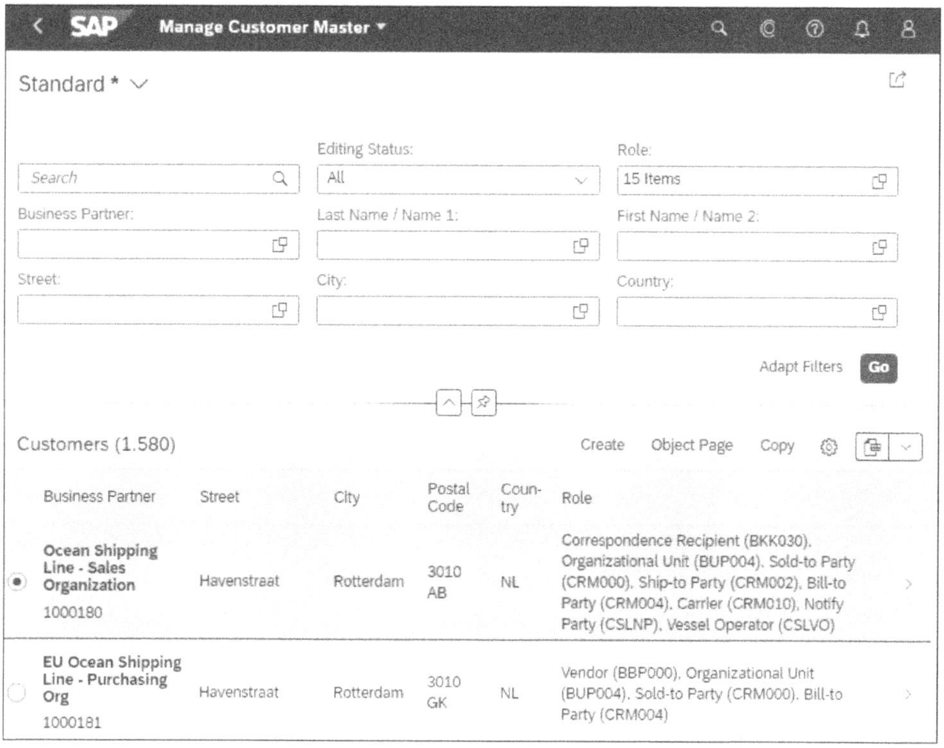

Figure 7.9 Customer List

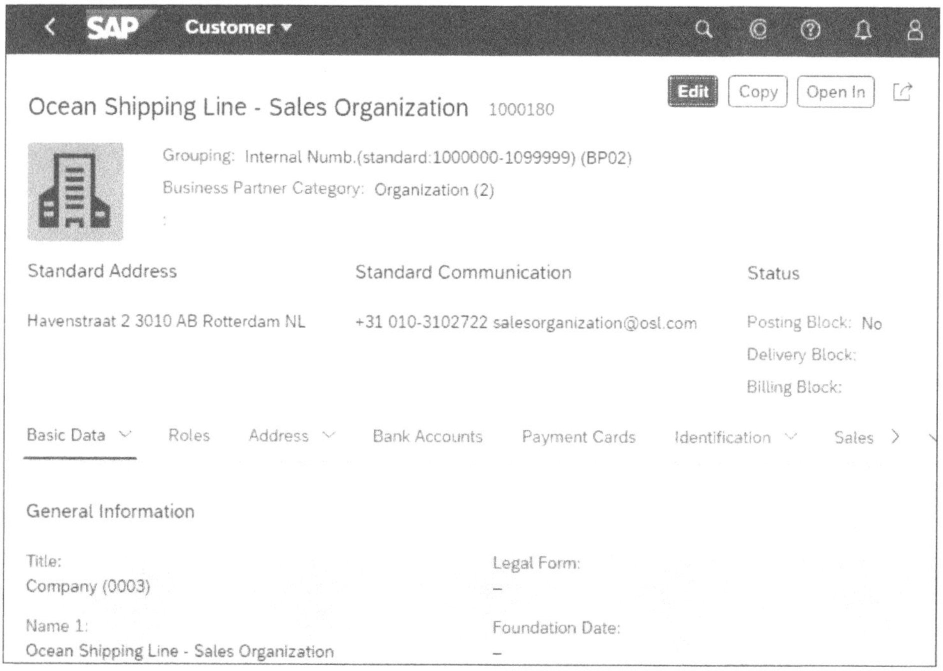

Figure 7.10 Customer Form with the Basic Data Tab

4. To exit the form, you have the following options in the shell bar:
   - The ❮ (**Back**) icon will take you back to the previous screen, which, in our case, is the list.
   - With the SAP icon from the shell bar, you immediately land on the start page.
   - The 🖻 (**Recent Activities**) and 🖺 (**Frequently Used**) menu items from the user menu, discussed in the previous sections, allow you to switch to another app in a direct way.
   - And, new for you is the small arrow ▼ to the right of the SAP icon or title of the initial screen. This navigation menu is only available in real SAP Fiori apps. You'll learn how to use it in the next steps.

5. In the shell bar, click ▼ on the navigation menu.

6. In the navigation menu, as shown in Figure 7.11, a dropdown list shows you the following options:
   - The previous stations of your journey, in our case, the **Home** and **Manage Customer Master** entries, and the current station **Customer**
   - The **All My Apps** menu item to switch to another app (for more on this, see the upcoming note box)

Figure 7.11 Navigation Menu

7. For the exercise, click on **Manage Customer Master** in the navigation menu. This will take you back to the list, as shown in Figure 7.12. Admittedly, clicking ❮ (**Back**) does the same thing and is faster. However, the navigation menu can be a help for apps with multiple screens.

8. Click again on a customer of your choice in the list. This will take you back to the form. Stay here, because we'll continue with the form in the next section.

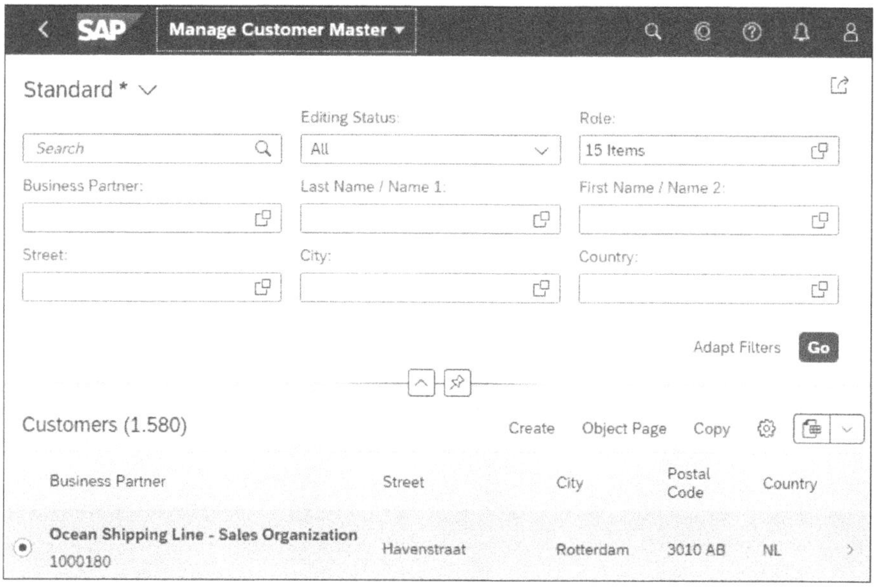

Figure 7.12 Customer List

Start Another SAP Fiori App with All My Apps

After clicking **All My Apps** in the navigation menu, you'll see a list with two parts, as shown in Figure 7.13.

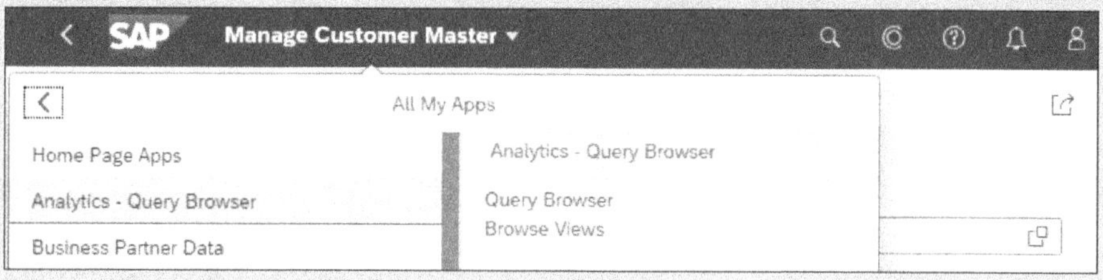

Figure 7.13 All My Apps Menu

In the left pane, you can see the catalogs, just like in the App Finder. In the right area, you'll find the corresponding apps for the catalog you clicked on. You can start an app by clicking on it.

7.3.2   Navigate in the Form

Do you still see the form with the customer master data? If not, switch back to this form using the instructions from the preceding section.

There are quite a few fields here, so we want to give you an overview of this app window first. The knowledge gained here can easily be transferred to other real transactional SAP Fiori apps.

There are three methods to make the data visible that is further down or up in a form and not shown on the screen:

- You scroll with the drag bar on the right edge of the window or with the mouse wheel/trackpad.
- You press one of the following keys: [ArrowDown], [ArrowUp], [PageDown], [PageUp], [End], or [Pos1].
- You select the appropriate tab at the top. For example, the form for the customer has the **Basic Data** and **Address** tabs. Click on a tab of your choice; in our example, click on **Address**, as shown in Figure 7.14.

7.3.3   Tab Bar and Tabs

A transactional SAP Fiori app is often divided into *tabs*, so that you can get to the desired data faster. In the tab bar, the tabs are in one line; here, you're offered navigation between the tabs (see Figure 7.14). As soon as you click on one of the displayed tabs, your browser automatically scrolls to the desired location.

Figure 7.14 Address Tab in the Customer Form

How Can I Realize Which Tab is Currently Open?

As soon as a certain tab is opened, its text is highlighted in the tab bar, and below the text there is also a colored line as an underline.

If a window is too small, you can display the hidden tabs by clicking ⟩ (**Scroll Right**) or ⟨ (**Scroll Left**) in the tab bar. These icons appear only as needed.

Instead, we recommend clicking on the ∨ icon on the far right because this displays all the tabs at a glance, as in Figure 7.15. Clicking on an entry in this dropdown menu will jump you to the corresponding tab.

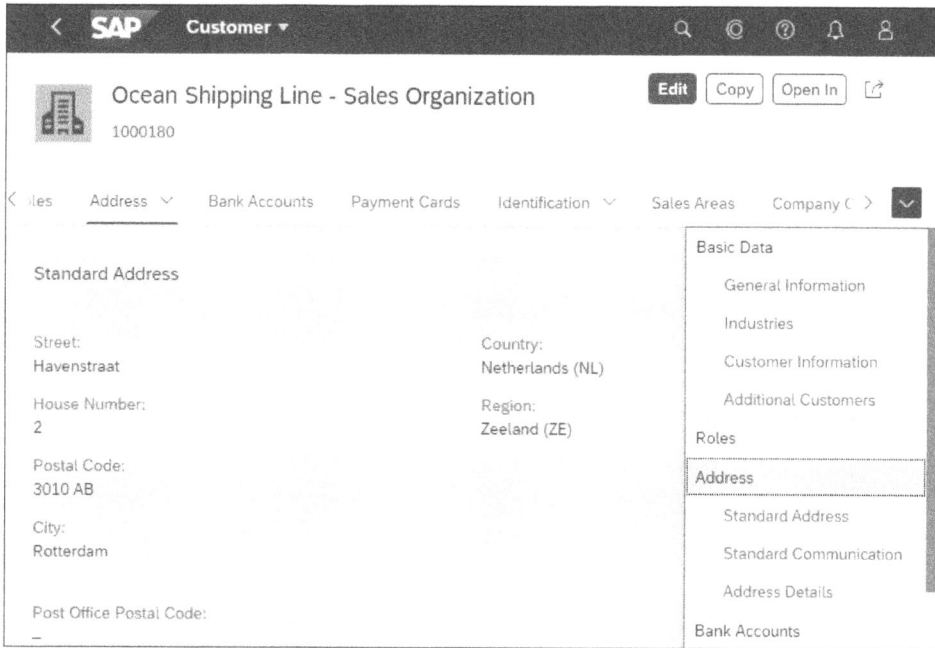

Figure 7.15 Overview of the Tabs Opened with the Icon on the Right Side of the Window

Find a Field or Field Content Faster

There are huge forms in SAP Fiori that span many, many screen pages. With a browser shortcut, you can quickly find the desired location in the form:

1. Press the key combination `Ctrl`+`F`. Depending on the browser, a search box will open above the shell bar as in Microsoft Edge (see Figure 7.16) or at the bottom of the screen as in Mozilla Firefox, for example.

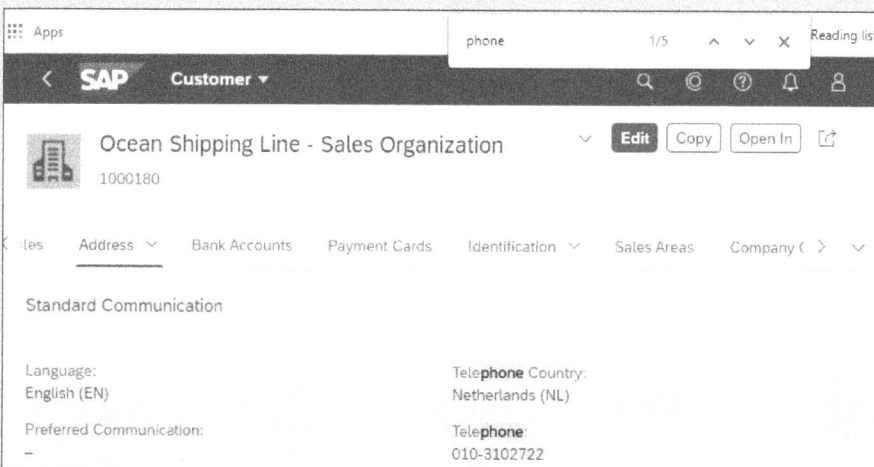

Figure 7.16 Opening a Search Box

2. Enter the search term in the browser search field. In our example, we searched for phone numbers using the search term "phone" as shown in Figure 7.16. The search term can be any text, for example, a field name or field content.

3. The browser jumps immediately during the input to the first occurrence and marks further places. With the `Enter` key, you jump to the next occurrence.

Now you may return to your SAP Fiori launchpad homepage for the next tutorial. The fastest way to do that is a click on the **SAP** icon in the upper-left corner.

7.4   Make a Change

Now we'll change a customer address and use the Customer Master app from the previous section again. And you'll see that it's a slightly different process than in SAP GUI because, this time, you first make a selection and only then change and save the data.

Follow these steps:

1. Open the Customer Master app by clicking the **Customer Master** tile.

2. You can click the **Go** button right away, that is, not filter and display all customers, as shown in Figure 7.17, or enter a customer number in the **Business Partner** field beforehand, for example, to pick out just the one customer.

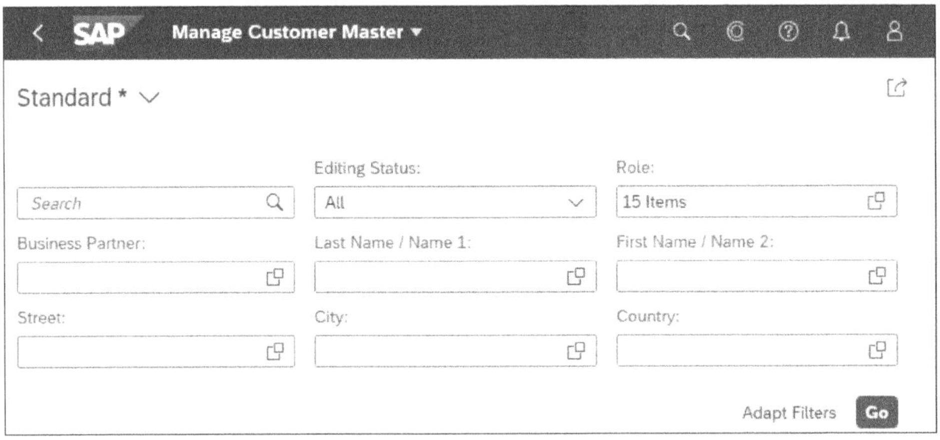

Figure 7.17 Customer Master App

3. In the list, as shown in Figure 7.18, click on the line of the customer you want to change, here the Dutch company "Ocean Shipping Line". This will take you to the corresponding form.

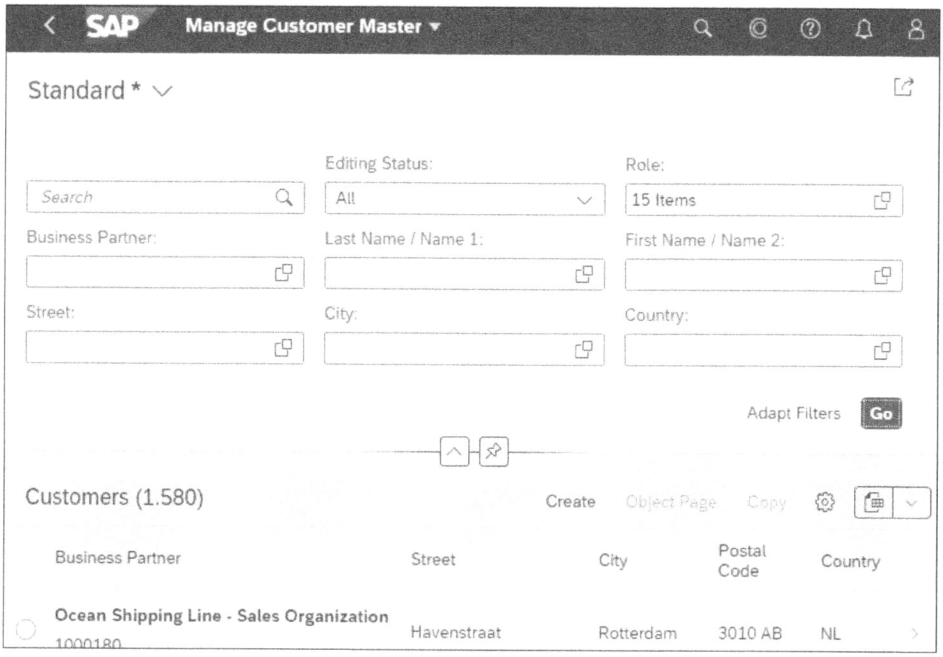

Figure 7.18 Clicking on a Customer in the Customer List

4. To change the address, first click on the **Address** tab in the tab bar, as shown in Figure 7.19.

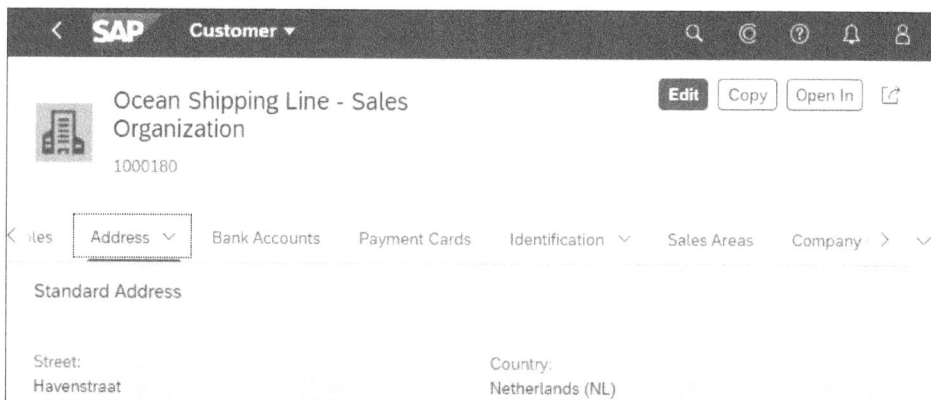

Figure 7.19 Address Tab

5. Now switch to edit mode by clicking the **Edit** button in the upper-right corner. As in SAP GUI, a record in edit mode is of course also locked in SAP Fiori against changes by other users.

And now for a little hidden object puzzle: What differences do you recognize between the form in display mode and the form in edit mode? There are three correct answers:

- The ⊡ **(Value Help)** icon in some fields such as **Country** or **Region**
- The boxes around the fields where data can be changed (fields without borders are protected against changes)
- The **Save** and **Cancel** buttons at the very bottom

6. You can now change the contents of the **Standard Address** field group as you wish, as shown in Figure 7.20.

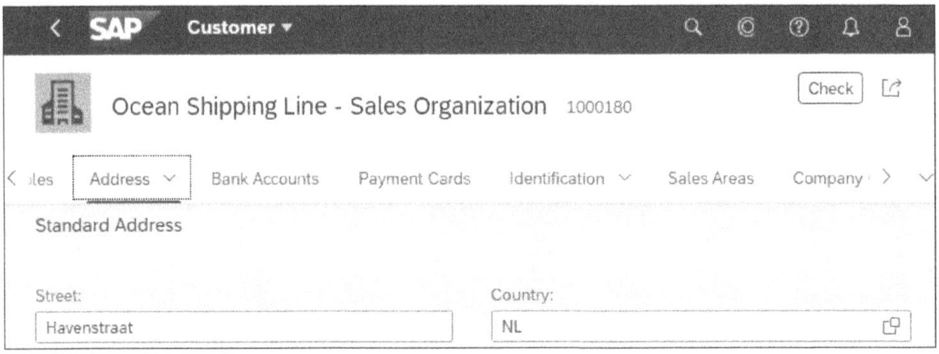

Figure 7.20 Change Standard Address

7. Afterward, save your changes by clicking the **Save** button at the very bottom. The app automatically switches back to display mode, as shown in Figure 7.21.

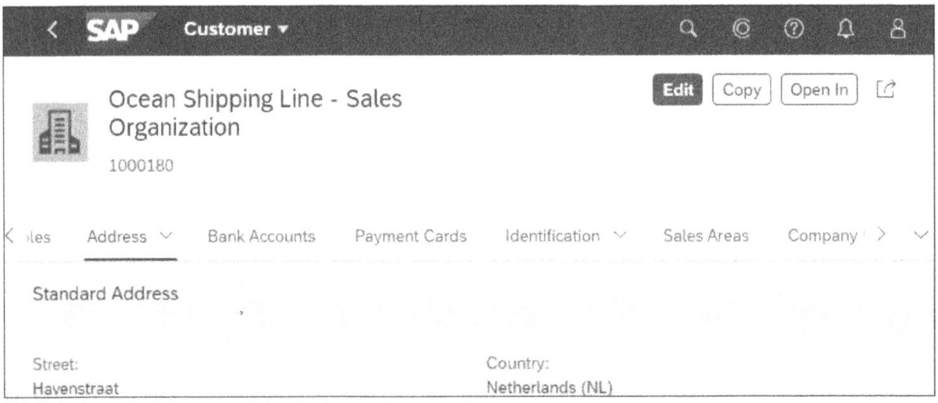

Figure 7.21 Display Mode after Saving Changes

8. To return to the SAP Fiori launchpad homepage, click **SAP** in the upper-left corner.

> **[!] Data Loss When Exiting the SAP Fiori App?**
>
> If you've made changes in an app and accidentally clicked **<** (Back) in the shell bar or launched another app from the navigation or user menu, SAP S/4HANA saves your change as a draft.

In other apps, a warning message saves you from data loss, as shown in Figure 7.22.

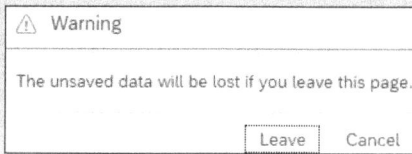

Figure 7.22 Warning Message about Data Loss

7.5   Create Data

In the previous section, you learned how to change the data of an existing customer in edit mode. To create a new customer, you use the same app as for viewing and modifying data.

In SAP GUI, there are usually *different* transactions for these operations, although many exceptions confirm this rule. In SAP Fiori, you usually have *one* real SAP Fiori app that you use to create, change, and re-create. Accordingly, with a real SAP Fiori app, three SAP GUI transactions are replaced, and, as a bonus, you also get a clear list with this app. Follow these steps:

1. Start the Customer Master app. The initial screen of the app appears, as shown in Figure 7.23.

Figure 7.23 Customer Master Data: Initial Screen

2. As in other real transactional SAP Fiori apps, you click the **Create** button in the initial screen to start a data entry process.

However, we'll stop this process here by clicking on the [SAP] icon and switch back to the SAP Fiori launchpad homepage because we'll discuss the creation of new records in great detail in Part III of the book.

[»]

> **Differences in Data Modification and Recording**
>
> There are some differences in data modification and entry between the SAP GUI and SAP Fiori apps:
>
> - Starting the acquisition with the **Create** button
> - Marking the mandatory fields with a red asterisk (*)
> - Using ⊡ (**Input Help**) for the value help, which is visible at all times and requires a click in the respective field in SAP GUI
> - Missing history as a list of all your inputs from the past in real SAP Fiori apps

7.6   Find Field Values with Value Help

Do you remember the picklists from SAP GUI? Of course, such a basic function also exists in SAP Fiori. However, it differs from the function in SAP GUI in more than just its name. (The procedures we cover in this section apply only to real SAP Fiori apps. The rules for picklists in legacy apps with "fiorized" transactions are the same as those in SAP GUI.) The values found via value help, such as country codes, are called *items*.

In this section, you'll learn how to use value helpers to find your input values, in the Customer Master app as an example. To train this function, we'll use the value list of the **Country** field from the Customer Master app and search for countries.

Follow these steps:

1. Launch the Customer Master app by clicking the **Customer Master** tile.
2. Click the **Go** button on the initial screen.
3. In the list, click in the row of any customer to get to the form.
4. In the form, first click the **Edit** button, and then click the **Address** tab in the tab bar, as shown in Figure 7.24.

   This point goes to SAP Fiori: In contrast to SAP GUI, here you can see immediately via ⊡ (**Value Help**) in which fields there is a value list. You don't need to click on a field to do this.

Figure 7.24 Address Tab

5. Click ⊡ (**Value Help**) in the **Country** field. (Alternatively, you can also click in the field and then press the ⌨ F4 function key as in SAP GUI.) The value help is displayed in a separate window, as shown in Figure 7.25.

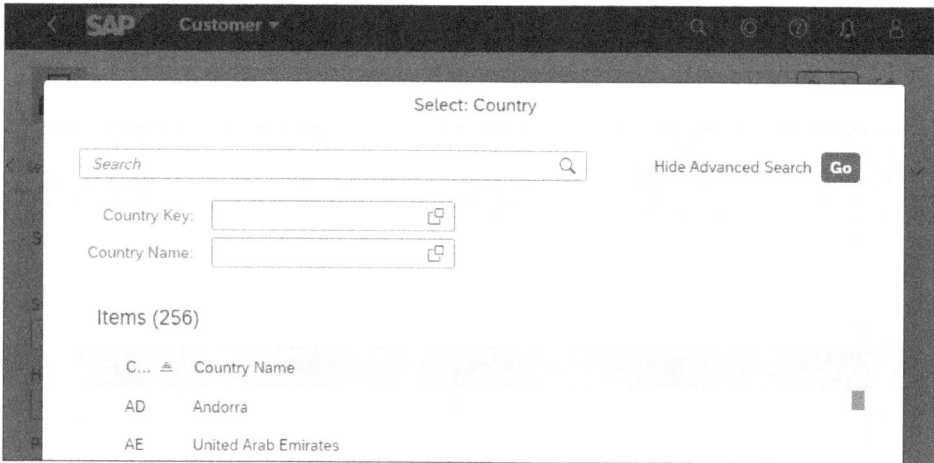

Figure 7.25 Value Help for the Country Field

6. In the upper part of the search window, you can now make specifications for the search, for example, enter the country you're looking for in the **Country Name** field, and then start the search by clicking on the **Go** button or by pressing the ⌨ Enter key.

   The items here are the country abbreviations (in SAP GUI they are called values). All countries in the world are displayed under the **Items** heading.

The Universal Search Box

A new SAP Fiori feature is only known in a few places in SAP GUI: a search field that searches not only the items of *a* column, but *all* columns. In our example, the **Country Key** and **Country Name** columns are searched simultaneously.

---

7. Enter "ME" in the **Search** field in the upper-left corner, as shown in Figure 7.26, and test the universal search. If you click the **Go** button, you'll see as items both a hit from the **Country Key** column (**ME** for **Montenegro**) and a hit from the **Country Name** column (**Mexico**).

8. The **Country Key** and **Country Name** fields under the **Search** field belong to the *advanced search*. You don't see these fields? In this case, the advanced search is hidden. You can make them visible by clicking on the **Show Advanced Search** button in the upper-right corner.

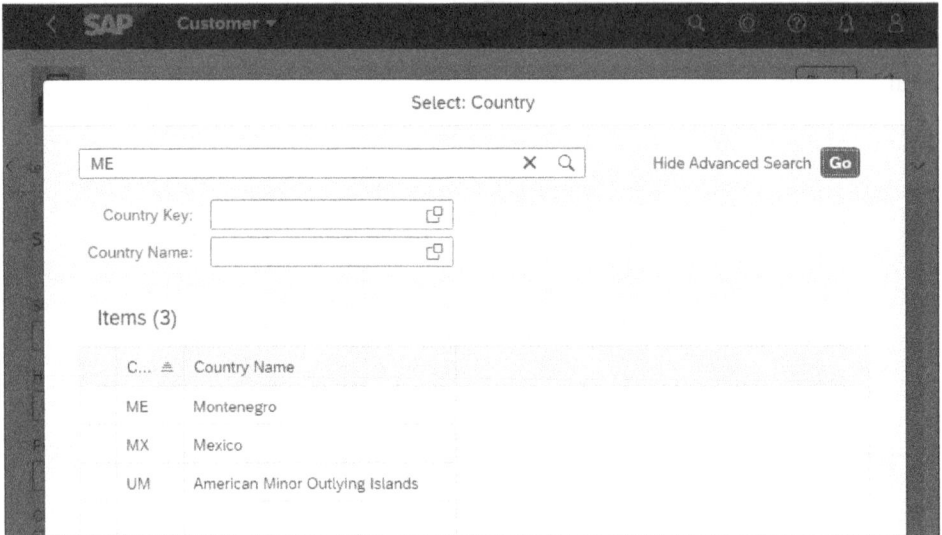

Figure 7.26 Universal Search

9. Now delete the "ME" entry in the upper-left search field, enter "ME" in the **Country Key** field instead, and click the **Go** button, as shown in Figure 7.27.

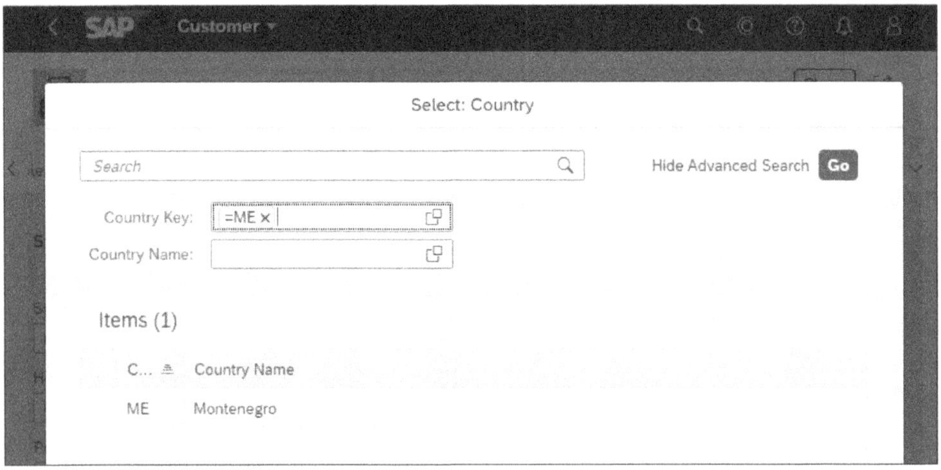

Figure 7.27 Searching with the Country Key Field

10. This time, you'll get only **ME** for **Montenegro** as a result because you've performed the search specifically for the country key in the **Country Key** column.

11. And as you may know from SAP GUI, with a double-click, you transfer an item into the input form.

You've now become familiar with the functions of this simple search.

> **Differences in the Input Aids**
>
> Do you miss some functions from SAP GUI here? We do too! In SAP GUI and in legacy apps, you can additionally create personal value lists. On the other hand, SAP Fiori also offers you advantages over SAP GUI with the universal search field and with the clear screen.

You may now close the value help by clicking the **Cancel** button at the bottom right, and then return to the SAP Fiori launchpad homepage with the by clicking **SAP**. You should be well rested for the next section because this it's about more complex searches in large data sets.

7.7   Find Master Data and Documents

In the previous section, you only searched for a country code. In this section, you'll learn how to search in larger data sets and find the matching information using the number of a business partner as an example. There are two methods for this:

- An automatic list that offers suggestions already as you type in the field in question
- The value help from the previous section, where you can use advanced search options when searching in extensive data sets such as master data and documents

7.7.1   Use the Automatic List

We'll show you the automatic list using the example of the Display Supplier Balances app. A supplier is a vendor, and based on the balance, you can see how much money you owe the vendor.

The method to search for a supplier is transferable to real SAP Fiori apps from other areas, no matter if you're in purchasing, sales, or anywhere else. Follow these steps:

1. Launch the Display Supplier Balances app, as shown in Figure 7.28.

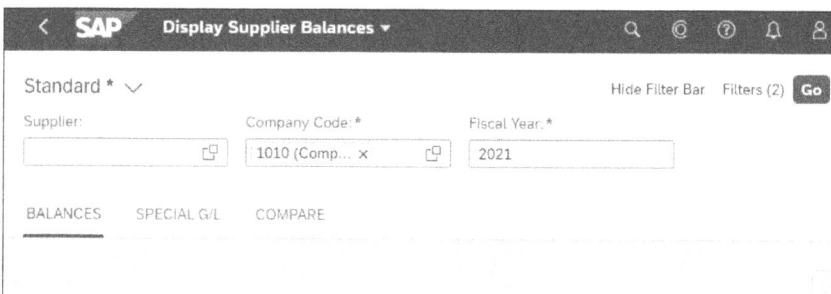

Figure 7.28 Display Supplier Balances App

2. To find the desired supplier, you might intuitively click (**Value Help**) in the **Supplier** field on the initial screen to start the search help—but don't do that! It's much easier to follow the next steps.
3. Enter "New" in the **Supplier** field, as shown in Figure 7.29. Even as you type, SAP Fiori shows you the search results below the fields.

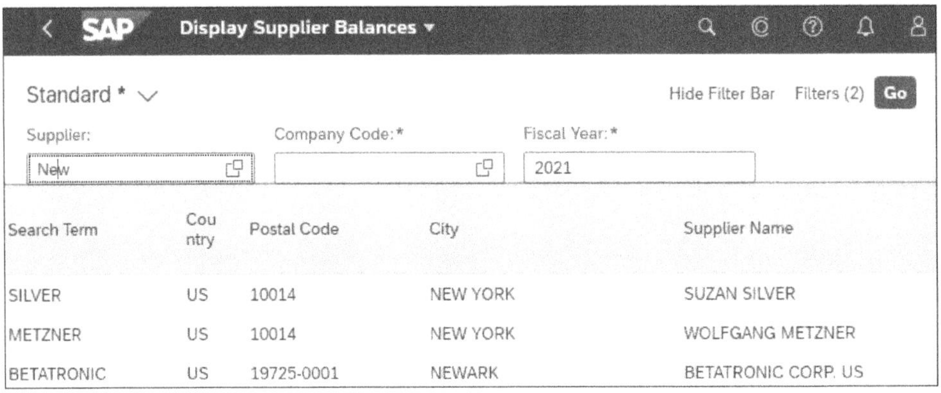
SAP
Display Supplier Balances
Standard *
Hide Filter Bar
Filters (2)
Go
Supplier:
New
Company Code:*
Fiscal Year:*
2021
Search Term
Cou ntry
Postal Code
City
Supplier Name
SILVER
US
10014
NEW YORK
SUZAN SILVER
METZNER
US
10014
NEW YORK
WOLFGANG METZNER
BETATRONIC
US
19725-0001
NEWARK
BETATRONIC CORP. US

Figure 7.29 Searching for the Supplier

4. Click on a line in the search result to transfer the supplier's name to the **Supplier** field. It's that easy!

Controllers, accountants, and, of course, all other interested parties, are welcome to click the **Go** button to view the balance and other information about this supplier.

The automatic list works in all apps, including "fiorized" transactions. However, it's available only for selected and frequently used fields such as **Material**, **Customer**, **Purchase Order**, or **G/L Account**.

Following are some more rules for automatic search in real SAP Fiori apps:

- The wildcard *, which you know from SAP GUI transactions, works with these entry helpers, but the wildcard + unfortunately doesn't.
- You've seen in our example that you can use contents from other fields of the supplier master record for the search in the **Supplier** field: we used "New", an entry from the **City** field. In principle, you may enter anything you think of for the item you're looking for in a field. The main thing is that it can be found somewhere in the data set.

7.7.2 Determine the Optimal Search Template

A *search template* is nothing more than a search help in SAP GUI or in a "fiorized" transaction, that is, a compilation of different fields. In SAP GUI and in "fiorized" transactions, there are different tabs for selecting search helps but not in real SAP Fiori apps. Therefore, it's best to compile a suitable search help yourself.

For our example, we again use the value help for the **Supplier** field. For the search, we specify a number range and then a postal code range. Follow these steps:

1. Start the Display Supplier Balances app if you got out after the last section. You'll arrive at the screen shown in Figure 7.30.

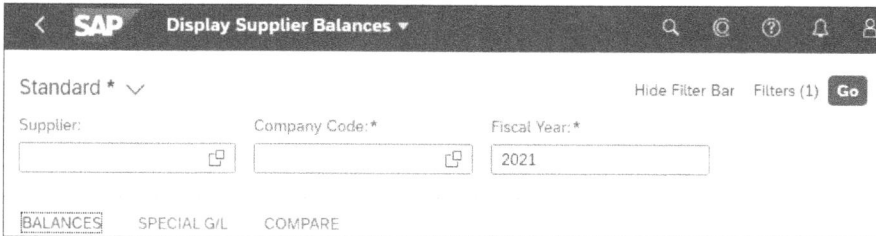

Figure 7.30 Display Supplier Balances App

2. This time, click ⊡ (**Value Help**) in the **Supplier** field. Now a similar value help appears as in the previous section for the **Country** field, as shown in Figure 7.31.

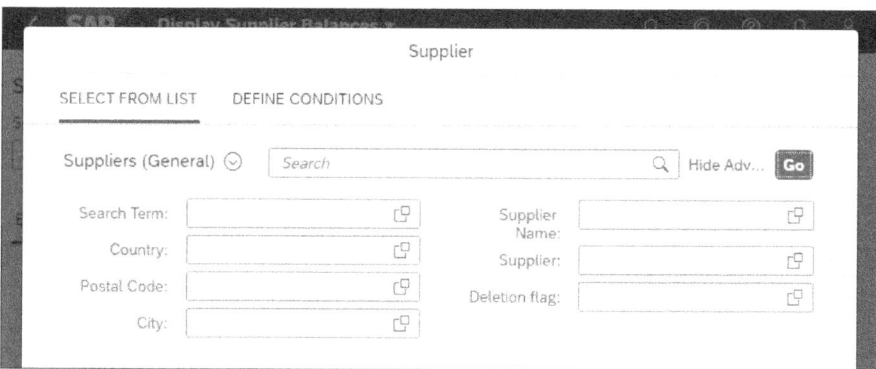

Figure 7.31 Value Help for Supplier

What differences does it show in the top half of the window compared to the automatic value help of the **Country** field from the previous section, except that there are more search fields? Here are the differences from top to bottom:

- You'll see an additional **Define Conditions** option at the very top, which allows you to additionally include or exclude values.
- To the right of the **Suppliers: (General)** entry, you'll find the ⊙ icon, which we'll discuss now.

3. Click on the ⊙ icon to see the possible search templates, as shown in Figure 7.32. The search templates correspond to the search helps on the tabs in SAP GUI.

4. If you select a different search template here, you'll get different search fields. With the **Suppliers: Purchasing** search template, for example, buyers can specify the purchasing organization.

In our example, we click on the **Suppliers by Country** search template.

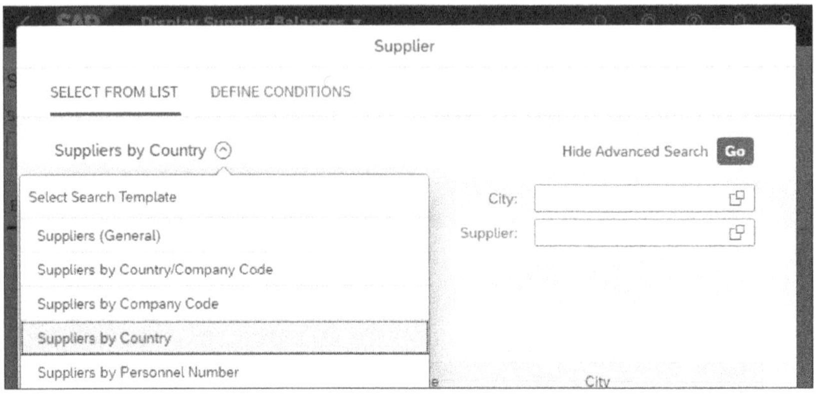

Figure 7.32 Search Templates

5. Now enter "GB" for Great Britain in the **Country** field, as shown in Figure 7.33. After clicking on the **Go** button, all corresponding suppliers will be displayed.

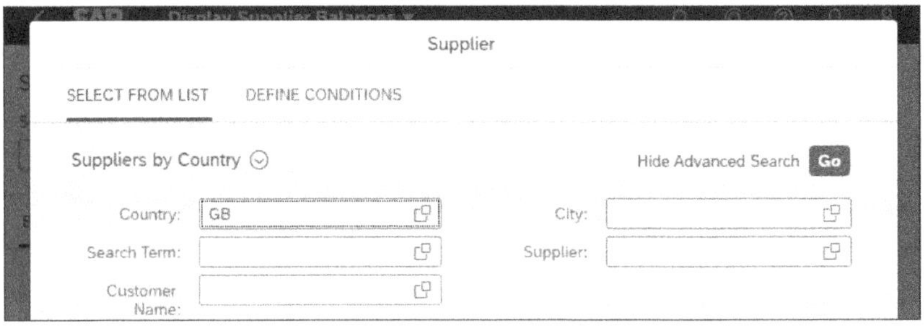

Figure 7.33 Country Search Template

6. Now you can select the desired supplier and apply it to the filter by clicking the **OK** button, as shown in Figure 7.34.

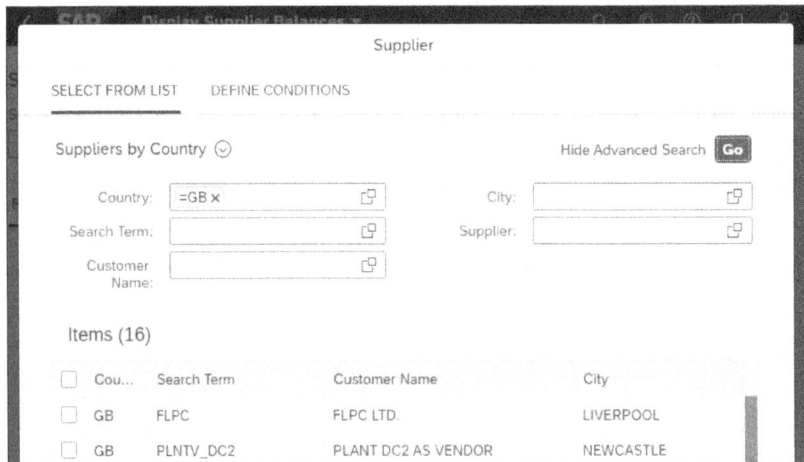

Figure 7.34 Appling the Supplier to the Filter

7. After that, you may close the app by clicking on the [SAP] icon.

Many other entry helpers of real SAP Fiori apps have only *one* search template, which is called **Standard**.

7.8   Enterprise Search for Global Search

Imagine the following common case in practice: A colleague or business partner calls and wants to get information about a certain order, a purchase order, or an invoice very quickly and tells you the corresponding document number. Depending on the performance and data volume of your SAP S/4HANA system, you conjure up the document on the screen before the person you're talking to has taken a breath. You don't know this from SAP GUI; instead, in SAP Fiori, you don't have to call up an app to display a specific order, for example. With a term or a number, you can search the entire SAP database via the enterprise-wide SAP search engine called *enterprise search*.

Before we get started, let's clarify the term *business object*, which you're about to encounter on the screen. A business object can be any record, such as a material or a supplier, or a document, such as a purchase order, an order, or a booking. A material master record is one business object, and a purchase order is another business object.

Now, let's walk through the steps to use the enterprise search functionality both with and without a business object type.

7.8.1   Search without Object Type

Here we want to search for purchase orders for a given material as an example. For the search, you can specify the type of a business object, that is, the *object type*. This can be, for example, the object type *purchase orders* or the object type *material*. Ideally, however, the search will also work without the specification of an object type, as you'll see in these instructions.

Follow these steps:

1. Click on [Q] (**Search**) in the shell bar. Two additional fields will appear to the left of this icon.
2. Enter the order number in the search field. In our example, we use the order number "4500000001" (hint: there are seven zeros between the 45 and the 1 at the end), as shown in Figure 7.35. Even as you type, SAP Fiori tries to preview the results list of your search. This includes all business objects matching your search.

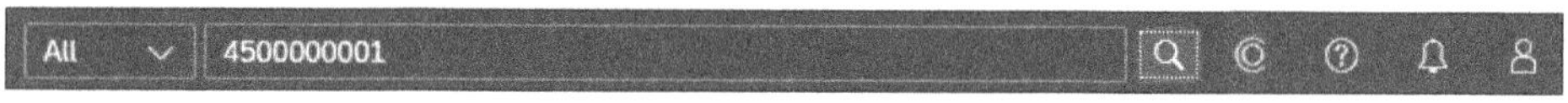

All ⌄	4500000001		Q	ⓒ	⑦	🔔	👤

Figure 7.35 Searching for the Purchase Order Number

3. After entering the number, click 🔍 (**Search**) again. (Alternatively, you can also press the ⌷Enter⌷ key.)

Now the found purchase order is displayed with the most important information in the *fact sheet*, as shown in Figure 7.36.

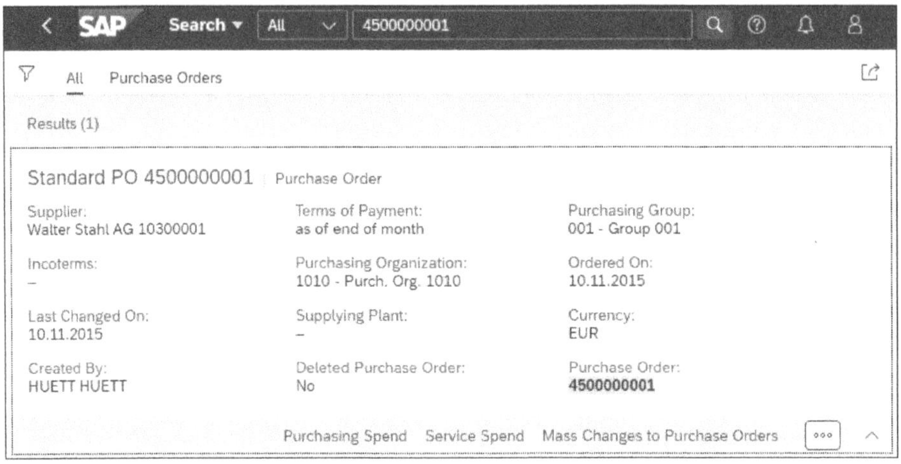

Figure 7.36 Fact Sheet

The information is at your fingertips so fast! If the information displayed here isn't sufficient, you have the following additional options:

- By clicking on the object number (in our example, the purchase order number), you switch to the corresponding app, which shows you more data in each case, or to another fact sheet.

- Under the information found (in our case, the order data), you'll usually find links to more apps that match the respective object, for example, the link **Purchasing Spend**. With the ●●● (**More**) icon, you get a further selection of matching apps.

[»]

Fact Sheets

As a result of the search, you've received a fact sheet , that is, an overview with important information about a business object. Fact sheets are a separate app category alongside transactional and analytical apps. They can't be accessed directly as a separate app, but only via enterprise search or links.

[+]

Context Navigation between Related Objects

If you click on data highlighted in color in apps, such as the order number in our example, you'll receive further information. From the order displayed after clicking on the order number, you can in turn call up further detailed information by clicking on the supplier or the ordered material. SAP also calls this *context navigation* between related objects.

7.8.2   Search with Object Type

Are you familiar with the hidden object book *Where's Waldo?* In this book, the task is to find Mr. Waldo, a man with glasses, in a very large crowd of people. You know exactly what he looks like, and yet it takes you a very long time to find him. Imagine how much more time and effort you would need if you didn't know his specific appearance, but only had the information "man with glasses"!

You can imagine the search in SAP Fiori in a similar way. Instead of searching in a huge amount of information with sparse information, it's usually better to narrow down the search to a specific object type beforehand. You save yourself time and save the system a lot of computing power!

In the following example, we're looking for purchase orders. We know the material number and want to search out the associated purchase orders in connection with a goods recall. To do this, proceed as follows:

1. Open the enterprise search by clicking ⧉ (**Search**) in the shell bar.

2. In the shell bar, click the ⌄ icon next to the **All** field. This will open the object list, as shown in Figure 7.37.

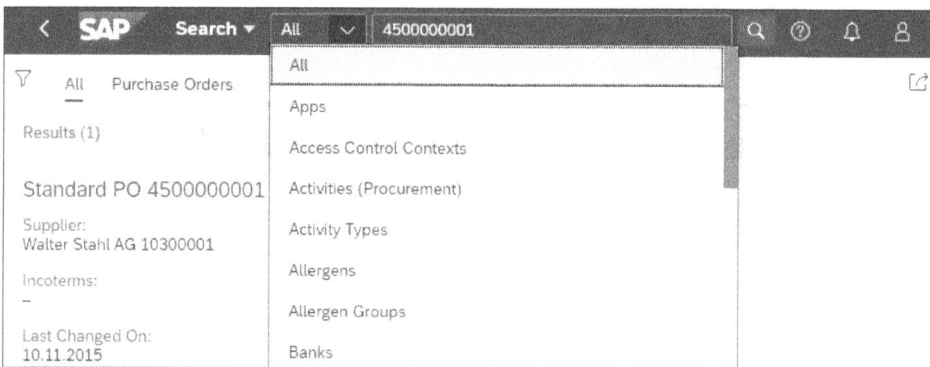

Figure 7.37 Opening Object List

3. Click the **Purchase Orders** entry in the list.

4. Enter the material number "TG0011" in the enterprise search field, as shown in Figure 7.38, and click the ⧉ (**Search**) icon again to confirm.

Figure 7.38 Entering the Material Number

Only the orders for the material number TG0011 will be displayed, as shown in Figure 7.39. The number next to an entry shows you the number of objects found.

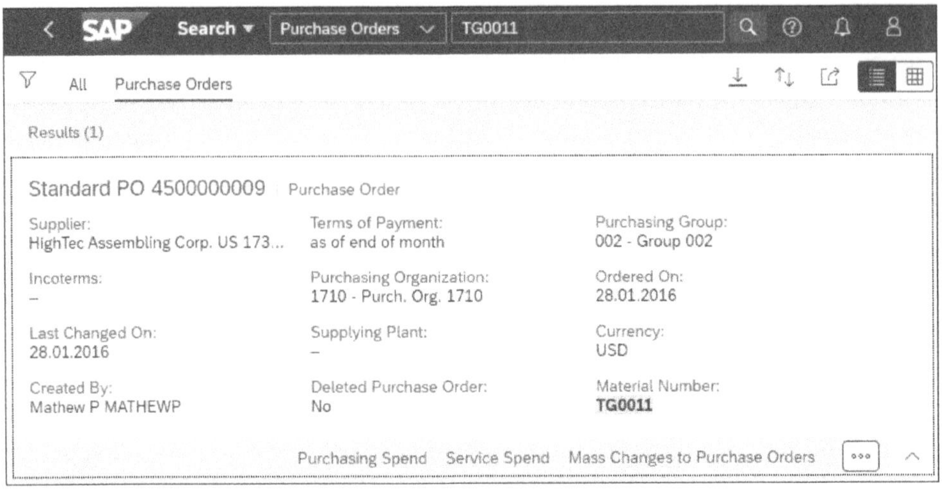

Figure 7.39 Display Order for the Material Number

With a little typing and two clicks you've called up a document overview. Of course, this also works with master data.

To get a better overview, click ⊞ (**Show as Table**) in the upper-right corner, and you'll see the search result as a table, as shown in Figure 7.40. To return to the initial state, click the SAP icon.

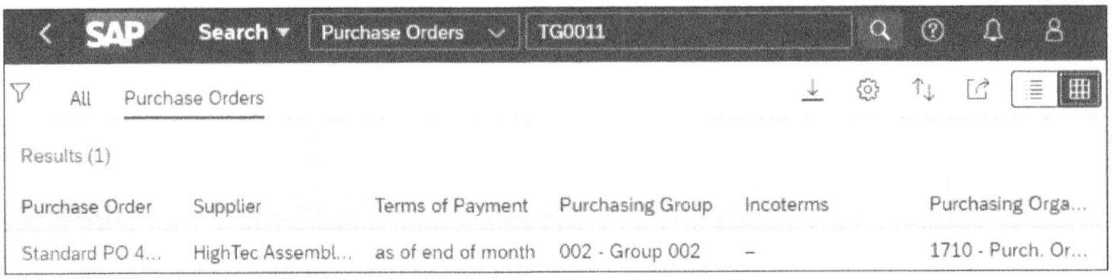

Figure 7.40 Search Results as a Table

Observe the following rules for enterprise search:

- SAP Fiori shows you only the data for which you have authorization.
- There is usually no distinction between uppercase and lowercase letters.
- You can also use multiple search terms; for example, in the supplier search, you can use the name of the supplier and its location.
- Not all fields are searched. For example, if you're searching for a supplier, you can create search templates for the **Search Term, City, House Number, Name, Postal Code, Street, Company Code, Country Key**, and **Purchasing Organization** fields. Other fields, such as **Telephone**, aren't searched.
- You can use the wildcard *, which you may know from the SAP GUI transactions. The wildcard + doesn't work in enterprise search.

7.9   Help for Self-Help

Are you somewhere in the depths of an app and don't know what to do? The usual "just Google it!" advice, shouldn't be your first choice in SAP Fiori because SAP Fiori offers its own comprehensive, intuitive, and reliable help for many questions. In this section, you'll learn how to access this help and use it optimally.

7.9.1   Help for a Field

Follow these steps to access **Help** for a particular field:

1. Start the Manage Sales Orders app.

2. Now direct your gaze to the shell bar. Here, click ⑦ (**Open Help**) to open the help (alternatively, press the F1 key). As you can see in Figure 7.41, some fields, icons, or functions on your screen will be surrounded by a colored circle. This signals that SAP Fiori offers help for these elements.

   The **Help** area on the far right is an interactive menu with help offers for the currently called app. By default, the **Help Topics** tab 🖵 is open here. Minimalist explanations of the items marked with a circle on your screen are displayed here.

   You can now either click directly on one of the colored circles or on an entry in the **Help Topics** area on the right.

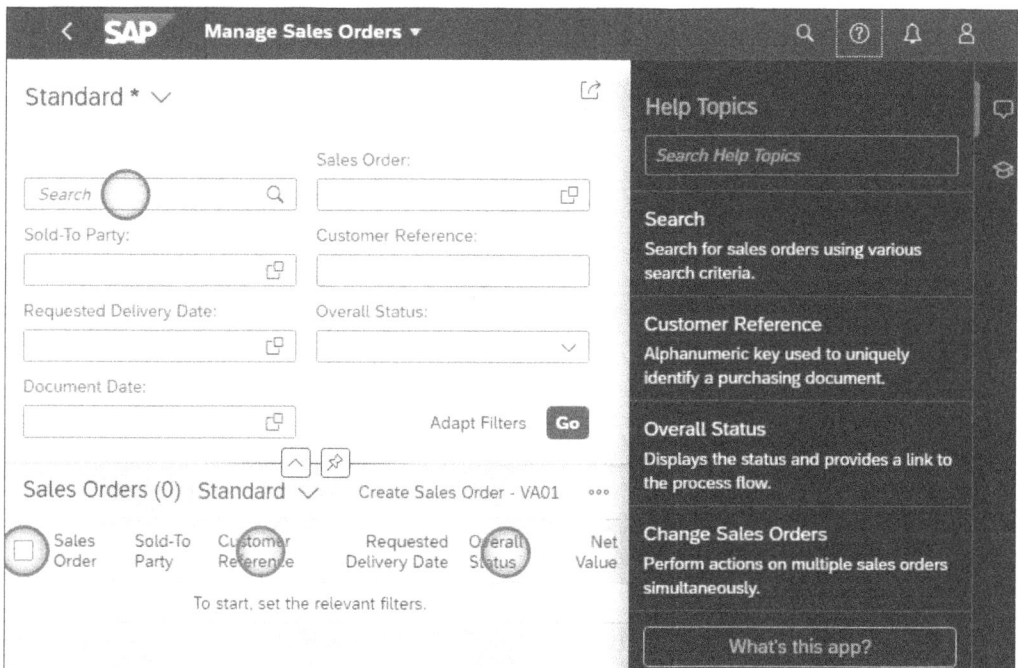

Figure 7.41 Open Help View

3. Click the marker around **Customer Reference**, and the requested help content for the **Customer Reference** field is displayed immediately, as shown in Figure 7.42. It includes a definition of the term and a description of its usage. Note that help content can also include videos as well as links to other content of interest to you.

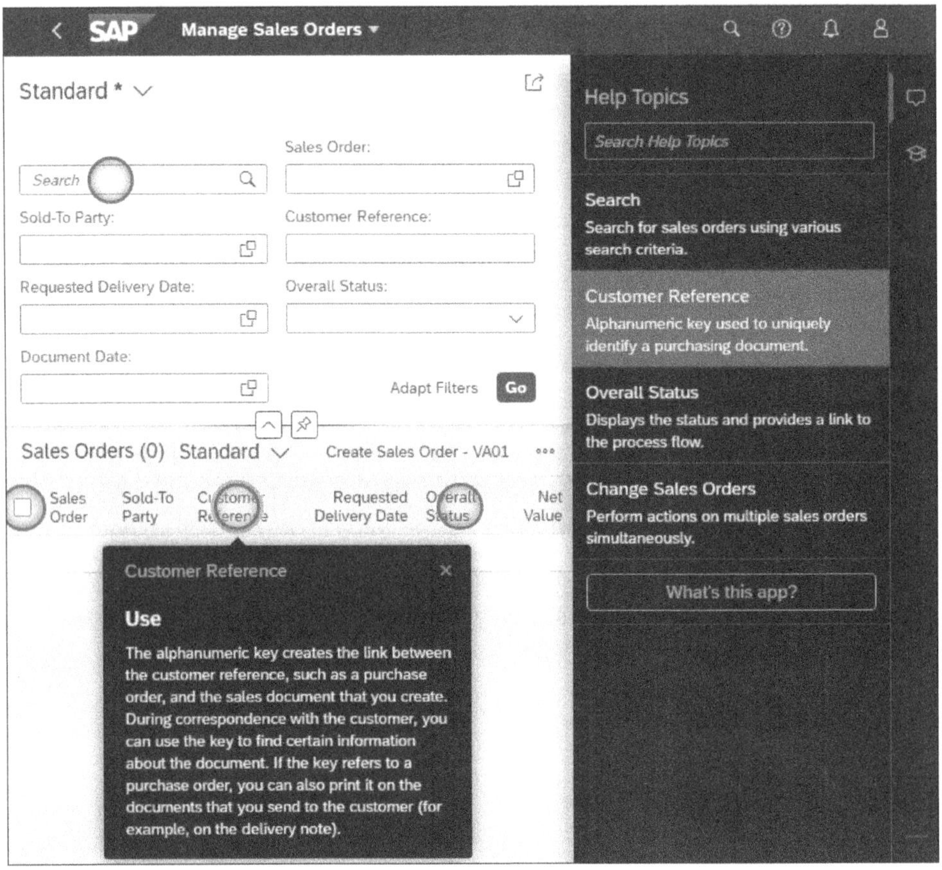

Figure 7.42 Help Content for Customer Reference

4. To hide the help again, just click on ⊘ (**Open Help**) in the shell bar again.

> [»]
>
> **Is the Open Help Icon Missing in the Shell Bar?**
>
> If the ⊘ (**Open Help**) icon is missing in the shell bar, the help isn't configured in your SAP system. Unfortunately, only an admin can assist you to get the help function.

7.9.2   Help with an SAP Fiori App

In addition to help on individual fields, icons, or functions within an app, SAP Fiori also provides extensive information on the apps themselves:

1. In the **Help Topics** pane on the right, click the **What's this App?** button.
2. A subpage of SAP Help Portal opens in a separate window in your browser, as shown in Figure 7.43. Here you'll find everything you need to know about the app.

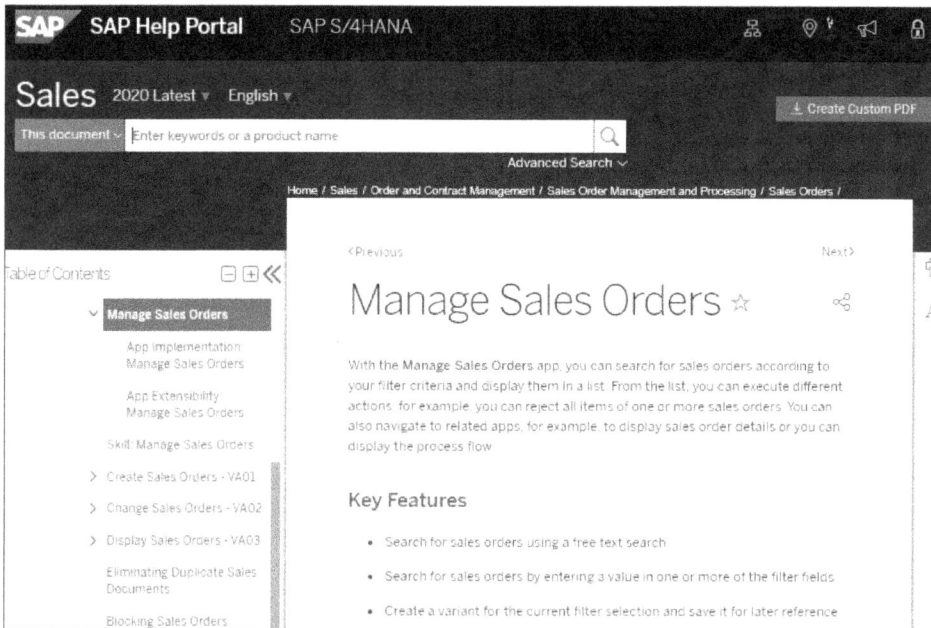

Figure 7.43 Help for the Manage Sales Orders App

Even More Help in the Learning Center

When you opened **Help**, you may have noticed that two icons are displayed in the **Help Topics** area on the right. So far, you've only seen ▣ (**Help Topics**).

When you click on ▣ (**Learning Content**), the **Learning Content** tab opens. In it, you'll be offered the **Learning Center** button. The learning center gives you access to a large amount of tutorial packages.

[+]

How Can I Get More Information?

To find even more information, you can use either the search or the table of contents contained in the left part of the window on the *https://help.sap.com* website.

[«]

7.10   Work Optimally with the Browser

Working with SAP Fiori always means working with the browser. In this section, you'll learn the necessary skills to create a "symbiosis" between SAP Fiori and your browser.

Mozilla Firefox, Microsoft Edge, and Google Chrome are the most commonly used browsers supported by the SAP system. The following instructions have the Microsoft Edge browser as a starting point (see Figure 7.44), which is the successor to Internet Explorer. Are you using a different browser? Don't worry, the following methods will work with other browsers in most cases.

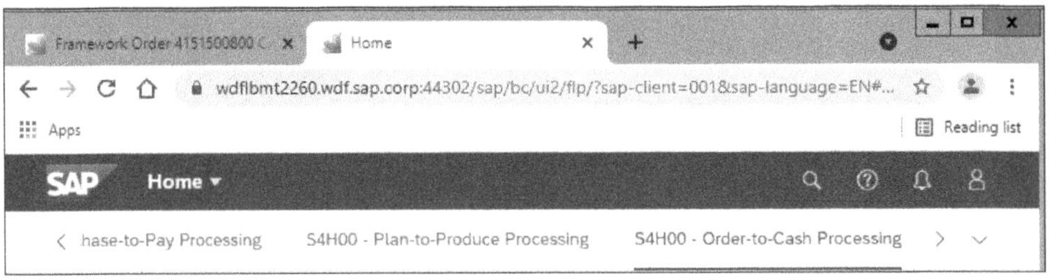

Figure 7.44 SAP Fiori Window in a Browser Window

7.10.1   Open a New SAP Fiori Tab

In Microsoft Edge, to create an additional SAP Fiori tab, simply use the shortcut Ctrl + Shift + K. This duplicates the already active tab.

On the SAP Fiori launchpad homepage, all you need to do is right-click a tile. In the context menu, you then select the appropriate browser command, for example, **Open Link in New Tab** or **Open Link in New Window**.

[+]
Switch Tab with Shortcut

With the shortcut Ctrl + #, you switch elegantly between different tabs. Just replace "#" with the number of the tab, to make it even easier: To switch to the first tab, press Ctrl + 1, with Ctrl + 2 you switch to the second tab, and so on.

7.10.2   Convert a Tab into a Window

In SAP GUI, you can open multiple SAP GUI windows and work with multiple SAP windows in parallel. This also works in SAP Fiori, or to be quite precise, it works with your browser. Follow these steps:

1. In Microsoft Edge, duplicate the tab with the shortcut Ctrl + Shift + K. A second tab has been created from the existing tab.

2. Now you want to open this tab in a new window. Click on the tab and hold down the mouse button.

3. Drag the tab to an area below the browser's navigation bar while holding down the mouse button, and then release. The tab has been "transformed" into a window, as shown in Figure 7.45.

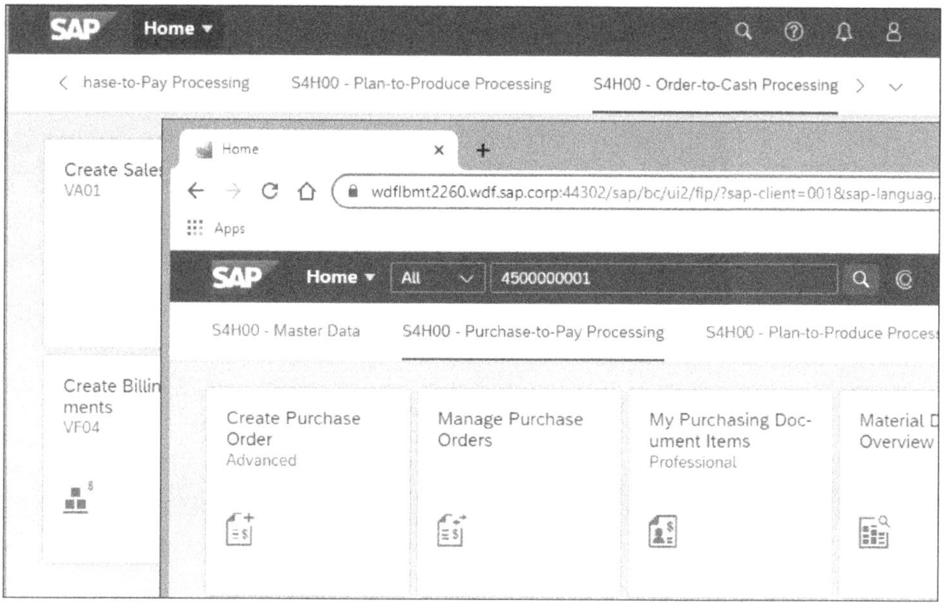

Figure 7.45 Creating a New Window

7.11 Summary

Done! Now, at the end of this chapter, you know the basics as well as some valuable tips on how to use real SAP Fiori apps to view, create, and change data. You can use filters and enterprise search to find your documents and master data, and you can get help resources for apps and fields. Stay tuned! Let's continue with the exciting topic of SAP Fiori reporting.

Chapter 8
Reporting with SAP Fiori Apps

One of our favorite topics is creating reports in SAP Fiori. In addition to tabular reports, similar to Chapter 3, you also design charts here.

Although the G in GUI stands for "graphical," the options for graphical displays are so limited that they are hardly ever used in practice. This is where SAP Fiori can score; the capabilities may surprise you!

What You'll Learn

- How to find and perform tabular evaluations
- How to filter the data you want from large amounts of data
- How to print evaluations or save them as PDF files
- How to download tables in Microsoft Excel
- How to create beautiful charts with analytical apps

We only cover "real" SAP Fiori apps in this chapter. "Fiorized" transactions work in SAP Fiori in the same way as you learned in Chapter 3 for SAP GUI.

8.1 Find and Launch SAP Fiori Apps with Tables

You've already seen an SAP Fiori app that allows you to create a table with the Customer Master app. What you know from SAP GUI under the term list is called a *table* in SAP Fiori. In Figure 8.1, we use a different app called the Manage Purchase Orders app. The structure of the two apps is pretty darn similar, and that's a good thing—know one app, know almost all!

So how do you find apps that contain tables? These apps have the word "manage" in common in their titles. So, if you search for the search term "manage" in the App Finder, you'll see the following apps, for example:

- Manage Business Partner Master Data (for customers and vendors)
- Manage Sales Orders or Manage Billing Documents (for the sales department)
- Manage G/L Account Master Data or Manage Journal Entries (for financial accounting)
- Manage Purchase Orders or Manage Purchase Requisitions (for purchasing)

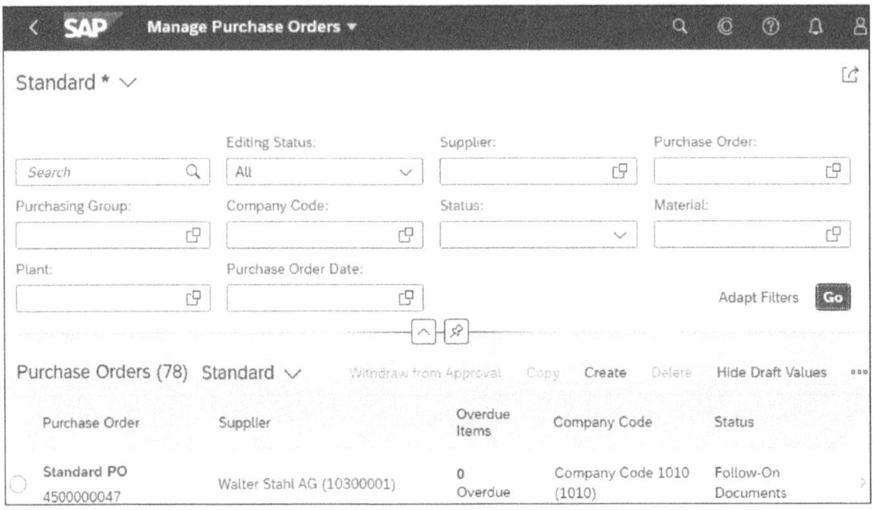

Figure 8.1 Table from the Manage Purchase Orders App

We'll show you the basic handling of all these apps using the Manage Purchase Orders app as an example. As always, you can also use another app with a table for this purpose.

Follow these steps:

1. Launch the Manage Purchase Orders app.

2. On the **Manage Purchase Orders** screen, you select your records (in our example in Figure 8.2, these are purchase orders) using filters, which we discuss in Section 8.2. This corresponds in SAP GUI to the selection in the selection screen.

3. However, to see all orders in the list right now, click the **Go** button on the far right.

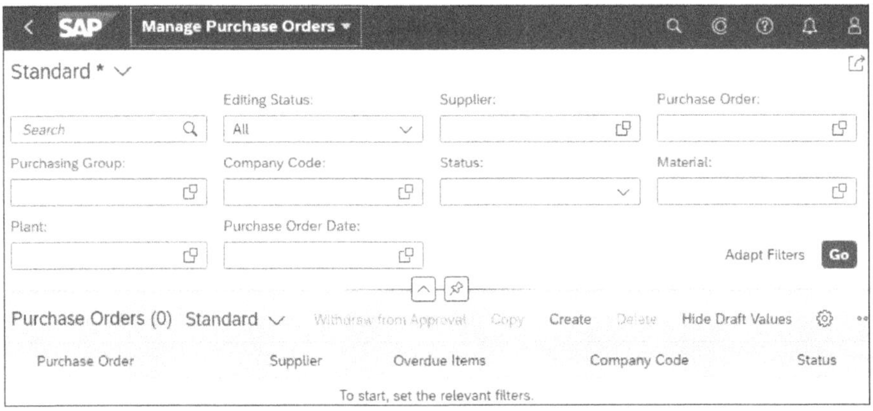

Figure 8.2 Manage Purchase Orders App

4. From the table, click on an order number in the **Purchase Order** column on the left to go to the order form, and then switch back using the ⟨ (**Back**) icon to return to the table.

In the row directly above the table, you'll see additional buttons, as shown in Figure 8.3, in our example for functions to copy, create, or delete an order.

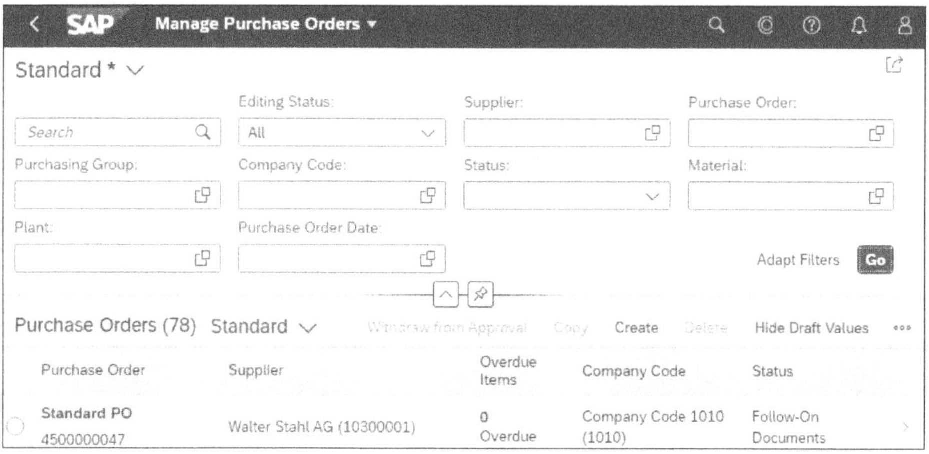

Figure 8.3 Table Options above the Table

No Real Table to See?

If several fields are displayed one below the other in your table, as shown in Figure 8.4, it's not a usable table. You'll have this effect if the window is too narrow.

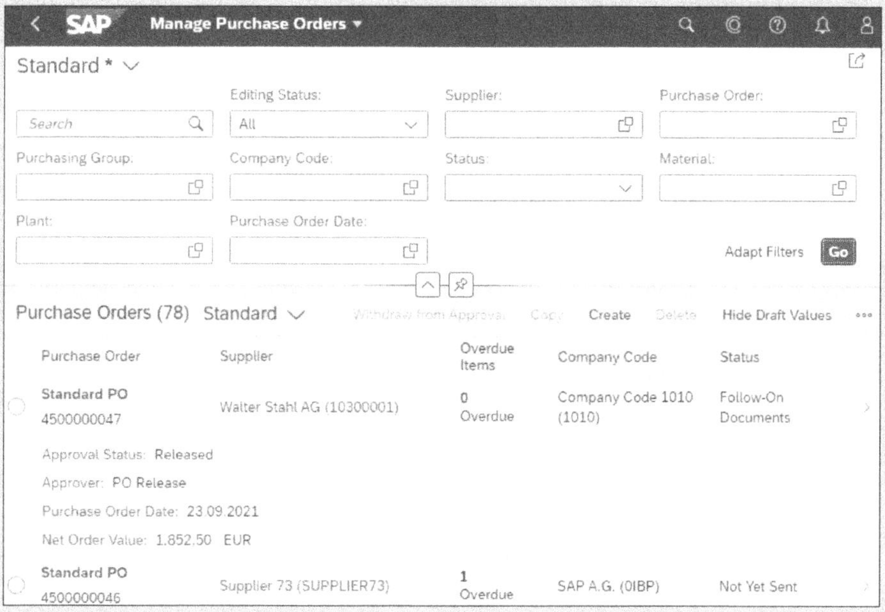

Figure 8.4 Table with Multiple Rows per Order

However, you usually want to see all the information in one line. There are two ways to do this: widen your window or reduce the zoom level in your browser.

8.2   Filter Tables in Header

For the most part, tables have very flexible and powerful filter functions for selecting data. In this section, we'll show you the filter options that you can use to select the desired documents. To do this, use the filters in the upper-part of the screen, which is also called the *header*. You must not confuse the header with the header line that contains the logo!

You've already become acquainted with some filter functions in Chapter 7, Section 7.6 and Section 7.7:

- The universal **Search** field in the upper-left corner, which searches not only the elements of one column, but several columns at once
- The ⧉ (**Value Help**) icon with which you can search for a *single value* (technical term: *element*) for a field
- The ⊡⌄⊡ (**Expand Header**) and ⊡⌃⊡ (**Compress Header**) icons, which are used to show and hide the filters

These techniques are also available in the filter functions of the table and won't be discussed again here.

8.2.1   Filter with Multiple Elements and Fields

Compared to the value help in forms, extended filter options are available in a table, which we'll present in the following instructions. Here we want to select all orders for two different purchasing groups. The purchasing group is used in practice either for individual purchasers or for purchasing teams.

Follow these steps:

1. Start the Manage Purchase Orders app to arrive at the screen shown in Figure 8.5.

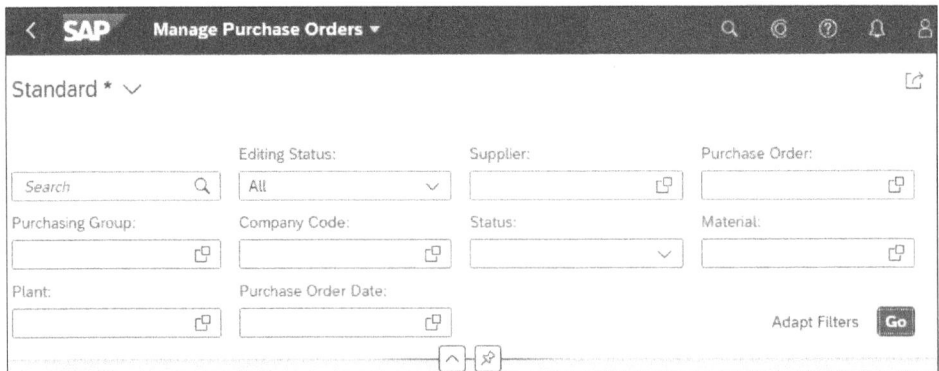

Figure 8.5 Manage Purchase Orders App

2. In the **Purchasing Group** field, click ⧉ (**Value Help**). As shown in Figure 8.6, a popup window appears with value help you can use to select the desired purchasing groups.

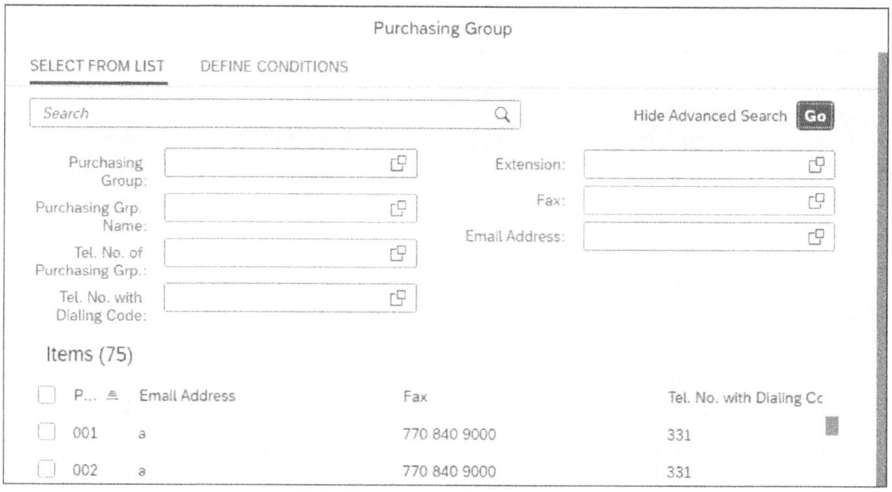

Figure 8.6 Purchasing Group Popup

3. If you click the **Go** button with any restrictive entries in the fields, purchasing groups will be displayed in the lower part of the screen under the **Items** heading.

4. Select the desired purchasing groups on the far left, as shown in Figure 8.7. In contrast to a value help in a form, you may select several elements at the same time here.

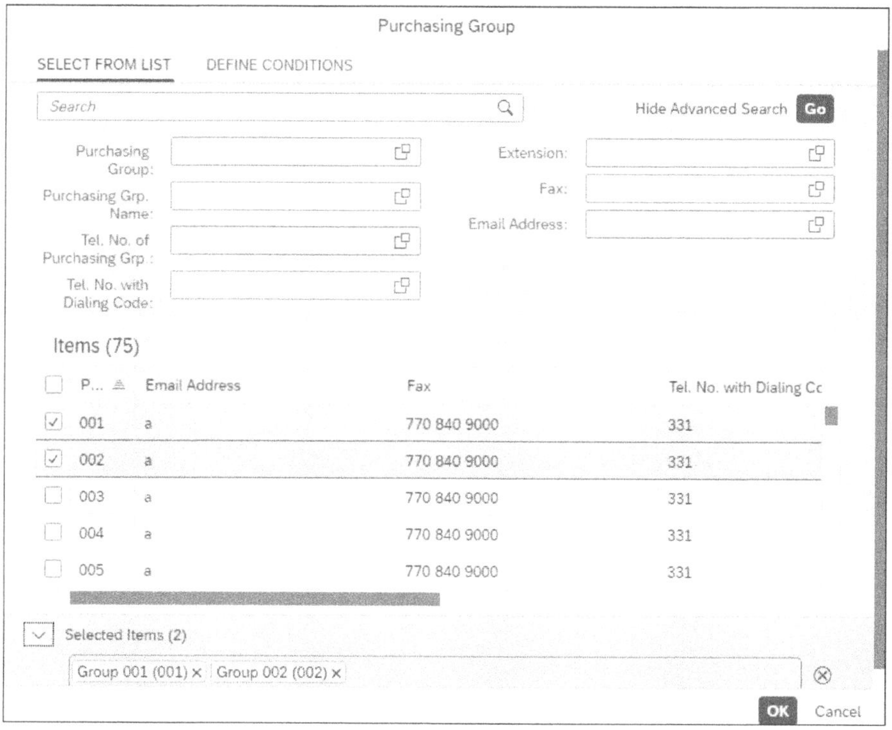

Figure 8.7 Selecting Purchasing Groups

[»]

Delete Selection of Items

Clicking ⊗ (**Remove**) in the **Selected Items** line deselects all selected items at once.

You can remove individual items by clicking the ✕ icon in the bottom row of the **Selected Items** area or by clicking the ☑ icon in the **Items** table.

5. Click on the **OK** button. This transfers the two selected purchasing groups to the **Purchasing Group** field in the initial screen of the app, and the value help disappears again.

6. Set another filter by entering your company code "1010" in the **Company Code** field, as shown in Figure 8.8.

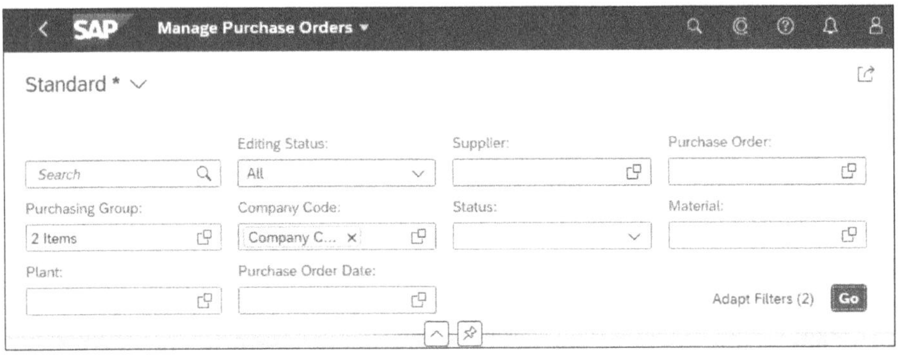

Figure 8.8 Entering the Company Code

7. Click the **Go** button on the right. Now you'll see all the orders corresponding to the filter in the table, as shown in Figure 8.9.

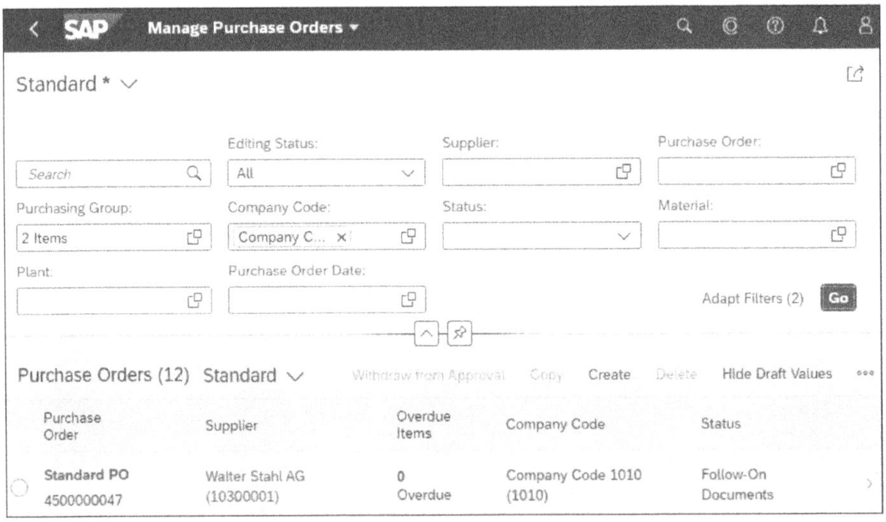

Figure 8.9 Orders for the Company Code and Purchasing Group Filters

In this section, you've used two filters: one for the **Purchasing Group** field and one for the **Company Code** field. SAP Fiori will then display in the table only the records that meet the conditions from both filters.

[«]

Delete Filter

In the header, click in the field for which you want to delete a filter. The set filter value(s) will be displayed. Then, to the right of the filter value you want to remove, click **X** (**Close**), as shown in Figure 8.10.

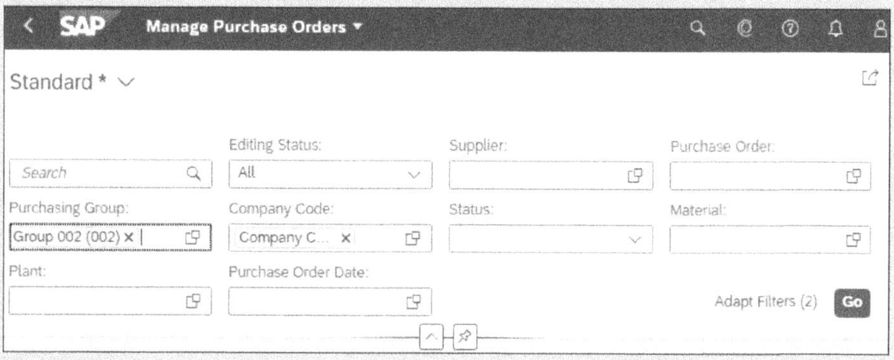

Figure 8.10 Display of the Filter Values for the Purchasing Group Field

[+]

Pin Filter Bar

If you have a long table in which you perform different filtering, you can click ⚲ (press on **Pin Header**) below the header to fix the filter bar directly below the shell bar. This way, when you scroll down through the data available for selection, you always have the header with its filters in view and can quickly adjust it.

8.2.2   Filter with Additional Fields

In the apps, the preselection of filters offered often isn't sufficient. SAP Fiori offers you the capability to customize the filters available for selection.

In our example, we need the **Created By** field to filter the orders by a username:

1. Start the Manage Purchase Orders app.
2. In the header, click the **Adapt Filters** button to the left of the **Go** button, as shown in Figure 8.11.

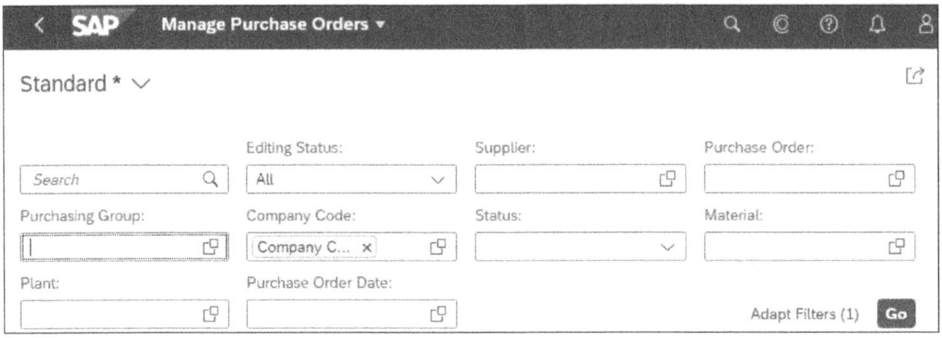

Figure 8.11 Adapt Filters Button

3. The **Adapt Filters** popup window opens, as shown in Figure 8.12. In the upper part of the window, you can see the fields that are visible in the header with the filters currently set. These fields are marked with ☑ on the far right. Further down, you'll find subject areas such as **Delivery and Invoice**, where more fields are waiting for you to use as filters. In the Manage Purchase Orders app, there are more than 40 fields.

To add more fields as filters in our example, scroll down further, and click **More Filters** in the **Purchase Order** topic area.

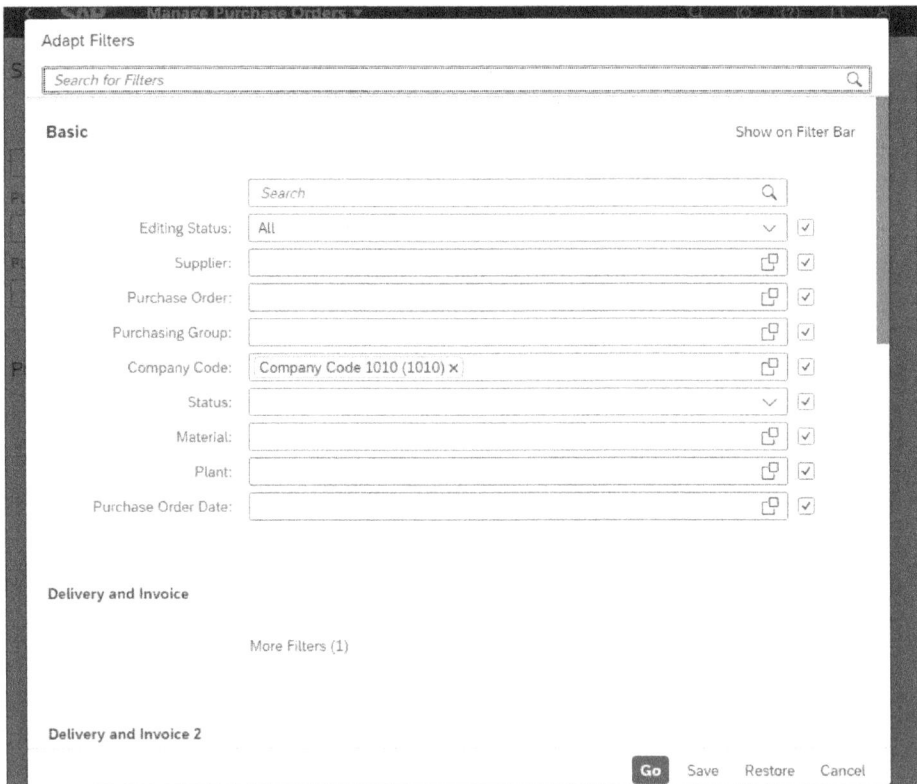

Figure 8.12 Adapt Filters Popup

4. In the **Select Filters** window that opens, as shown in Figure 8.13, the fields that belong to the subject area are displayed. In the case of the **Purchase Order** topic, these are **Created By** and many other indicators.

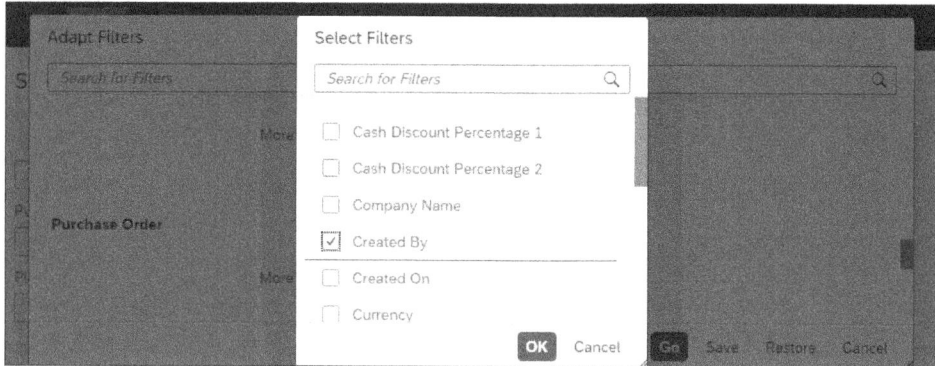

Figure 8.13 Select Filters

Search Filter

Do you know the field names? If yes, the search is faster using the topmost field, **Search for Filters**. As soon as you enter the first letters of the field name you're looking for there, the subject area(s) containing the corresponding field will be displayed below. Unfortunately, the wildcard * doesn't work here. Try this in our app, for example, with the field names **Created By** or **Postal Code**.

---

5. Check the box on the left side of the field you want to apply, for example, **Created By**, and click **OK**.

6. Now you can enter a username in the **Created By** field.

7. Click the **Go** button, the **Adapt Filters** popup window will close (see Figure 8.14), and you can put the newly added filter directly into operation.

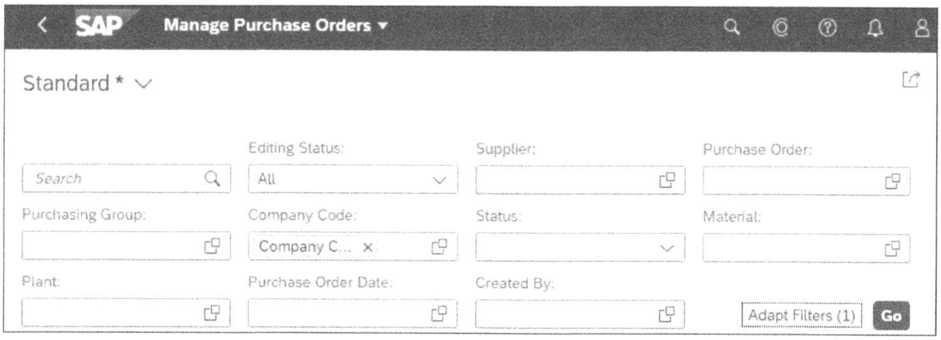

Figure 8.14 Adding a Filter: Created By

8.3   Print Tables or Save Them as PDF Files

In Chapter 3, Section 3.4, we showed you how to output or print the lists as PDF files in SAP GUI. Have you already discovered the 🖨 (**Print**) icon in real SAP Fiori apps? No? Neither have we!

But you don't need it either. For output, just use the corresponding function in your browser. Right-clicking on the table will bring you to the context menu, as shown in Figure 8.15, where you'll find the **Print** menu option.

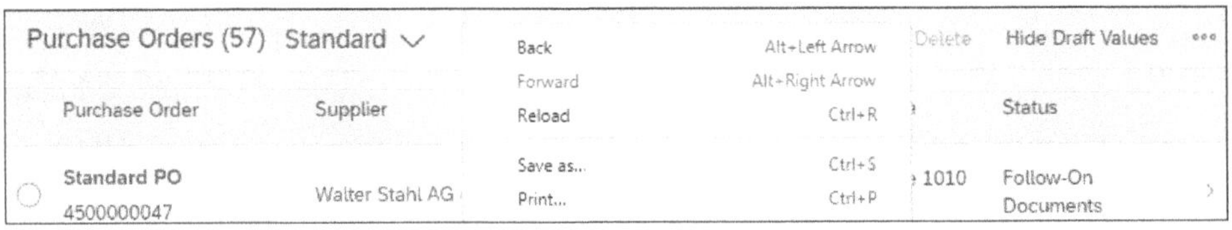

Purchase Orders (57) Standard ⌄		Back	Alt+Left Arrow	Delete	Hide Draft Values	•••
		Forward	Alt+Right Arrow			
Purchase Order	Supplier	Reload	Ctrl+R		Status	
		Save as...	Ctrl+S			
Standard PO	Walter Stahl AG	Print...	Ctrl+P	1010	Follow-On	⟩
4500000047					Documents	

Figure 8.15 Context Menu with the Print Menu Option

After selecting the **Print** menu option, as shown in Figure 8.16, you can also choose PDF output instead of a printer (**Save as PDF** in the **Destination** field) and set whether the printout should be in portrait or landscape format, how wide the margins should be, and how many copies you need.

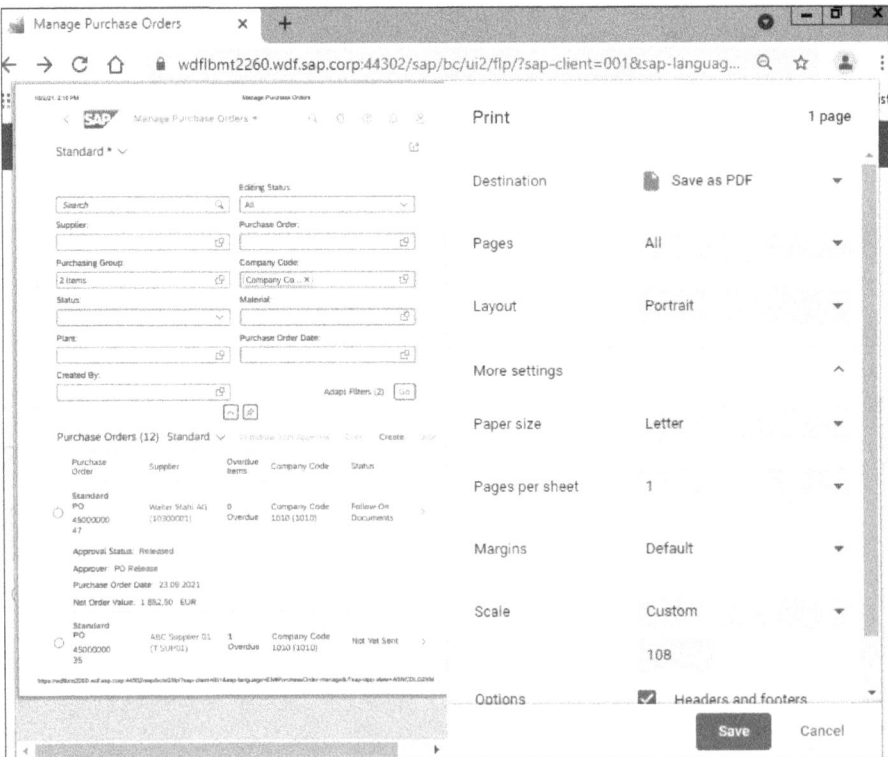

Figure 8.16 Printing Options

A big advantage over SAP GUI in the browser is the possibility of scaling, with which you determine the font size and thus also the print area. Another advantage is that you can output any partial areas of the table by first selecting the area to be output in the table and then selecting the **Custom Area** option in the **Pages** field directly before output.

8.4   Download Tables

You want to process the table in a spreadsheet program such as Microsoft Excel? We're not big fans of the download per se because SAP Fiori offers very useful functions for designing evaluations. Still, there can be the following legitimate reasons for downloading a spreadsheet:

- You want to make data available to colleagues who don't have SAP access.
- You want to create charts or perform other complex calculations in Microsoft Excel. (Before you use Microsoft Excel for any other reason, you should finish reading this chapter first. In Section 8.5, we describe the capabilities of analytical SAP Fiori apps.)

In the following instructions, you'll trigger a download for the table of orders:

1. Start the Manage Purchase Orders app.
2. Click on the **Go** button. You'll now see the table with all orders below, as shown in Figure 8.17.

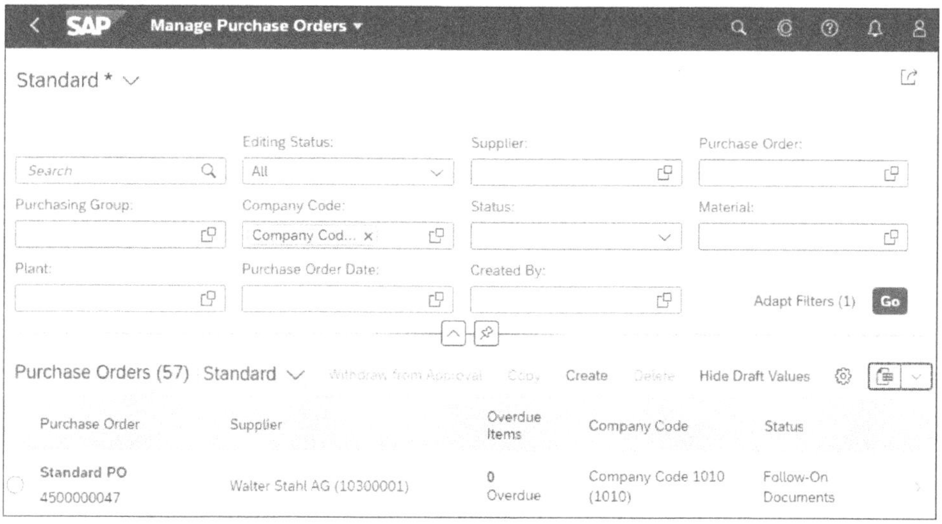

Figure 8.17 Display Table with Purchase Orders

3. Click [icon] (**Export to Spreadsheet**) on the far right of the screen. This will start exporting the data for a spreadsheet program .

Another popup window opens in which SAP Fiori shows you the progress of exporting. With a large amount of data, it can sometimes take a little while.

[+]

Too Much or Wrong Data Selected?

If you realize only after clicking 📇 (**Export to Spreadsheet**) that you're exporting too much data or the wrong data, simply click the **Cancel** button. SAP Fiori will then finish the export.

4. Depending on the settings of the browser you're using, you may have to allow the download of the file separately after the export has finished. If necessary, do this.
5. Depending on the browser, a question about how to proceed appears at the end, as shown in Figure 8.18. For our example, click the **Open** button.

| Do you want to open or save **Purchase Orders.xlsx** (5.30 KB) from **is53004.t4t.biz**? | Open | Save ▾ | Cancel | × |

Figure 8.18 How to Proceed

In Google Chrome, for example, you'll find a Microsoft Excel icon in the bottom-left corner that you can click on. In both cases, you immediately open the corresponding Microsoft Excel spreadsheet, as shown in Figure 8.19.

Have fun with further processing with Microsoft Excel!

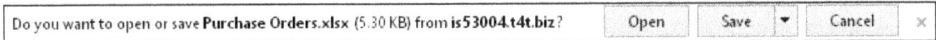

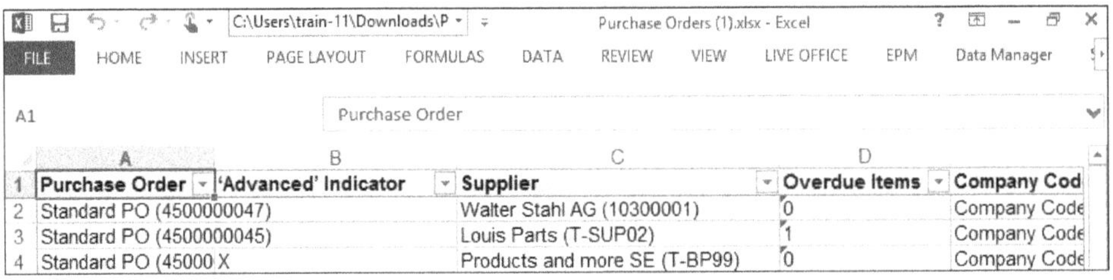

Figure 8.19 Microsoft Excel Spreadsheet Opens

[+]

The Disadvantage of Downloading

In SAP Fiori apps, you have the luxury of working with real-time data. This means that the data you evaluate is always up to date! As soon as you start a download, it only captures a snapshot.

It's therefore smarter to check whether you can't achieve the same or perhaps even a better result with the onboard tools of SAP Fiori than with post-processing in Microsoft Excel. In addition, you'll save yourself a lot of time with the corresponding presets from the next chapter.

Send a Table by Email

Say a colleague with SAP access needs the data from your spreadsheet. You want to download the data in Microsoft Excel and email the Microsoft Excel spreadsheet to your colleague. It's much more convenient and faster to follow these steps:

1. Click in the upper-right corner on ⬀ (**Share**), and select the **Send E-Mail** command, as shown in Figure 8.20.

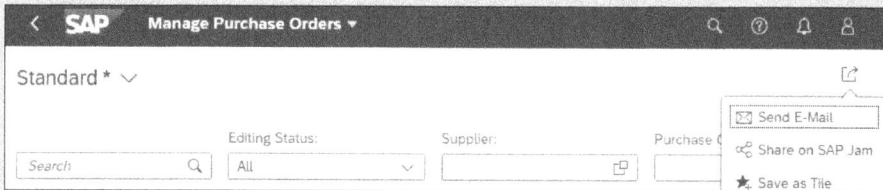

Figure 8.20 Send E-Mail Option

2. With this, you open, if installed, an email program such as Microsoft Outlook, whereby the link for the table is taken over into the automatically created email, as shown in Figure 8.21. Now just enter the recipient and off you go!

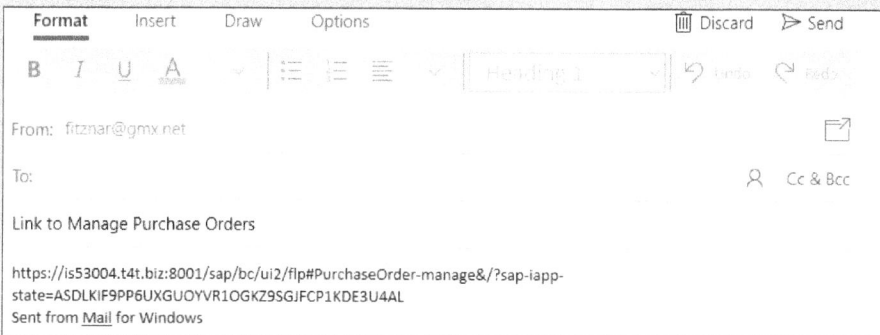

Figure 8.21 Email Created in Microsoft Outlook

After opening the email and clicking on the link, your colleague will see the table from which you created the link.

8.5   Visualize Data with Analytical SAP Fiori Apps

Fiori, as you already know, is Italian for *flowers*. But SAP Fiori hasn't been really colorful so far, and that's a good thing. It's not just we authors who sometimes sit in front of the computer for days and nights on end and enjoy the calm and relaxed design of the SAP Fiori interface. Not for nothing did its design concept win the Red Dot Award in 2015.

From now on, however, it's going to be colorful here. Analytical apps offer you the possibility to display data graphically. These apps form their own app category (besides the already presented categories of transactional apps and info sheets). We'll walk through their reporting features in the following sections.

8.5.1   Charts and Measures in Tiles

On some tiles, you've probably already spotted small charts and colorful metrics as the only splashes of color on the otherwise largely monochrome SAP Fiori launchpad homepage. The charts in tiles are also called *micro charts*, as shown in Figure 8.22.

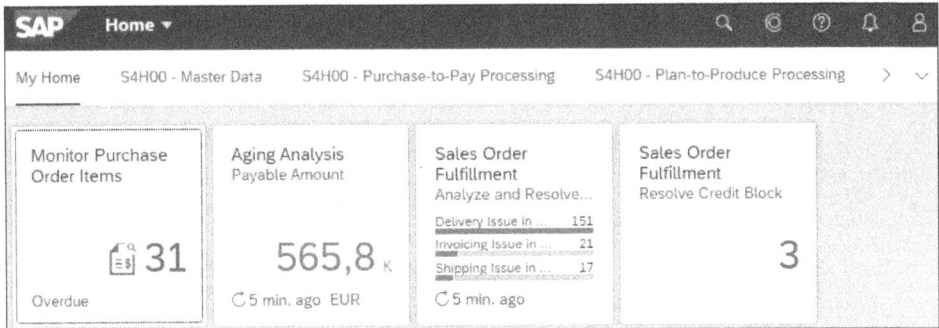

Figure 8.22 Tiles with Measures and Micro Charts

Analytical apps are usually hidden behind the tiles with micro charts and measures. You may also recognize them by the following features:

- The colored display of the analysis content in the tile
- The C (**Refresh**) icon in the tile, which can be used to update the measures contained in the tile
- The display of the three icons ▦ ▦ ▦ (**Chart and Table View, Chart View, Table View**) after launching the app

You don't get this in SAP GUI, but in SAP Fiori, the *key performance indicators* (KPIs) on the tiles are calculated in real time from large data sets. Starting on the SAP Fiori launchpad homepage, you get up-to-the-minute KPIs. After that, the KPIs are automatically updated at preset, regular intervals. If you don't want to wait for the automatic update, you can click C (**Refresh**) in the tile, and you'll see the current status.

Different Colors of the Measures

The color displayed is based on predefined thresholds. The color green stands for OK, yellow for a warning, and red for a critical condition that needs special attention. As a rule, a key user can also set threshold values and create KPIs.

8.5.2   Launch an Analytical App

Find an analytical app on the SAP Fiori launchpad homepage or use the App Finder. In our example, we use the Monitor Purchase Order Items app.

Follow these steps:

1. Launch an analytical app, such as Monitor Purchase Order Items.

2. In the initial screen, make the selection in the header, and click the **Go** button. In our case, we specify the currency in the mandatory field **Display Currency** with **USD**. After clicking the **Go** button, you'll usually see the preset chart and—as in our example shown in Figure 8.23—a table of the corresponding data.

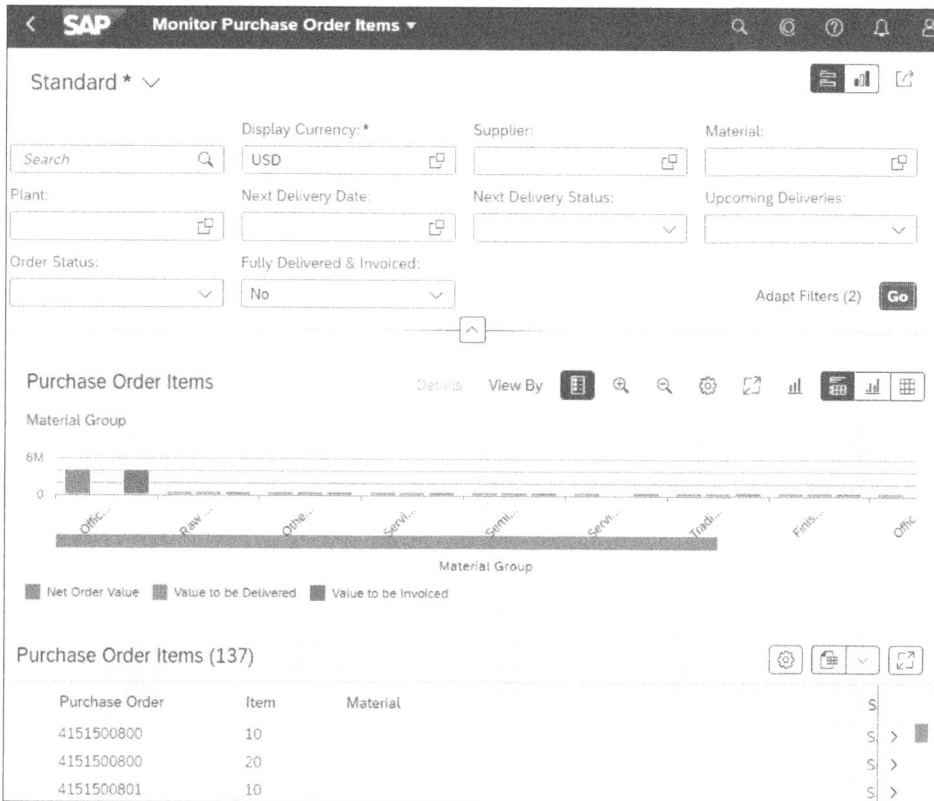

Figure 8.23 Monitor Purchase Order Items App: Chart and Table

What if I See Only a Table or Only a Chart?

Don't worry. Analytical apps can contain three different views: *chart view* [icon], *table view* [icon], and *chart and table view* [icon], which consists of both a chart and a table.

The word *can* used in the preceding sentence means that an analytical app can contain either one, two, or all the views mentioned. If the app you selected contains only a table view, use a different analytical app for this tutorial, that is, an app that contains at least one of the ▯ (Chart View) or ▯ (Chart and Table View) icons.

Your first analytical app? No problem! We'll give you an overview of the complete screen first:

- At the top, you can see the header with the filter options. You can expand this area as with tables by clicking ▭ (Expand Header) and ▭ (Collapse Header). Here you select the data for which a chart is to be created.
- The number shown on the left above the table or above the chart corresponds to the measure that was already displayed to you from the corresponding tile on the SAP Fiori launchpad homepage. In our example, this is **Purchase Order Items (137)**. This measure is sometimes also located in the upper-left corner directly below the header line.
- Further down, depending on the default settings, you'll see a table, a chart, or a combination of both in the main area.

So much for the brief overview—now let's take a closer look at everything! Click ▭ (Collapse Header).

8.5.3   Switch between Views

Don't get confused now. The term *view* is used more often in the SAP system, so you can switch between different views in the filter bar. But now, you switch between different views for the main area of the analytical app. For this purpose, there are the following icons under the filter area, which are sometimes also hidden behind the ▯ (More) icon:

- ▯ (Chart and Table View)
- ▯ (Chart View)
- ▯ (Table View)

You can recognize the currently selected view by the colored icon. In our example shown in Figure 8.24, we're currently in the chart and table view. To switch to the pure chart view, click ▯ (Chart View).

After the click, the table positioned under the chart disappears. In addition, the ▯ (Chart View) icon is highlighted. The ▯ (Chart and Table View) icon again loses its color marking.

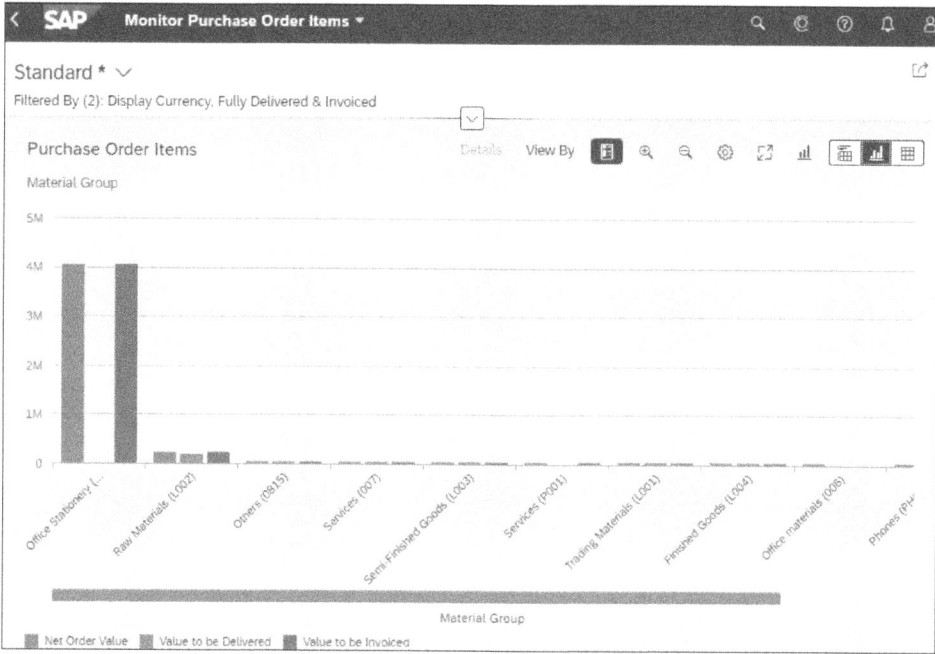

Figure 8.24 Chart View for Purchase Order Items

8.5.4   Change Settings for the Chart

Now we want to edit the chart. We'll show you the setting options using a specific example. In Figure 8.25, you can see our current state on the left and the state we want to achieve on the right.

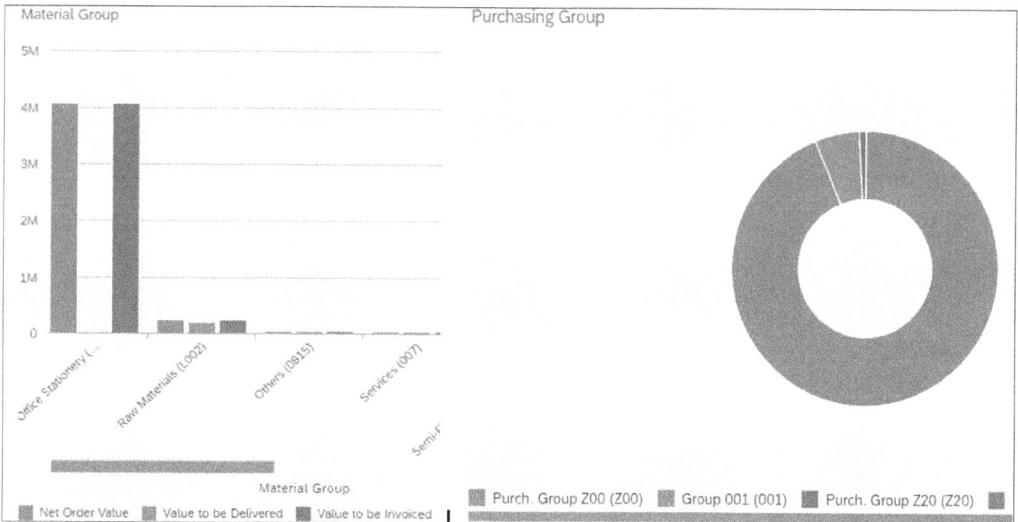

Figure 8.25 Current Chart and Desired Chart

What changes are these, and what do we want to achieve?

- Currently, the order value is displayed by material group such as office stationery or raw materials, but we intend to display the order value as a percentage by purchasing group.

- Currently, three columns are displayed side by side for each material group, showing the net order value, the value of orders to be delivered, and the value of orders that haven't yet been invoiced. We want to have only the net order value displayed.

- Instead of the **Column Chart** type of chart, we want to use a **Donut Chart**.

To change the display by material groups to a display by purchasing groups, switch to another dimension. This is necessary because the material group is one dimension, and the purchasing group is another dimension. At the moment, you can see the **Material Group** dimension below the bar graph in Figure 8.25.

In addition, in the next steps, we'll reduce the measures and change to the **Donut Chart** dimension type. All this is done in a window that you can access via ⚙ (**Settings**). Follow these steps:

1. To change the dimension and make further settings, click ⚙ (**Settings**). The **View Settings** window opens, as shown in Figure 8.26.

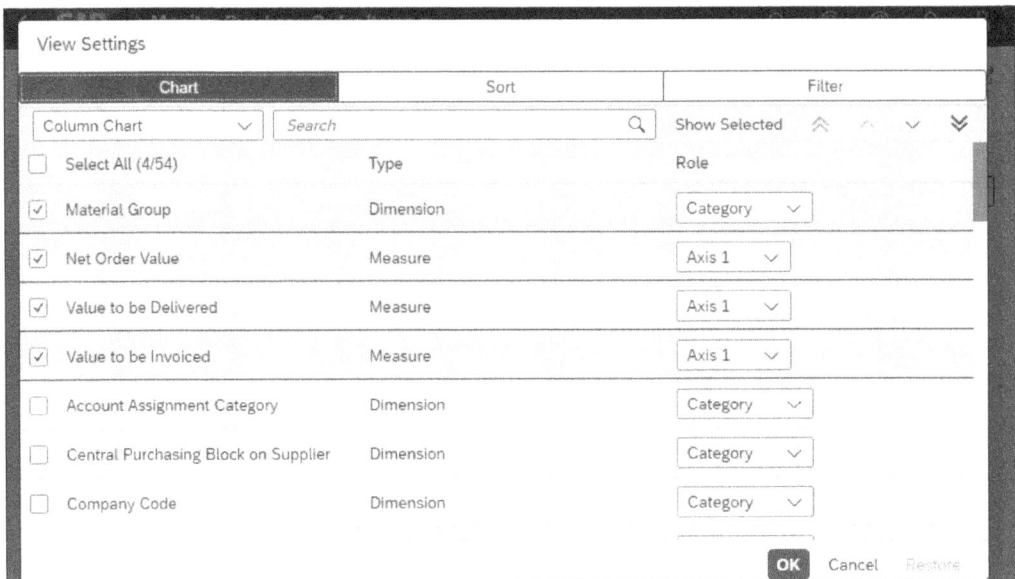

Figure 8.26 View Settings

The **View Settings** window often shows a huge number of fields, and this gives you a lot of flexibility for your charts. With the ≫∧∨≫ icons, you bring the fields that are of interest to you to the top of your field of view.

The window allows you to make adjustments to the chart such as adding and removing dimensions and measures or changing the chart type. Moreover, you can add sorting and filters.

2. Remain on the **Chart** tab. Here you see a table with a total of 54 entries and 4 of them are already selected. In the **Type** column, you can see that an entry is either a dimension or a measure.

   The effect of a setting in the **Role** column depends on the selected chart type. Here is an example: in a bar chart, if you change the measure setting from **Axis 1** to **Axis 2**, you tilt the bars 90 degrees. For a donut or pie chart, this setting has no effect because of the missing axes.

   On the far left, you can recognize the selected entries by the fact that they each have a checkmark in the box. The total number of selected entries can be seen from the **Select All (4/54)** text in the upper-left corner of the **View Settings** window.

3. So now proceed like this:
   - Remove the checkmark ☑ for the **Material Group** field.
   - Further down, place a checkmark for the **Purchasing Group** field. This will change the dimension.
   - Remove the checkmarks ☑ for the **Value to be Delivered** and **Value to be Invoiced** fields. This will reduce the number of measures and at the same time the number of bars from three to one.

4. Now you want to change the chart type. To do this, click in the upper-left corner directly under the **Chart** tab on the ☑ icon. A list of possible chart types will open, as shown in Figure 8.27. Select **Donut Chart** from the list.

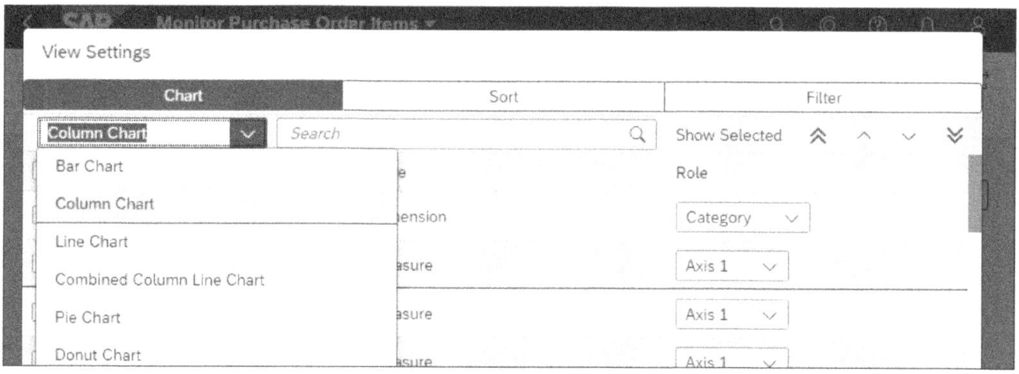

Figure 8.27 Changing the Chart Type

5. Click the **OK** button in the **View Settings** window, and voilà! You'll see the redesigned chart, as shown in Figure 8.28.

245

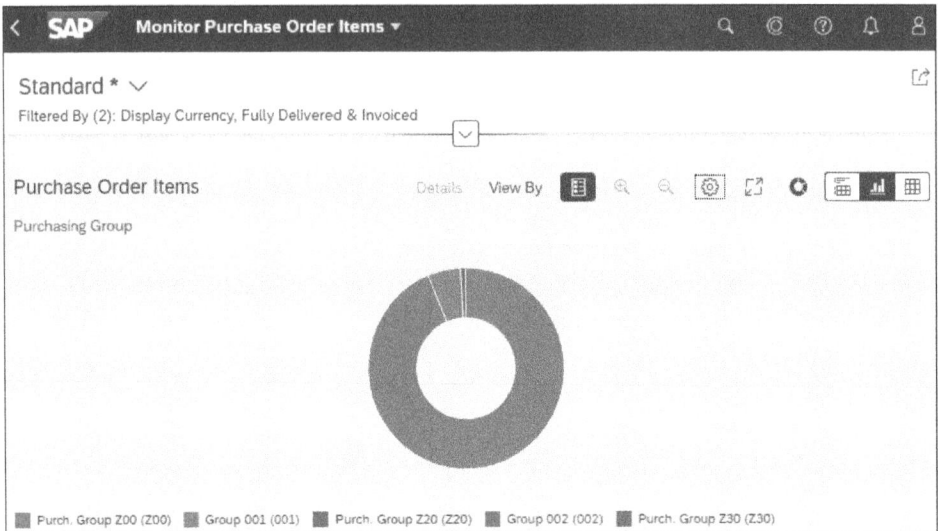

Figure 8.28 Redesigned Chart

[»]

What Happens if I Don't Select a Dimension or Measure?

Interesting idea! Because a dimension divides the total value of a chart according to the dimensions, the total value just isn't distributed if you don't select a dimension. Accordingly, a donut chart in this case shows you only a closed ring without further subdivisions.

And what if you haven't selected a measure? Then there are no numerical values, and no chart can be created without numerical values. In this case, after confirming with the **OK** button, you'll receive the following elegantly worded message: **Size does not meet the specified minimum or maximum number of feeds definition.**

[»]

Are Multiple Dimensions Possible in One Chart?

Yes, you can use two dimensions in a chart, such as **Material Group** and **Purchasing Group**. In a bar chart, this means that there is a separate bar for each combination of material group and purchasing group.

8.5.5   Functions in the Chart View

We recommend the **View Settings** window from the preceding section if you're launching an analytical app for the first time and want to get an overview of the fields offered, or if you want to perform a "complete renovation" of your chart. However, you can also change some of these settings faster and more directly in the chart view. There are some icons and buttons for this purpose, as shown in Figure 8.29.

| Purchase Order Items | Details   View By | ▦ | ⊕ | ⊖ | ⚙ | ⤢ | ◐ | ▦ | ▦ | ▦ |

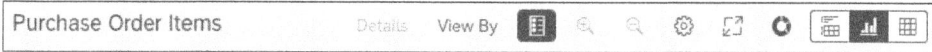

Figure 8.29 Icons in the Chart View

In Table 8.1, we've listed these icons and buttons individually and described their functions.

Button	Description
Details	This button only works if you've previously marked at least one element in the chart, such as a bar or a pie slice, with a click. Then, after clicking the **Details** button, you'll see more information, in our case, the order value of the respective pie slice.
View by	Use this button to add more dimensions.
▦ (**Toggle Legend Visibility**)	Use this icon to show or hide the legend in a chart.
⊖ (**Zoom Out**), ⊕ (**Zoom In**)	Use these icons to change the display of the chart size.
⤢ (**Maximize**)	Use this icon to see the displayed chart on full screen width in a separate window.
◐ (**Donut Chart**) ▦▦▦ (**Chart and Table View, Chart View, Table View**)	This icon shows the current chart type (in Figure 8.29, ◐ is shown). If you click on it, SAP Fiori will show you a list of possible chart types from which you can choose to right of that icon.

Table 8.1 Functions in the Chart View

8.5.6   Mini Charts

The headline doesn't lie—there are not only large charts but also mini charts in the header! These *mini charts* have two advantages: they provide you with quick information, and you can filter with them.

Expand the header with the filter options via ⸺⌄⸺ (**Expand Header**). To look at the visual filters now, click ▥ (**Visual Filter**) in the upper-right corner of the screen. This replaces the classic filter fields with mini charts, as shown in Figure 8.30.

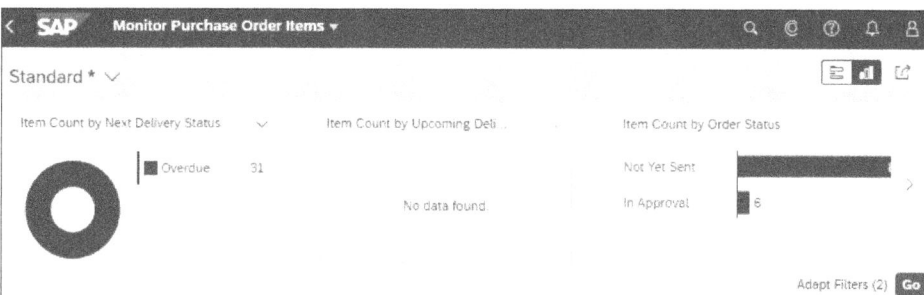

Figure 8.30 Mini Charts

If you want to set a filter, just click on an element, such as a bar or a pie slice in a mini chart, and eventually click again on the **Go** button. The chart and the table below the filter bar will adapt to the new filtering.

You can recognize the active visual filters by the highlighted area in the mini chart. And you can see the number of filters in the upper-right corner of a mini chart, which currently shows **(1)**.

Customization of the visual filters is possible via the **Adapt Filters** button at the bottom right. After clicking this button, you can select mini charts for display (see Figure 8.31) via the **Show on Filter Bar** field, set other chart types, and change the sorting. After clicking the **Go** button, you'll see the result.

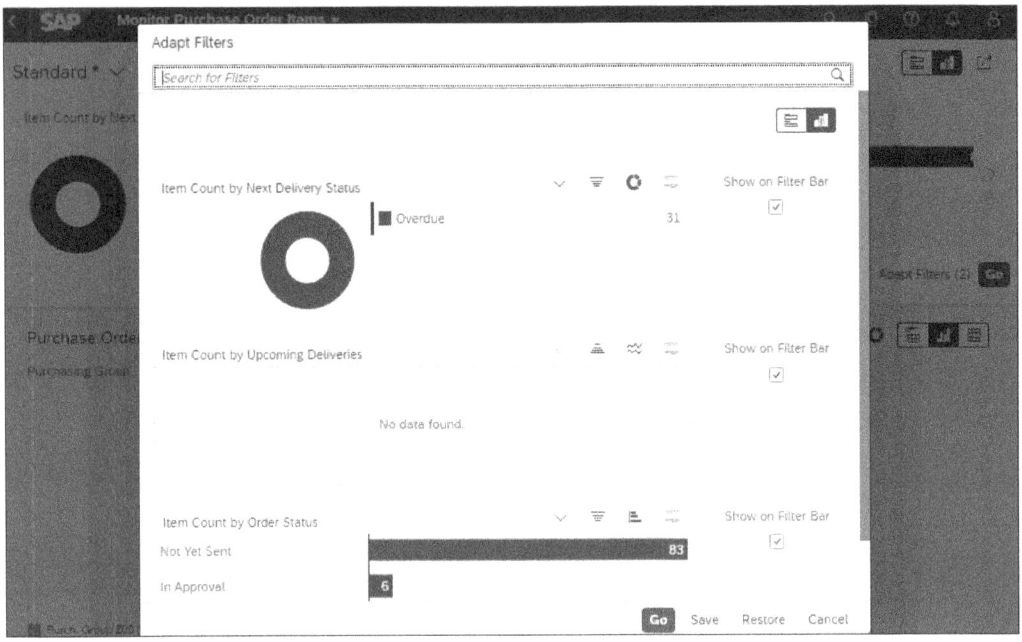

Figure 8.31 Adapting Filters

8.6 Summary

You've learned about key SAP Fiori reporting features, including finding, filtering, printing, and downloading tables. You've also seen how to visualize data using analytical apps.

Now, let's move on to customizing your SAP Fiori reports.

Chapter 9
Customizing SAP Fiori Reports

In SAP GUI, you've seen with the report variants and the layouts that your reports can be adapted very flexibly without the help of key users or programmers. In the same way, these techniques are available to you in "fiorized" transactions.

Fortunately, there are also similar techniques in "real" SAP Fiori apps, some of which are even easier to implement. These have been given new designations in real SAP Fiori apps: both report variants and layouts are called *views* here.

> **What You'll Learn**
>
> - How to save filter settings as views so you don't have to retype the filter values or reassemble the filter fields each time you start a report
> - How to create table settings such as sorts, collations of columns, and groupings to give you a better overview of your data
> - How to permanently save table settings as views
> - How to make and save special presets for tables in analytical apps
> - How to create your own tiles for your preset reports in no time

9.1 Views for Filters

You already know the procedure for setting filters in real SAP Fiori apps from Chapter 8, Section 8.2. You performed the following actions there:

- Determined single and multiple *values* (other term: *elements*) for filtering via the 🔲 (**Value Help**) icon
- Searched and defined additional fields for filtering with the **Adapt Filters** button

What you don't know yet is that these filters can also be saved as a *view*! But what is that? Basically, a view for filters corresponds to a report variant from SAP GUI. The following settings are saved:

- Filter values such as the number of a plant
- All settings you make with the **Adapt Filters** button

Later, you can call up this package of presets again in no time at all and apply the presets directly. This procedure is especially worthwhile if you need certain filters frequently. This procedure can save you a lot of time in your everyday work.

We'll walk through how to use and customize views in the following sections.

9.1.1   Save Filter as View

In this section, we'll create one view. In practice, you can create as many views as you want for an app.

We'll use the Manage Purchase Orders app as an example. You're welcome to use any other real transactional SAP Fiori app.

Follow these steps:

1. Launch the Manage Purchase Orders app, as shown in Figure 9.1.

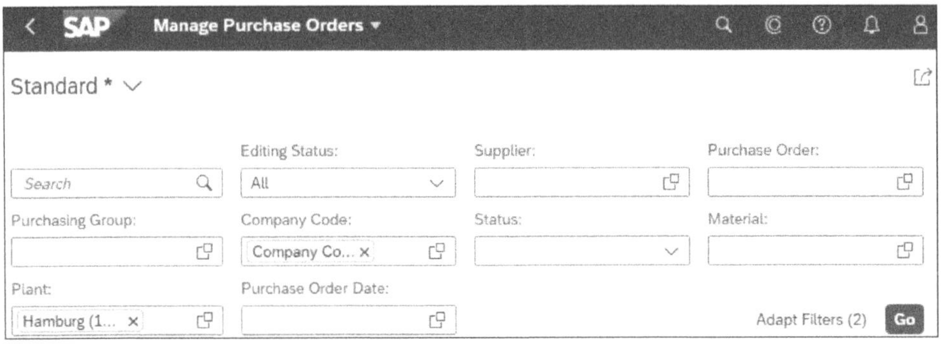

Figure 9.1 Manage Purchase Orders App

2. Filter using one of the **Plant** or **Company Code** fields, and, if necessary, perform additional filtering using the **Adapt Filters** button (refer to Chapter 8, Section 8.2.2).

3. In the upper-left corner, directly below the shell bar, click ☑ (**Select View**). A window will open showing the views that are already available (see Figure 9.2). In our case, there is only the **Standard** view. This was created by SAP; it's protected against changes and can't be deleted.

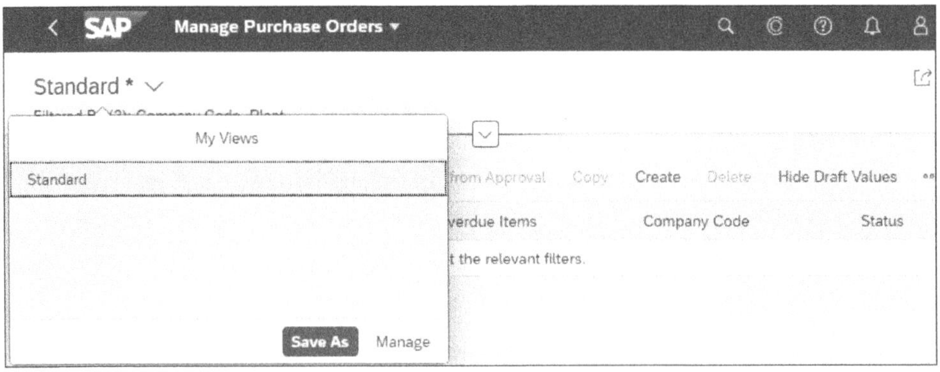

Figure 9.2 Selecting the View

4. In this window, click the **Save As** button. The **Save View** window appears.

5. Enter the label "My first view" in the **View** field.

The selection fields shown in Figure 9.3 offer you the following exciting functions:

- **Set as Default**: The filters you set will always be set automatically right after you launch the Manage Purchase Order app. Great! We would like to see this feature in SAP GUI as well because the corresponding default setting only works with a special trick. (By the way, we've compiled this and other tricks in our German language book, *SAP for Users – Tips & Tricks*, for advanced users [Rheinwerk Publishing, 2020].)

- **Public**: This option allows a key user to use this view across all users. For our example, we leave this option disabled so that only you can see and use the view.

- **Apply Automatically**: With this option, the filtering is performed immediately when the app is launched. This saves you from having to click the **Go** button. This option is normally used when no adjustments need to be made to the filters. Unfortunately, this helpful function also doesn't exist in this form in SAP GUI.

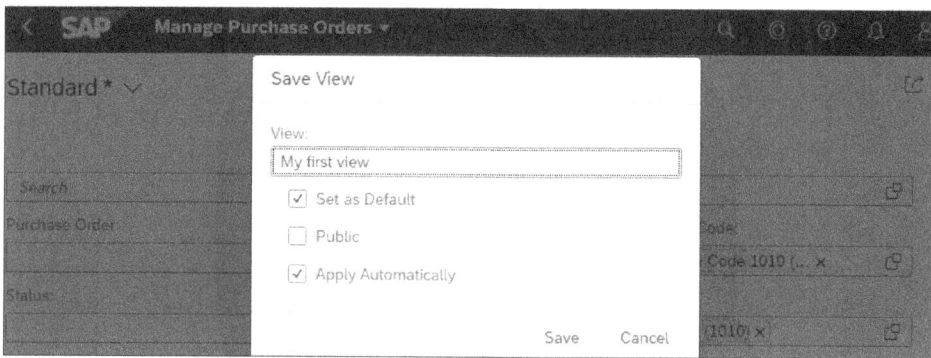

Figure 9.3 Save View Popup

6. Select the **Set as Default** and **Apply Automatically** checkboxes.

7. Confirm by clicking the **Save** button. You'll then be taken directly to the view you created, as shown in Figure 9.4. The name of the current view is always displayed at the top left.

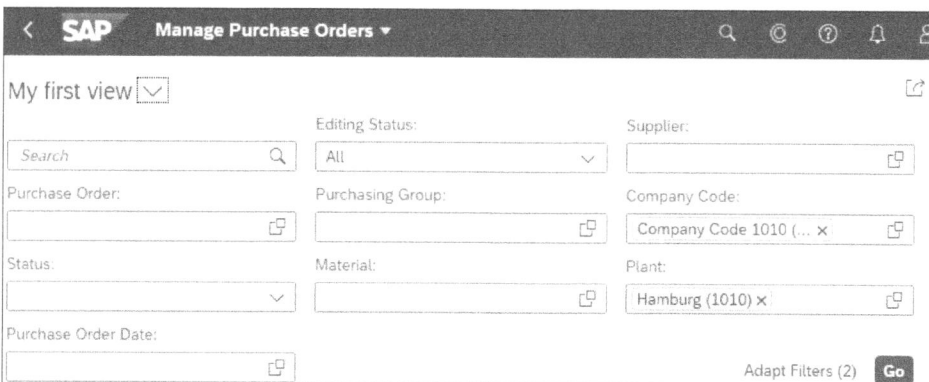

Figure 9.4 Your New View

8. To return to the SAP Fiori launchpad homepage, click ◀ (**Back**) in the shell bar.

[»]

Sometimes a View Can Also Be Saved in the Adapt Filters Window

You've already used the **Adapt Filters** button in Chapter 8, Section 8.2.2, to define additional fields as filters. Also in this window, in some apps such as Manage Purchase Orders, you can use the **Save** button to save the new settings as a view. For this button to be active, you must have made at least one change for the filters.

9.1.2   Call Up a View

If you've implemented the previous tutorial, you now have at least two views to choose from. In addition to the view you've created or the views your colleagues have shared as public, there is also always the **Standard** view.

Often, in practice, you need multiple views. For example, you use one view to display only the orders with **Draft** status and another view to display orders with **In Approval** status.

This requires that you can switch between different views, which you'll learn how to do now:

1. Start the Manage Purchase Orders app. The app now starts immediately with **My first view** due to the **Set as Default** option activated in the instructions in the previous section, if the standard view is not set as default. Thanks to the **Apply Automatically** option that is also enabled, you've saved yourself clicking on the **Go** button, and the purchase orders are displayed immediately.

2. Click ☑ (**Select View**) next to the **My First View** line. The **My Views** popup window opens, as shown in Figure 9.5.

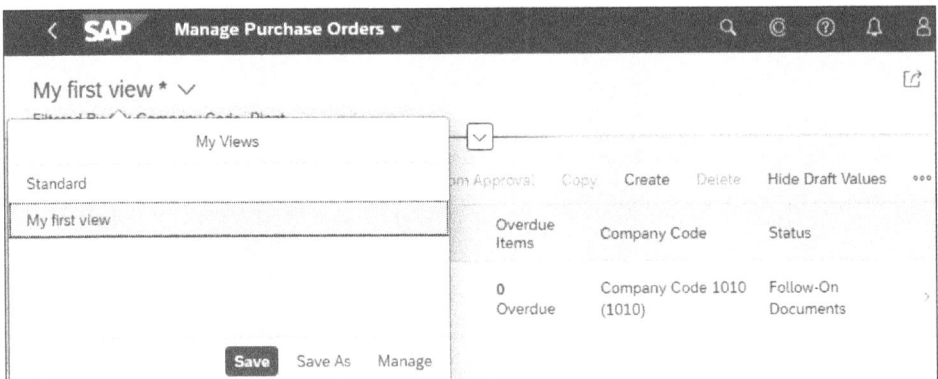

Figure 9.5 My Views Popup

[+]

How Can You Recognize the Currently Selected View?

Directly below the shell bar, the name of the currently selected view is displayed just to the left of the ☑ (**Select View**) icon. In addition, the currently selected view is highlighted in the **My Views** popup window.

3. To select a different view, click on the desired view. In our example, we switch to the **Standard** view. The **My Views** window disappears, and the settings maintained in the selected view are set.

4. Click the **Go** button, and you'll see the filtered result in the list if the **Apply Automatically** checkbox isn't selected.

9

9.1.3 Change Filter for a View

What you love, you should care for. To adapt the views of your filters to changing requirements, you should adjust them. We'll show you how to do this in the next steps:

1. If necessary, start the Manage Purchase Orders app.

2. Call **My First View**, and expand the header, if it is collapsed.

[«]

Do I Need to Have a View Selected to Change It?

Before you can change a view, you must select it. It works like report variants in SAP GUI; that is, when you make a change, you save over an existing view with the updated status.

3. Make a change. For example, as shown in Figure 9.6, set an additional filter for the **Status** field, such as the **Follow-On Documents** status, or change a setting using the **Adapt Filters** button.

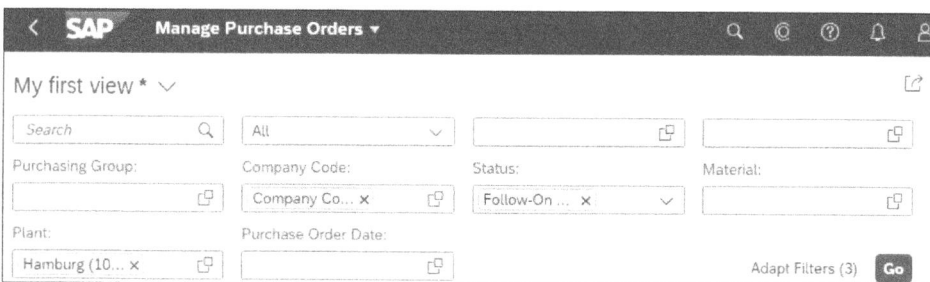

Figure 9.6 Changing Filters

How Can I Tell That I've Made a Change?

Take a look at the view name on the left below the shell bar. An asterisk (*) will be added to it after you've made changes to the currently selected view. When you save your changes, the asterisk disappears again.

4. Click ☑ (**Select View**) next to the **My First View** line. As usual, the **My Views** window opens.

5. Click the **Save** button in the **My Views** window. The next time you call up the view, your change will be taken into account.

Help! The Save Button Doesn't Exist for Me!

If you don't see a save button, there are two possible reasons for this:

- You haven't made any changes before, so there's nothing to save and therefore no corresponding button.
- For a change of the **Standard** view, only the **Save As** button appears because this view is protected against changes and therefore can only be saved under a new name.

9.1.4   Change the Name of a View

Maybe you already had the thought that the name **My First View** isn't really practical. Therefore, in this section, you'll learn how to change the name of a view. Follow these steps:

1. If necessary, start the Manage Purchase Orders app.

2. Click in the filter bar on ☑ (**Select View**) next to the **My First View** line.

3. Click the **Manage** button in the **My Views** popup, as shown in Figure 9.7. The **Manage Views** window opens, as shown in Figure 9.8.

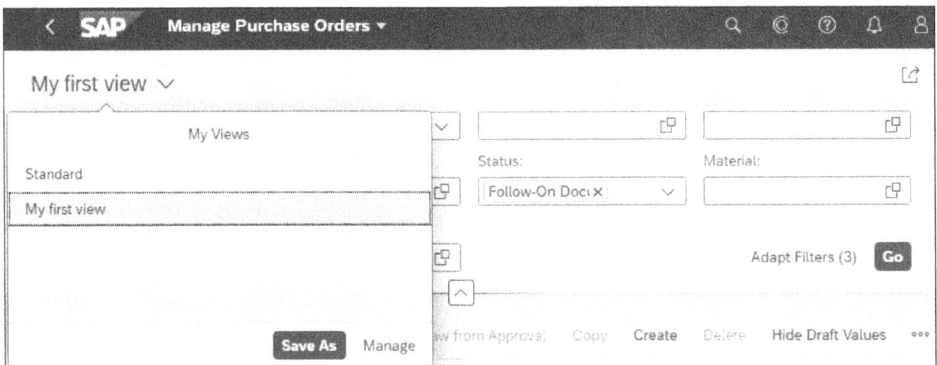

Figure 9.7 My Views Popup

In this window, you can see all the views you've created and, if necessary, other views that your key user has created and released, for example.

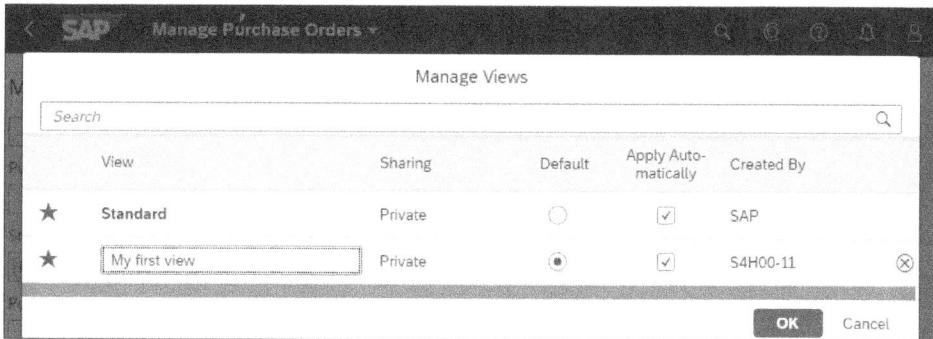

Figure 9.8 Manage Views

4. Click in the field with the content **My First View**. Overwrite the existing text with, for example, "Follow-on document check", and confirm your entry with the **OK** button.

9.1.5   Show/Hide View

Say your key user has created several views that you don't need or there's a view that you no longer need for the time being, but don't want to delete. To solve these problems, you can hide views that you don't need. Follow these steps:

1. If necessary, start the Manage Purchase Orders app.
2. Open the **My Views** window by clicking ✓ (**Select View**) icon on the left. Then click the **Manage** button in this window.
3. In the table of the **Manage Views** window, before each view, you'll see the ★ (**Remove from Favorites**) icon. Click on this icon. As you can easily see in Figure 9.9, the previously filled star is now displayed "empty" ☆.

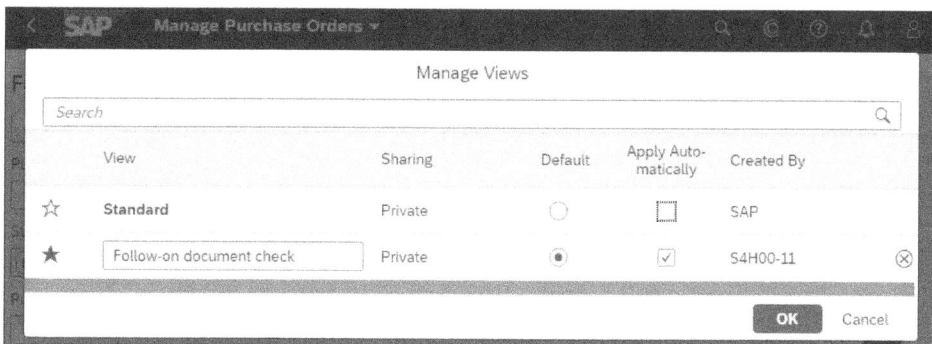

Figure 9.9 Remove from Favorites

255

4. Confirm your change by clicking the **OK** button.

5. Now when you click ☑ (**Select View**), the previously hidden view is no longer listed.

To show a view again, proceed as described in the previous steps, but finally click the ☆ (**Add as Favorite**) icon.

[+]
> **Why Does the Switch to the Icon Not Work?**
>
> If the view in the **Default** column is checked, it can't be hidden. In such a case, before clicking ★ (**Remove from Favorites**), activate another view by clicking its radio button in the **Default** column.

9.1.6   Delete View

It happens that certain views become permanently redundant. To keep the overview, you clean up and delete views that are no longer needed. As a normal user, you can usually only delete the views you've created yourself. The **Standard** view generally can't be deleted. Follow these steps to delete a view:

1. If necessary, start the Manage Purchase Orders app.

2. Open the **My Views** window by clicking ☑ (**Select View**).

3. Click the **Manage** button in this window to arrive at the screen shown in Figure 9.10.

4. To delete a view, click ⊗ (**Delete View**) on the far right in the respective line.

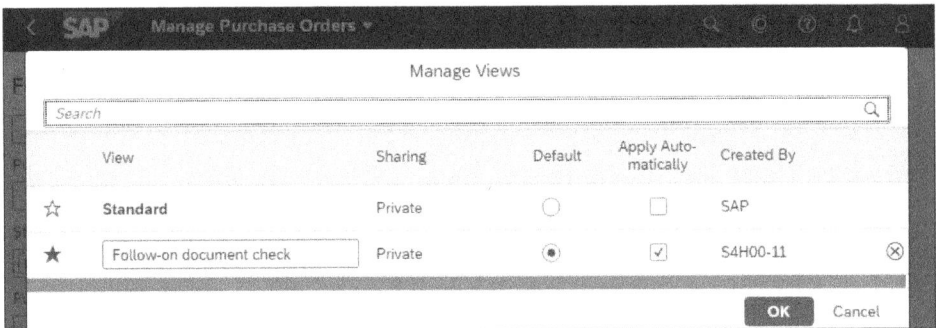

Figure 9.10 Manage Views

5. To confirm the deletion, you could now click the **OK** button. (You should be really sure because there is no possibility to restore the view after deletion!)

Aren't sure? Then click **Cancel**. (In addition, in our example, click **Cancel** because you'll need the sample view again later.)

9.1.7   Functions Overview

In Table 9.1, you'll find again all the functions from the **Manage Views** window that you can use to maintain your views.

Function	Impact
🔍 **(Search)**	Filters the available views via the input field
★ **(Remove from Favorites)**	Removes the view from the **My Views** selection window
☆ **(Add as Favorite)**	Adds the view to the **My Views** selection window
Default	Determines which view should be used immediately after the app is launched
Apply Automatically	Determines whether the list of records should appear immediately without clicking the **Go** button
⊗ **(Delete View)**	Allows you to permanently delete a view

Table 9.1 Functions in the Manage Views Window

9.2   Customize Tables in Transactional Apps

The default settings for tables defined by SAP or within your company aren't enough for you? You need a different sorting of the contents, a different grouping, a different order of the columns, and maybe one column is superfluous, but the other one is missing, and so on? This is typical for most businesses, and you have it in your own hands to design your own optimal table.

In this section, you'll learn about the extensive table design capabilities of SAP Fiori. These functions only apply to tables from real SAP Fiori apps. For legacy apps, use the SAP GUI functions (see Chapter 4).

> **Can This Personalization Be Saved?**
>
> Of course, you can create, maintain, release, and delete views not only for filters but also for tables or diagrams. The procedure here corresponds to the general work with views.

For our example, we'll again use the Manage Purchase Orders app, which you're already familiar with. However, you're also welcome to use another real transactional SAP Fiori app.

Follow these steps:

1. Launch the Manage Purchase Orders app, as shown in Figure 9.11, and then click the **Go** button in the filter bar.

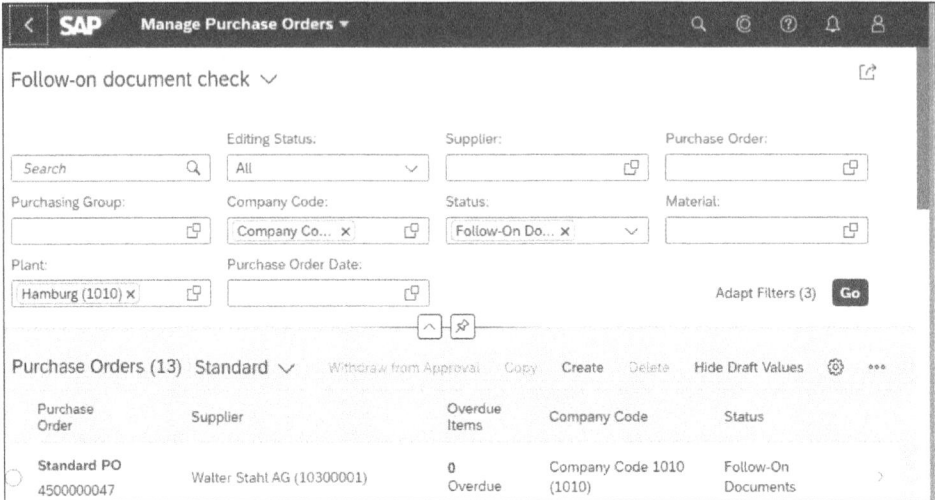

Figure 9.11 Manage Purchase Orders App

2. To start editing the table in the lower part of the screen, click ⚙ (**Settings**). In the following instructions, we'll show you how to move columns of the table, show and hide columns, sort the table, and group records using the **View Settings** popup window that opens now (see Figure 9.12).

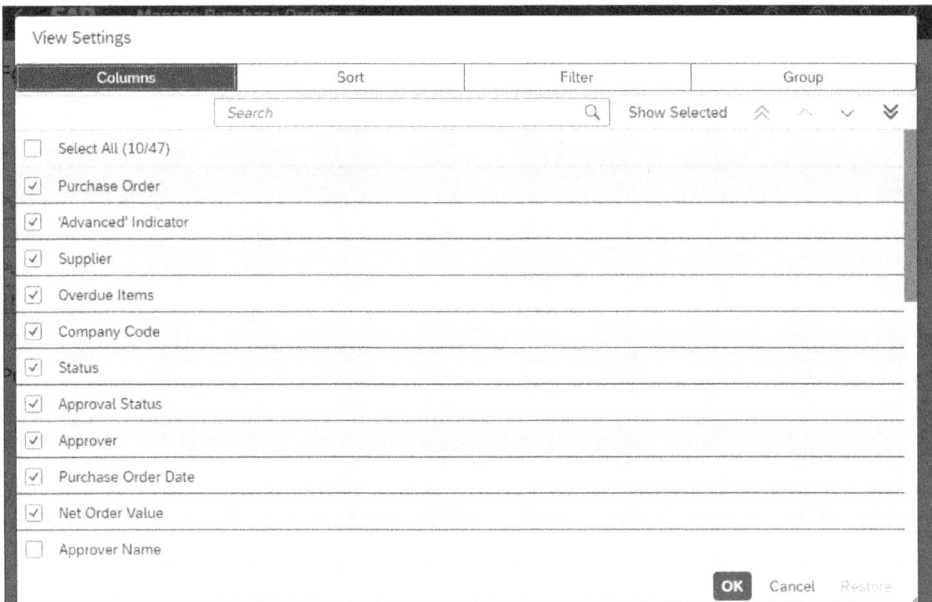

Figure 9.12 View Settings

9.2.1   Move Columns

Do you have a table that scrolls horizontally at the bottom of the window? There is no such thing in our example Manage Purchase Orders app, but there is in other apps, such as the Manage G/L Account Master Data app. You can set the optimal column order here to avoid scrolling. Frequently used columns belong to the left, so you have to scroll less! In addition, for our app, you can now use this function to optimize the table.

Follow these steps:

1. In the **View Settings** window, switch—if necessary—to the **Columns** tab, as shown in Figure 9.13. The entries shown in the **Columns** tab each represent one column of the table. Thus, the order of the entries also reflects the order of the columns. In our example, the **Purchase Order** column is shown first and the **'Advanced' Indicator** column is shown second.

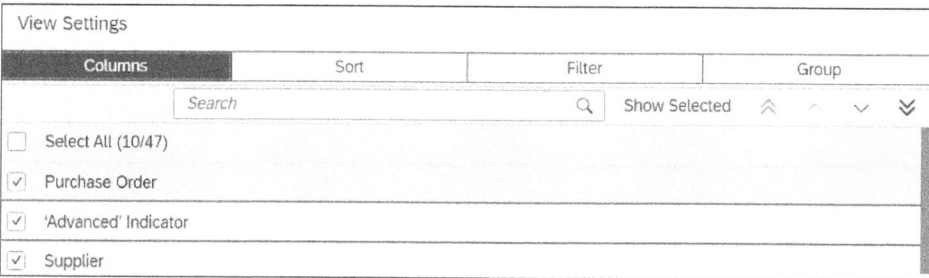

Figure 9.13 View Settings: Columns

2. To move a column, first click on the column name, in our case, on the entry **Supplier**, as shown in Figure 9.14. You can recognize the marker set with it by the colored background.

3. Because you want to see the **Supplier** column as the first column, click ⩓ (**Move to Top**) in the upper-right corner, as shown in Figure 9.14. With ⌃ (**Move Up**) and ⌄ (**Move Down**), move the columns one position at a time.

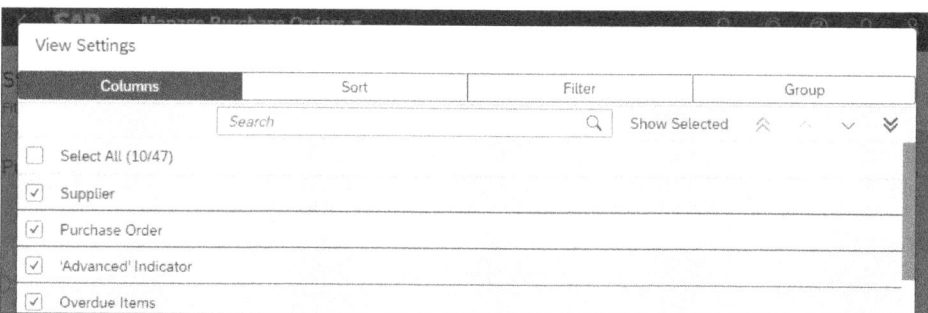

Figure 9.14 Supplier Column at the Top

4. To view the changes, click the **OK** button in the lower-right corner of the **View Settings** window. You'll then see the new display of the table in the app, as shown in Figure 9.15.

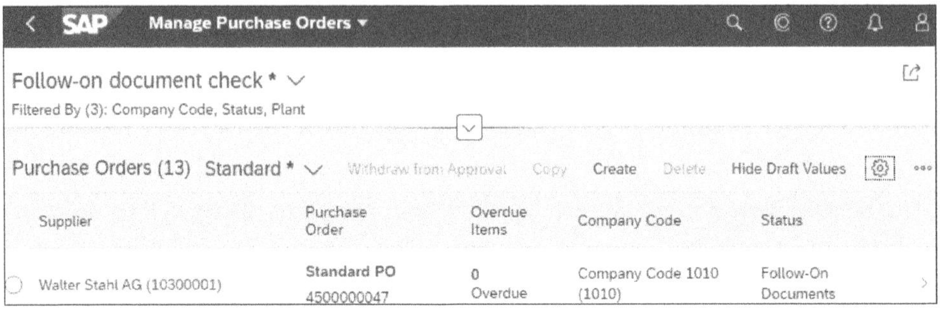

Figure 9.15 New Table View Settings

Table 9.2 provides an overview of the functions of the **Columns** tab in the **View Settings** window.

Function	Impact
🔍 (Search)	Allows you to browse and quickly find fields
Show Selected	Reduces the display to the selected entries
☐ (Select All)	Allows you to select all columns for display
⌃ (Move to Top)	Places the selected column to the far left at the top of the table
⌃ (Move Up)	Moves the selected table column one position to the left of the table
⌄ (Move Down)	Moves the selected table column one position to the right of the table
⌄ (Move to Bottom)	Places the selected column to the far right at the end of the table

Table 9.2 Functions of the Columns Tab

9.2.2   Show and Hide Columns

The optimal table displays exactly as many columns as necessary. To do this, you simply hide superfluous columns. On the other hand, many apps offer valuable information that is hidden by default and is just waiting to be displayed.

In the following instructions, you'll show the **Created On** column and hide the **Overdue Items** column as an example:

1. You're in the table of the Manage Purchase Orders app (with the state of the previous section, as shown in Figure 9.16). Click 🔧 (**Settings**).

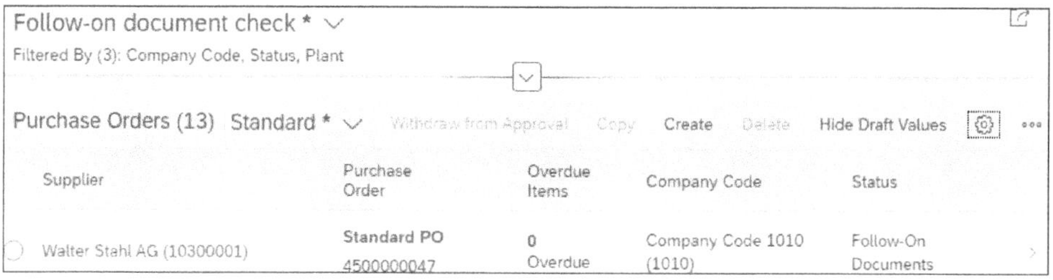

Figure 9.16 Manage Purchase Orders Table

2. In the **Columns** tab of the **View Settings** popup window, click the corresponding checkbox to display the **Created On** column ☐. The checkbox is filled with a checkmark ☑.

3. To hide the **Overdue Items** column, click on the corresponding checkbox that is currently checked ☑. After the click, the checkmark is removed from the checkbox ☐.

4. To view the changes, click the **OK** button in the lower-right corner. Now you see the **Created On** field in the report, as shown in Figure 9.17.

Figure 9.17 Manage Purchase Orders Table with a New Field: Created On

---

Cherry-Picking

[+]

You're in a transactional app and see many, many fields in the **View Settings** window that aren't displayed. Now you're puzzling over several field names, such as **Doc. Type Descript.**, and what content is hiding behind them.

The quickest way to see what you can display in the table is to select the ☑ (**Select All**) checkbox, and then click the **OK** button. Now the table shows you all the available content. You go "cherry-picking" by noting the columns you want to see in your optimal table. Then you show them in the **View Settings** window. With this method, you won't miss any columns, and you'll find many hidden treasures.

9.2.3 Sort

For sorting, there is a separate tab in the **View Settings** window. You use this to sort the table by the **Status** field. Follow these steps:

1. You're in the table of the Manage Purchase Orders app. Click ⚙ (**Settings**), and switch to the **Sort** tab in the **View Settings** popup window, as shown in Figure 9.18.

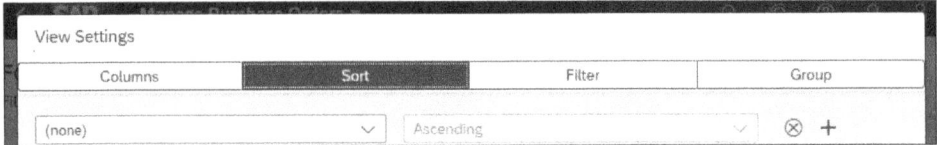

Figure 9.18 Sort Tab

2. To add a new sorting by the contents of a column, click the ☑ icon in the field with the content **(none)**, as shown in Figure 9.19. You'll see an alphabetically sorted list of all fields, including the hidden ones.

Figure 9.19 Adding a New Sorting

3. For example, select the **Status** field. To search the offered fields, enter the word "Status" in the field where the content **(none)** is displayed so far, as shown in Figure 9.20. After each entry of another letter, the displayed fields are filtered accordingly.

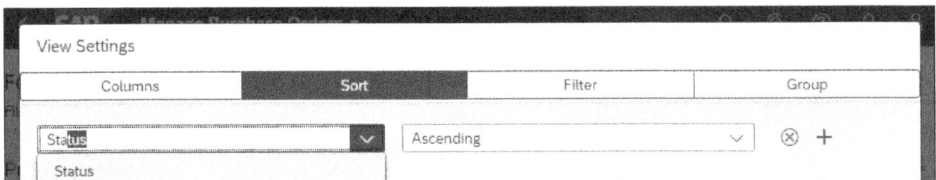

Figure 9.20 Search for Status Field

4. To choose between ascending or descending sorting, change the default in the second field to **Ascending** or **Descending**.

5. Finally, click the **OK** button in the lower-right corner to view the changes.

You can see where and which sort order is set in a table by the icon placed in the column header, as shown in Figure 9.21. The ⊒ icon stands for a descending sort, and the ≜ icon stands for an ascending sort.

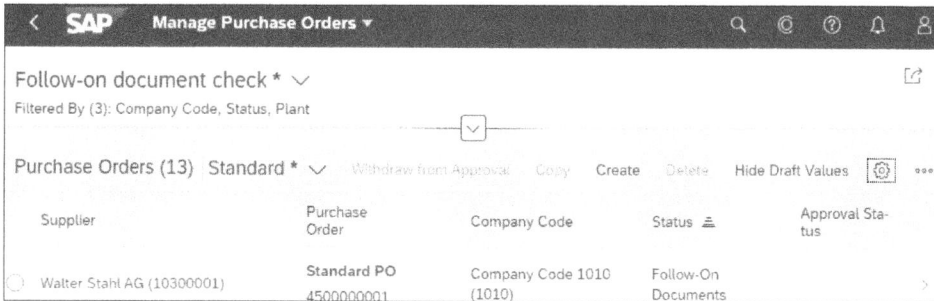

Figure 9.21 Status Column Header with Sorting Column Marker

In Table 9.3, you'll find further functions of the **Sort** tab.

Function	Impact
+	The **Add Sort Criterion** icon allows you to add more sort fields for multiple sorting.
⊗	The **Remove Sort Criterion** icon allows you to remove existing criteria.

Table 9.3 Other Functions of the Sort Tab

[+]

Sort Faster and Save Clicks

Do you want to sort by only *one* column? For this, you don't need the ⚙ (**Settings**) icon to switch to the **View Settings** window. Just click on the column header in the table, as shown in Figure 9.22. This will open a small window for most of the columns.

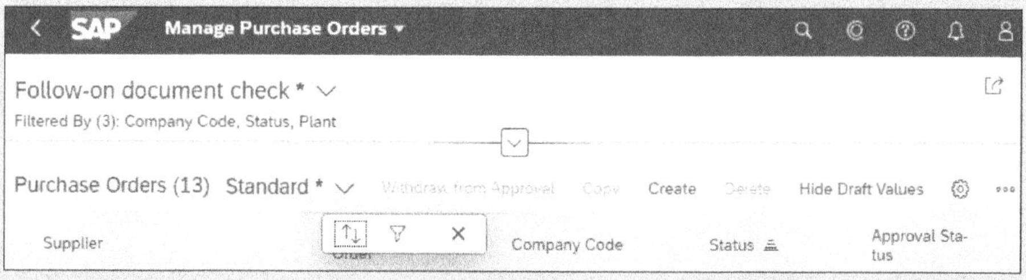

Figure 9.22 Window That Appears after Clicking on a Column Heading

After clicking on ↑↓ (**Sort**), sort the clicked column in descending order ⊒. In the same way, you'll get an ascending sorting afterwards.

Do you need multiple sorting in different columns? That doesn't work with this method. For this, you need the **View Settings** window after all.

9.2.4   Group

A grouping corresponds to a summary of related records. To do this, proceed as follows:

1. You're in the table of the Manage Purchase Orders app. Click ⚙ (**Settings**), and in the **View Settings** popup window, switch to the **Group** tab, as shown in Figure 9.23.

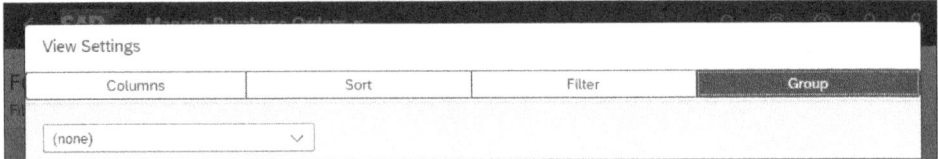

Figure 9.23 Group Tab

2. To see the offered fields for groupings, click the ☑ icon in the field with the content **(none)**. Click the **Created On** field in the list, as shown in Figure 9.24.

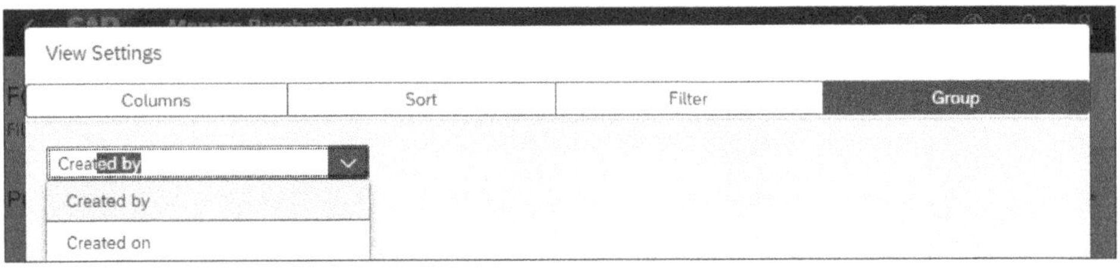

Figure 9.24 Select the Created On Field

3. To view the changes, click the **OK** button in the lower-right corner of the **View Settings** window.

What happened? The list you see in Figure 9.25 is sorted by the **Created On** grouping field, and the corresponding grouping field no longer appears as a column. Instead, you'll find the date values of the **Created On** grouping field as subheadings for one group at a time.

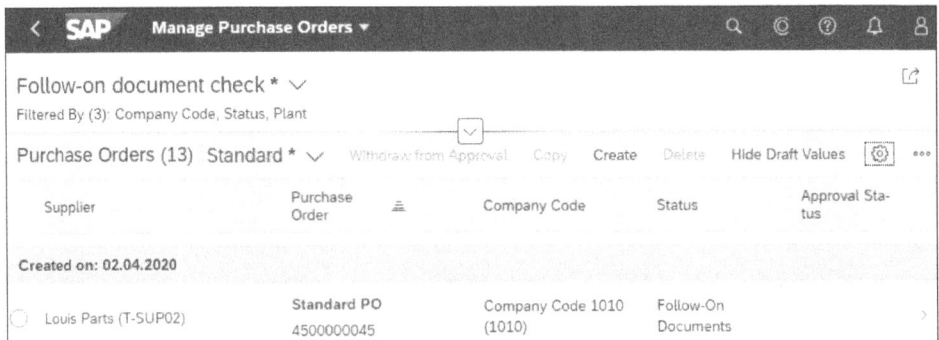

Figure 9.25 New Grouping

9.2.5   Save and Manage Table Settings

You may now save the many settings you've made up to this point as a view. The methods for saving, accessing, and managing these views are the same as those we showed you earlier in this chapter in Section 9.1. Therefore, you'll only get a quick guide at this point.

Follow these steps:

1. You're in the table of the Manage Purchase Orders app. Click directly above the table (not above the filters!) on ☑ (**Select View**). A window opens showing the views that are already available (see Figure 9.26).

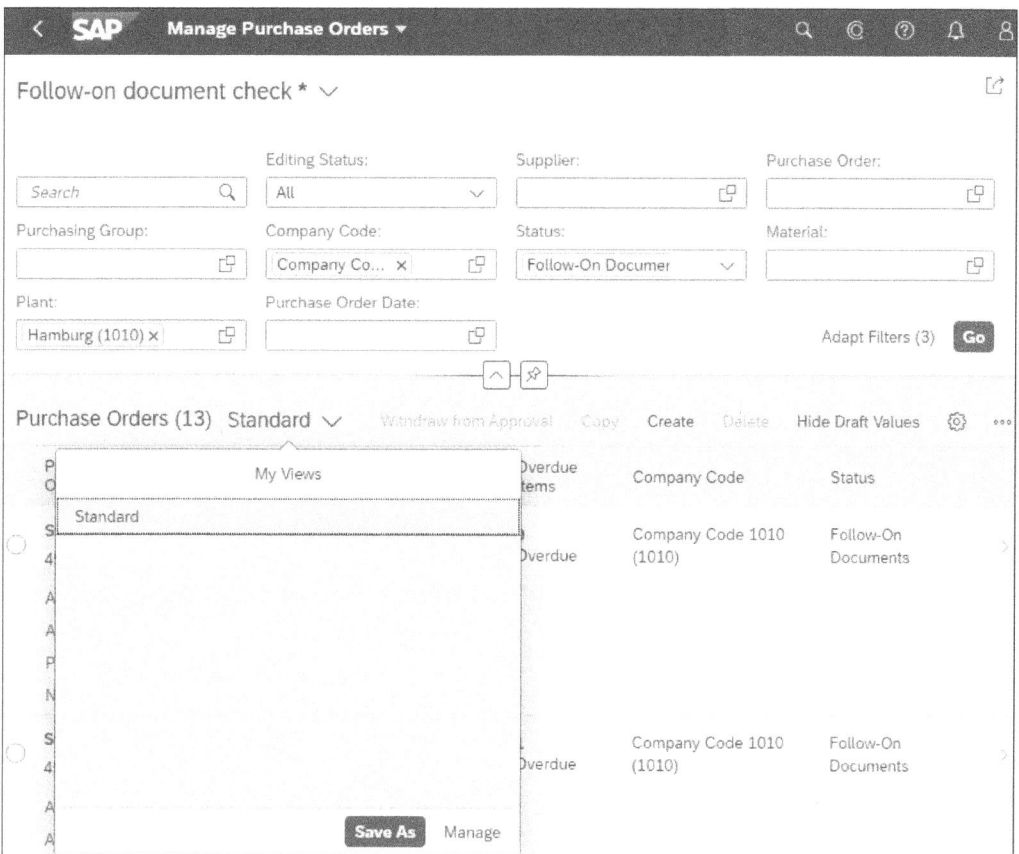

Figure 9.26 Select Views Popup

2. In this window, click the **Save As** button at the bottom right. The **Save View** window will appear, as shown in Figure 9.27.

3. Enter a name in the **View** field, and optionally select the **Set as Default** checkbox if you want this to be your new default.

Figure 9.27 Save View

4. Confirm your entry by clicking the **Save** button.

Calling and managing the views for table settings is done in the same way as for the views for filters.

[»]

Are You Missing Functions?

Especially if you're already familiar with reporting with SAP GUI, you'll miss some features. Compared to SAP GUI, real transactional SAP Fiori apps still lack some features, for example, fixing columns, customizing column widths, and much more.

Just read on. What you can't "get right" here in the transactional SAP Fiori app will work in many cases in the tables of the analytical apps. Of course, we'll be happy if SAP soon delivers these functions for the transactional apps as well.

9.3   Customize Tables in Analytical SAP Fiori Apps

Why do we now revisit the topic of tables? Well, we're not just hitting the repeat button here! We're here because some functions of tables in analytical apps differ from those of transactional apps.

Basic functions are identical to those from real transactional SAP Fiori apps such as Manage Purchase Orders. And other functions evoke déjà vu experiences from SAP GUI. As already announced, SAP Fiori offers you more possibilities in analytical apps than in transactional apps. Of course, the already-familiar options of the **View Settings** window are available here as well. Because their use in analytical apps doesn't differ from their use in transactional apps, we'll disregard them here and refer you to Section 9.2.

To follow the instructions in this section, pick a nice analytical app. In our example, we use the Monitor Purchase Order Items app.

Follow these steps:

1. Start an analytical app such as Monitor Purchase Order Items.

2. If necessary, make a selection using the filters, and click the **Go** button. For the Monitor Purchase Order Items app, you must at least make the entry "USD" in the **Display Currency** field, as shown in Figure 9.28.

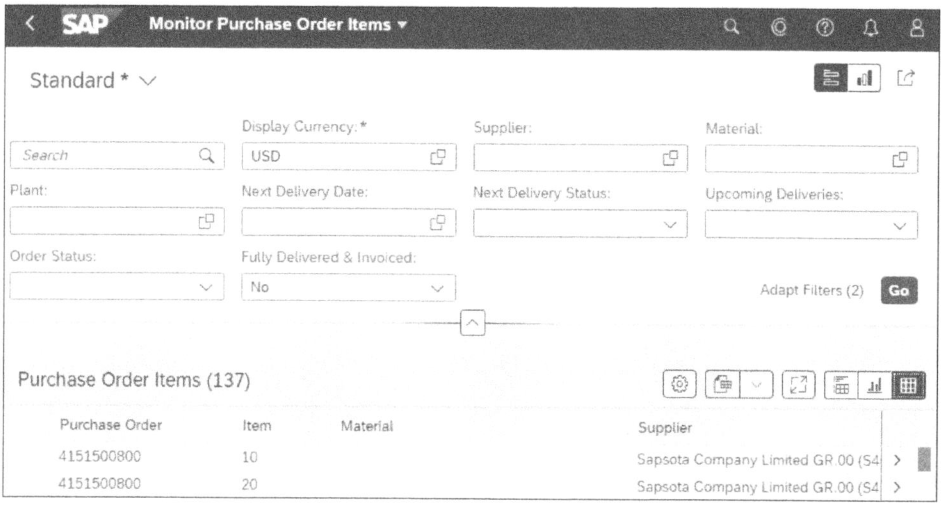

Figure 9.28 Monitor Purchase Order Items: Header and Table

3. To switch to the pure table view, click ⊞ (**Table View**).

9.3.1   Move Columns

The **Supplier** column with the supplier names should be on the far left. For this purpose, you don't need the ⚙ (**Settings**) icon. Instead, you may use the drag and drop technique as in SAP GUI. Follow these steps:

1. To move a column, click its column header (in our case, **Supplier**), and hold down the mouse button.

2. Drag the selected column to the desired position while holding down the left mouse button. Once there, release the mouse button. In Figure 9.29, you can see the current state.

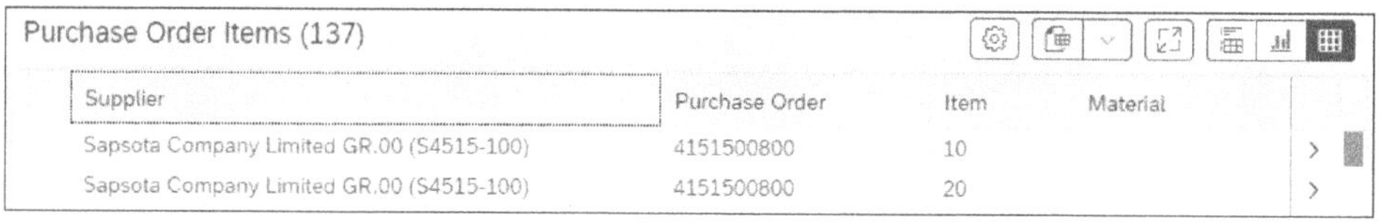

Figure 9.29 Columns after Drag and Drop

9.3.2 Adjust Column Width

Columns are often unnecessarily wide. With the mouse button pressed, drag the separator between two column headers to the left to reduce the column width or to the right to increase column width.

[+]

Optimal Column Width

In analytical SAP Fiori apps, the following elegant method from SAP GUI (also from Microsoft Excel) works: double-click on the hyphen between two column headers to get the optimal column width, matching the longest entry.

9.3.3 Sort Columns

Sorting works in analytical apps, just like in transactional apps, usually via ⚙ (**Settings**). However, when sorting a column directly in the table, there is a small deviation that can also be found in some transactional SAP Fiori apps:

1. In our example, click on the **Supplier** column heading. Now a small window appears offering you several options, as shown in Figure 9.30.

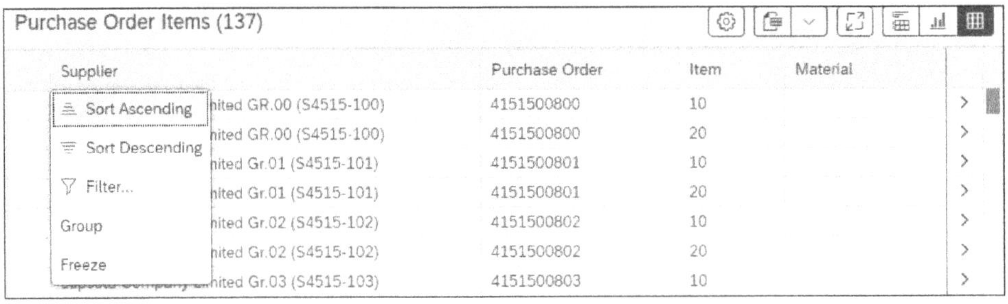

Supplier		Purchase Order	Item	Material	
≞ Sort Ascending	hited GR.00 (S4515-100)	4151500800	10		>
≡ Sort Descending	hited GR.00 (S4515-100)	4151500800	20		>
	hited Gr.01 (S4515-101)	4151500801	10		>
▽ Filter...	hited Gr.01 (S4515-101)	4151500801	20		>
Group	hited Gr.02 (S4515-102)	4151500802	10		>
	hited Gr.02 (S4515-102)	4151500802	20		>
Freeze	hited Gr.03 (S4515-103)	4151500803	10		>

Purchase Order Items (137)

Figure 9.30 Sort Options

2. For an ascending sort, click on ≞ (**Sort Ascending**), and for a descending sort, click on ≡ (**Sort Descending**).

9.3.4 Group

Now we group properly; not just with subheadings like in transactional SAP Fiori apps, but with a drilldown function. You don't know drilldown? No problem, just follow these instructions:

1. In our example, click on the **Supplier** column heading. A small window will appear offering several options, as shown in Figure 9.31.

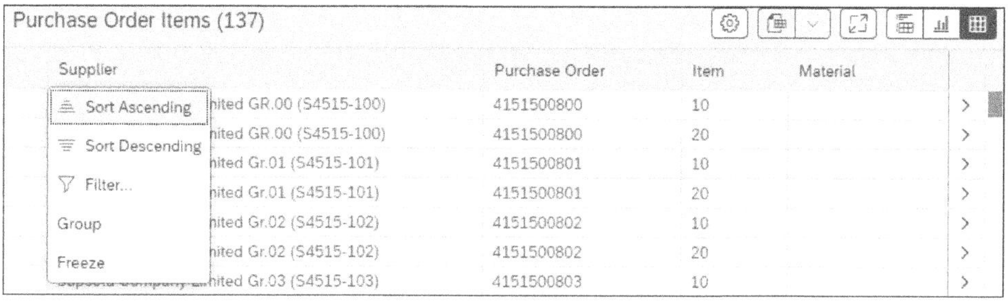

Figure 9.31 Options for a Column

2. Click the **Group** option. SAP Fiori will then group all purchase order items that contain the same supplier, as shown in Figure 9.32.

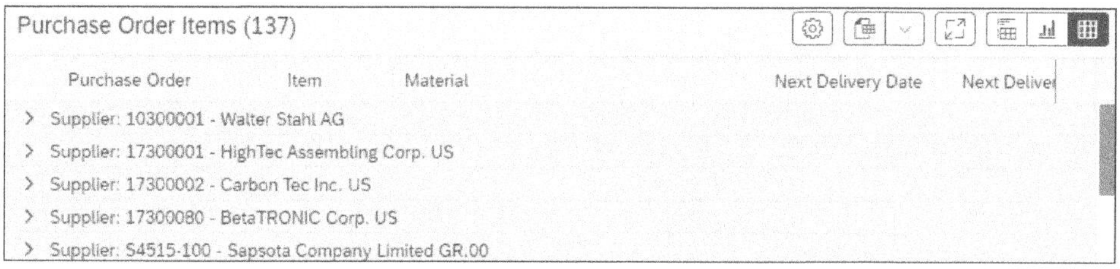

Figure 9.32 Grouping All Purchase Order Items

3. For example, to view all order items with the supplier **HighTec Assembling Corp. US**, open the corresponding grouping by clicking on the ⟩ icon, as shown in Figure 9.33. This technique is called *drilldown*.

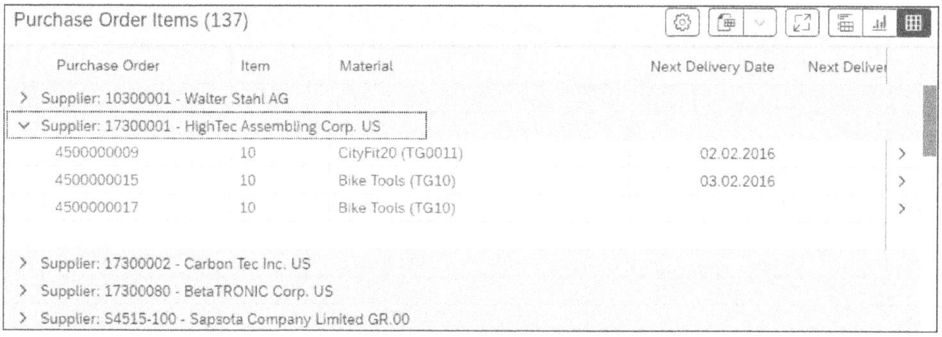

Figure 9.33 Drilldown to a Supplier

4. If you scroll further to the right, you'll see more details, such as the **Net Order Value** field, as well as the totals.

5. You can also create multilevel drilldowns by repeating steps 1 to 3 of these instructions for additional columns.

6. You can see the groupings after clicking ⚙ (**Settings**) and clicking the **Grouping** tab. There you have the following options for a grouping:
   – Remove a grouping by clicking ⊗.
   – Add another grouping by clicking ＋.
   – Set whether the field should be displayed as a column.

[»]

> ### Can I Save These Settings?
>
> Fortunately, you can save the table settings as a view, but only as a package together with the filter settings. To do this, click ☑ (**Select View**) at the top left, directly below the shell bar. The procedure is the same as saving a view for a filter (see Section 9.1.1).

Do you have about 10 minutes left? Then leave the Monitor Purchase Order Items app. There's another highly exciting topic in the next section.

9.4   Create Your Own SAP Fiori Apps

Welcome to the supreme discipline of personalization! Creating your own apps? In a fundamentals book? Something doesn't add up, does it? But don't worry, there's no coding or programming here.

Basically, the headline is an exaggeration, and we apologize for it right here. In reality, you create your own tile with which you can, for example, create a personalized evaluation with your own filter list and your own table settings—simple, fast, and comfortable! And the result will feel like your own app.

Follow these steps:

1. Launch a real SAP Fiori app of your choice, such as the Manage Purchase Orders app.
2. Click on the **Go** button. Is the correct view displayed with the desired filter and table settings? If so, you may skip step 3 and proceed directly to step 4.
3. Do you need special filter and table settings? Then make them now and save them as a view (refer to Section 9.1.1 and Section 9.2.5).
4. Click ⤴ (**Share**). A small menu appears in the upper-right corner, as shown in Figure 9.34.

[»]

> ### You Can't Find the Share Icon?
>
> The ⤴ (**Share**) icon isn't at the top of some transactions, but at the bottom right of the window. Or have you accidentally launched a "fiorized" transaction and not a real SAP Fiori app? The **Share** function isn't available for legacy apps.
>
> However, there are also various real SAP Fiori apps, such as Post Incoming Payments in which the ⤴ (**Share**) icon isn't available.

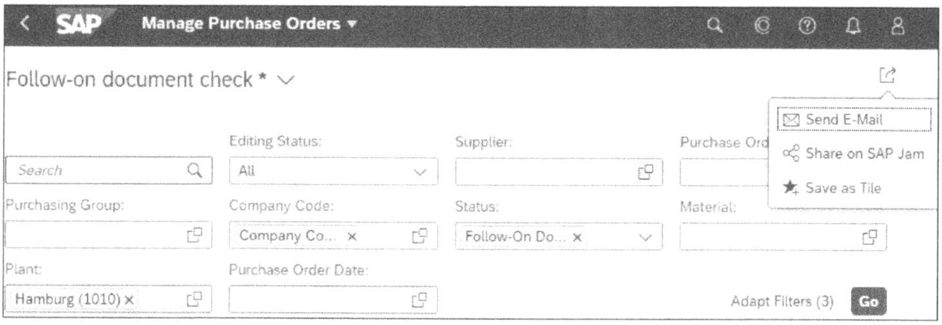

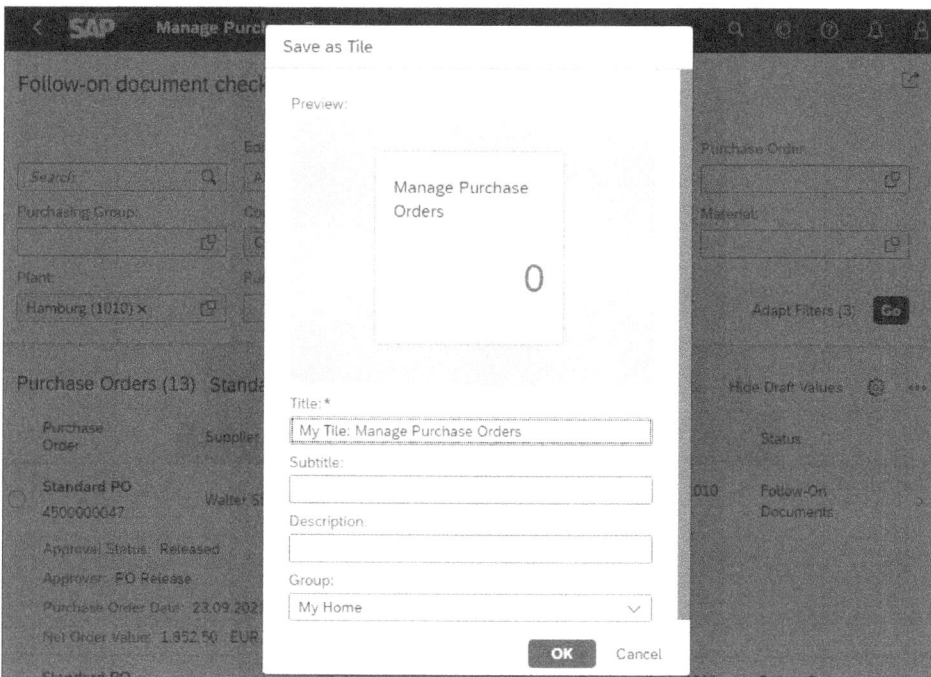

Figure 9.34 Share Options

5. In this menu, click 🟊 (**Save as Tile**). The **Save as Tile** popup window appears, as shown in Figure 9.35.

Figure 9.35 Save As Tile

6. In the **Save as Tile** window, specify at least the title and, at the bottom, the tile group in which you want your new tile to appear. Click **OK** when you're finished.

Other Options in the "Save as Tile" Window

In analytic SAP Fiori apps, additional options for KPI metrics are usually available in the **Save as Tile** window. Transactional apps such as Manage Purchase Orders have only **Title**, **Subtitle**, **Description**, and **Group** options besides **Preview**.

---

7. If you now return to the start page of your SAP Fiori launchpad and switch to the respective group, you'll find the tile you created there, as shown in Figure 9.36.

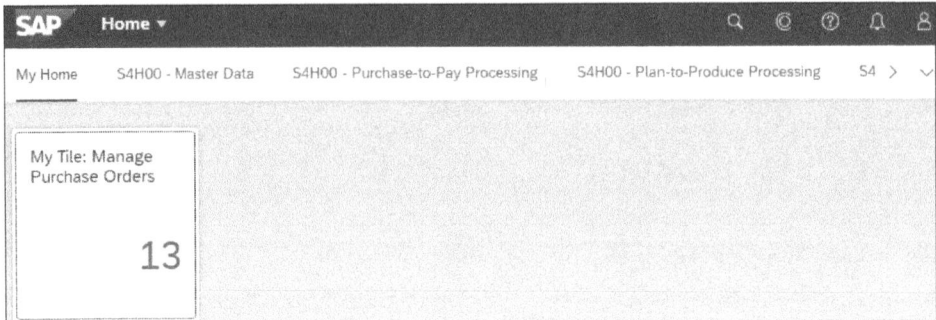

Figure 9.36 Your New Tile with a KPI

You can also create multiple tiles for one SAP Fiori app this way. This way, you can always get to exactly the information you need with a single click from the SAP Fiori launchpad homepage. What a fantastic feature! We liked it so much that we would like to invite the inventor(s) of this feature to a Munich beer garden or even to the Ace of Clubs restaurant in St. Leon-Rot.

9.5 Summary

You've seen with the views that your reports can be adapted very flexibly to your desires without the help of key users or programmers. In the selection screen, you save the default values for reports in report views.

In the list, the prefabricated standard reports often don't fit. Then you sort by columns, move and hide columns, change the columns widths, or show additional columns. Then you save these settings as views so you can quickly recall them when you need them. Analytical apps offer you additional functions.

You can also create your own tiles with which you can, for example, create personalized reports with your own filter list and your own table settings.

To conclude our discussion of the SAP Fiori user interface, we'll cover personalization in the next chapter.

Chapter 10
Personalizing and Optimizing SAP Fiori

Until now, you learned step by step how to walk with the SAP Fiori user interface. Now it's time to optimize and personalize SAP Fiori and thus turn walking into jogging. Because with personalization, you'll accelerate your work pace.

You've already made the first preliminary adjustments in Chapter 6, remember? So here comes an overview of our fitness training from Chapter 6:

- Create an SAP Fiori shortcut on the desktop.
- Customize a theme.
- Add, move, and remove tiles from the homepage.

But there is much more to it! You can find out the "what" and "how" now.

What You'll Learn

- How to set up your homepage optimally by creating your own tile groups to shorten the path to app launches
- How to convert tiles into space-saving links and change tile texts
- How to define cross-app default values and save yourself lots and lots of typing

10.1 Optimize the Homepage with Tile Groups

You already know what the homepage is and where it's located. In Chapter 6, Section 6.5, you also added a new tile to it from the App Finder. In this section, you'll learn how to use edit mode in the homepage in a practical way to optimize it.

For example, you can group related tiles into custom tile groups and place frequently used tile groups at the beginning of the homepage to reduce scrolling. In SAP GUI, this is similar to setting up the **Favorites** menu using folders.

10.1.1 Create a Group

The first step to a clearer homepage is to create your own groups in edit mode. We'll explain all the functions step by step in the following sections using examples.

Follow these steps:

1. Start SAP Fiori using the browser you trust. On the homepage, in the shell bar, click ⌷ (**User Menu**), as shown in Figure 10.1.

2. Select the ⌷ (**Edit Home Page**) menu item.

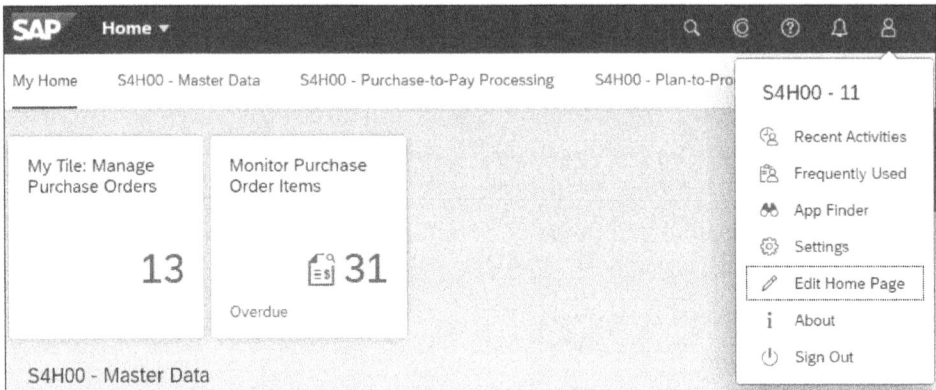

Figure 10.1 Opening the User Menu

3. This takes you to edit mode in the homepage, as shown in Figure 10.2. Tiles and tile groups can be edited here, but no apps can be launched.

4. Below each existing group, you'll see the **+ Add Group** button. Now click on this button, and SAP Fiori will add a new group.

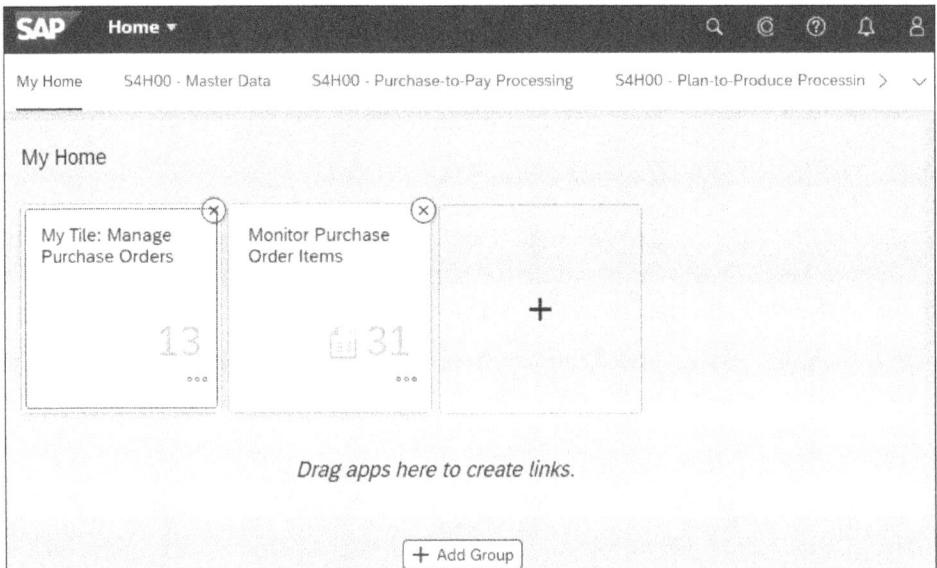

Figure 10.2 Edit Mode

What about the My Home Group?

You may have already seen the **My Home** group on the SAP Fiori launchpad homepage. You see this group, set up by default, on the launchpad homepage only when it contains tiles or links.

In edit mode, you'll always see this group in the tab bar and at the very beginning of all groups. The **My Home** group can't be renamed, moved, or deleted in edit mode.

Is this group displayed on your SAP Fiori launchpad homepage, and you want to get rid of it? To do so, drag all the apps from the **My Home** group to other groups, and close edit mode. Then the group will disappear by itself.

5. Click on the **Enter Group Name** text, as shown in Figure 10.3. With this click, an input field is displayed in which you can now define the new name of your group, for example, "My first group". Press ⌷Enter⌷.

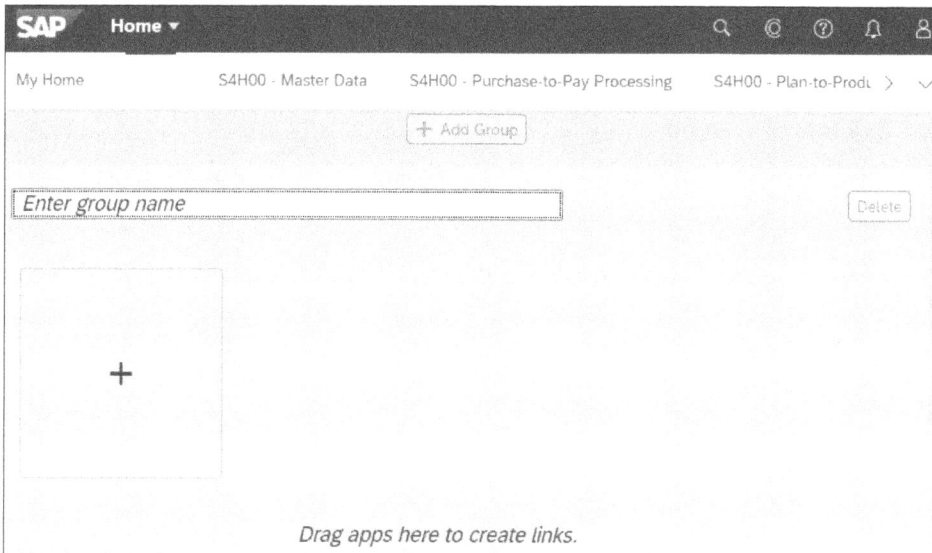

Figure 10.3 Enter Group Name in Edit Mode

6. To complete the process successfully, click **Close** in the footer bar. Because your new group doesn't yet contain any tiles or links, only the name of the group is displayed, as shown in Figure 10.4.

Figure 10.4 My First Group in the Group Selection Bar

You can now drag tiles to the new group on the homepage, as already described in Chapter 6, Section 6.5. You'll learn how to add a new tile in edit mode in the homepage in Section 10.2.1.

10.1.2   Rename a Group

Of course, the name "My first group" isn't exactly practical. That's why we're now renaming this one. To do so, follow these steps:

1. On the homepage, in the shell bar, click 8 (**User Menu**), and select the ✎ (**Edit Home Page**) menu item. This will put you back in edit mode.

2. Position your mouse pointer exactly on the heading of the group you created. You can tell if you're in the right place when the name of the group is outlined in color.

3. Click on the group name, in our case, **My First Group**. The displayed name of the group turns into an input field, as shown in Figure 10.5.

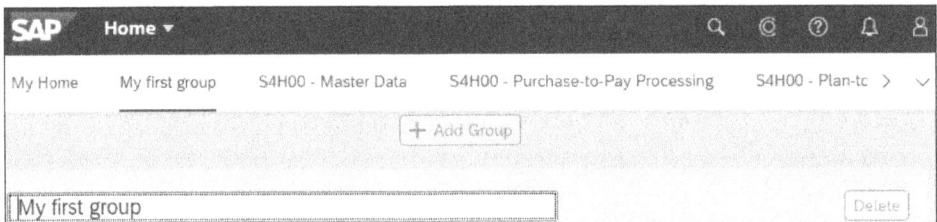

Figure 10.5 Renaming the Group in Edit Mode

4. Delete the existing content, enter "Reports for orders", and confirm your entry with the Enter key.

5. Exit edit mode by clicking **Close** in the footer toolbar.

10.1.3   Move a Group

In contrast to moving tiles, moving a group only works in edit mode. The procedure there is the same as moving a tile:

1. On the homepage, in the shell bar, click ⟨8⟩ (**User Menu**), and select the ⟨✎⟩ (**Edit Home Page**) menu item. This will put you back in edit mode.
2. Keeping the left mouse button pressed, drag a group up or down to the desired location and—once there—release the mouse button.
3. Click **Close** in the footer bar.

Note that you can't move a group in front of the first **My Home** group.

Because a group always uses the entire image width, moving the groups naturally only works vertically.

10.1.4   Delete a Group

Unnecessary groups have accumulated on your home page and you feel like cleaning up? Then delete the groups you no longer need! However, this only works for groups that you've created yourself. Follow these steps:

1. On the homepage, in the shell bar, click ⟨8⟩ (**User Menu**), and select the ⟨✎⟩ (**Edit Home Page**) menu item. This will put you back in edit mode.
2. Scroll to the group you created, and click the **Delete** button, as shown in Figure 10.6.

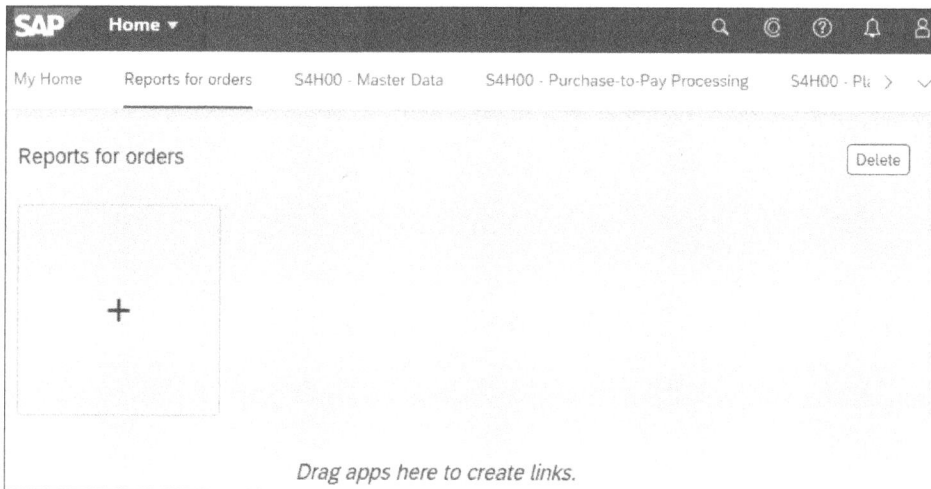

Figure 10.6 Delete Button in Edit Mode

3. In a separate window, as shown in Figure 10.7, you can confirm or cancel the operation. If you're sure, click the **Delete** button.

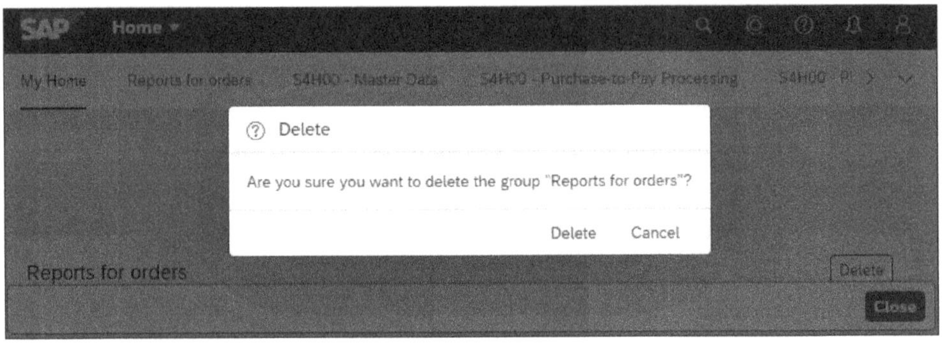

Figure 10.7 Deleting a Group

4. By clicking **Delete**, the group will disappear permanently. The successfully performed operation will be confirmed in a separate window.

5. Now you can click **Close** to display the result.

> [!]
> **You Should Be Really Sure!**
> Deleting groups can't be undone! It also automatically removes all the tiles contained in the group. If you accidentally delete a group with tiles that are important to you, you'll have to manually add those tiles to your SAP Fiori launchpad homepage again. Even closing the browser window won't save the group.

10.1.5   Reset a Group

You've deleted one or more tiles from a group. Oops! You didn't create the group yourself and want to undo this? You can, by following these steps:

1. As described previously, start edit mode in your homepage.

2. Delete any tile from a group that you didn't create by clicking the ⊗ (**Remove**) icon on the tile, as shown in Figure 10.8.

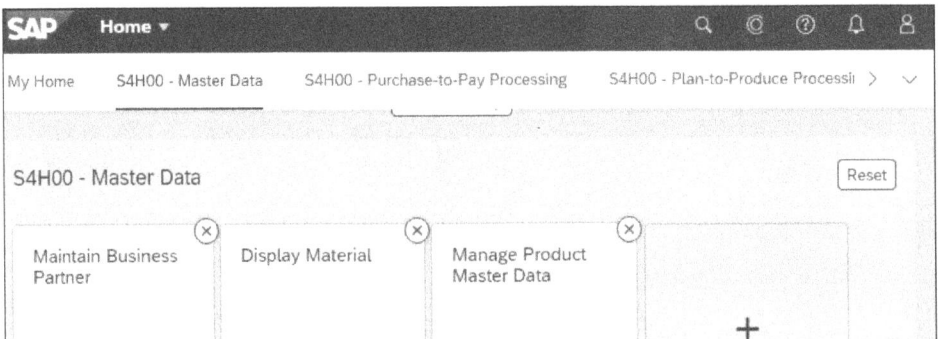

Figure 10.8 Tile Group in Edit Mode

3. Click the **Reset** button.

4. A query appears, as shown in Figure 10.9. Confirm the operation by clicking the **Reset** button.

5. Click **Close** in the footer.

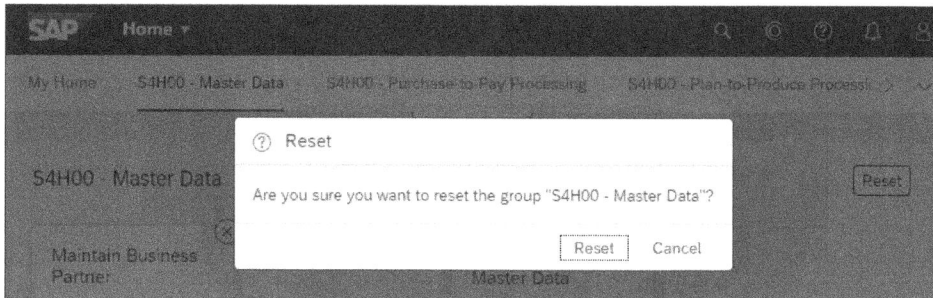

Figure 10.9 Resetting a Tile Group

The tile deleted in step 2 is back. Just a few clicks, and everything is back to normal!

What Is Reset?

The **Reset** button restores all. Not only are deleted tiles back, but the following actions are also undone:

- Renaming a tile
- Converting a tile into a link
- Renaming a tile group
- Moving a tile within a group

If you've moved a tile to another group, it remains there and also appears in the group that has been reset.

10.2 Edit Tiles

You can do more with tiles than you might think. Besides adding, moving, and deleting tiles, tiles can be converted into space-saving links, and tile texts can be changed. We'll discuss the steps to edit tiles in the following sections.

10.2.1 Add a Tile in Edit Mode

Now let's fill the new group you created with a little life. You already know how to add tiles with the App Finder from Chapter 6, Section 6.5. The App Finder can also be launched in edit mode.

Follow these steps:

1. On the homepage, in the shell bar, click ⌂ (**User Menu**), and select the ✎ (**Edit Home Page**) menu item. This will put you in edit mode.

2. To add a new tile, navigate now to the respective group. We'll use the top group, **My Home**. Click on the indicated tile with the ➕ icon, as shown in Figure 10.10. This is the last tile in each group.

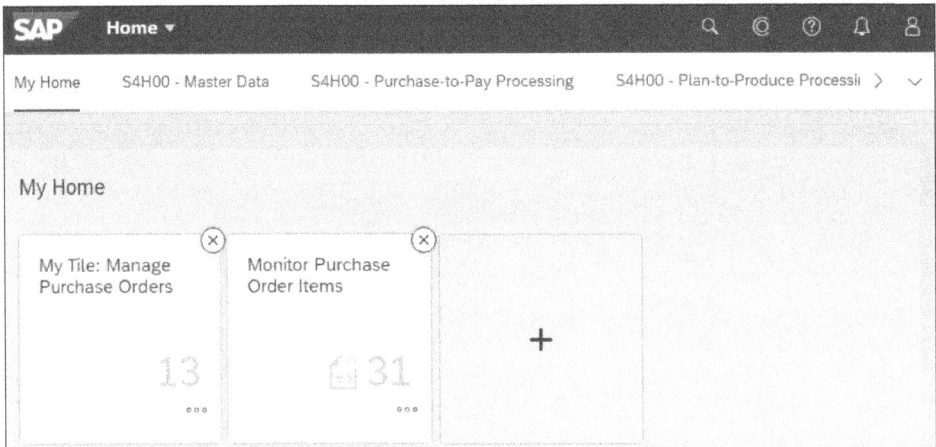

Figure 10.10 Clicking the Plus Icon in Edit Mode

3. After clicking on the ➕ icon, the image should already look familiar to you (see Figure 10.11). That's right—it's the App Finder! That was to be expected because, after all, before you add an app, you have to find it first.

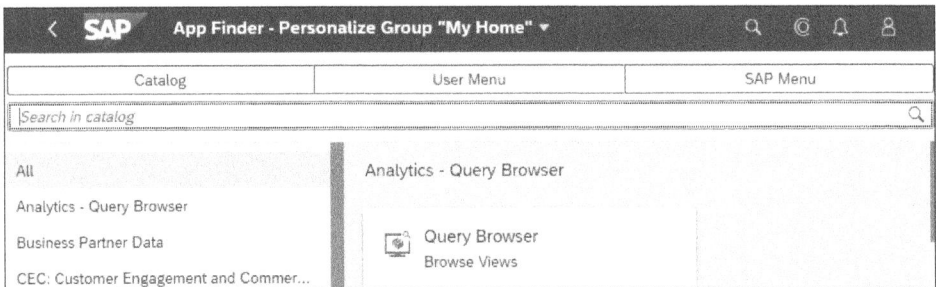

Figure 10.11 App Finder

4. As already described in Chapter 6, Section 6.4, search for an app of your choice using the 🔍 (**Search**) field for an app of your choice, such as Manage Purchase Orders.

> **Search in the Catalog**
>
> You start the search after entering a text such as "manage purchase orders" either by clicking 🔍 (**Search in the Catalog**) or by pressing the Enter key.

5. Don't click hastily on the displayed result! Clicking on the complete tile only launches the app, but doesn't add it to your homepage.

   If you instead hover the mouse cursor over the 📌 (**Add to Group "My Home"**) icon, as shown in Figure 10.12, you'll see a small hint after a couple of seconds. This will tell you the particular SAP Fiori launchpad group to which the app will be added. In this example, the destination is the **My Home** group.

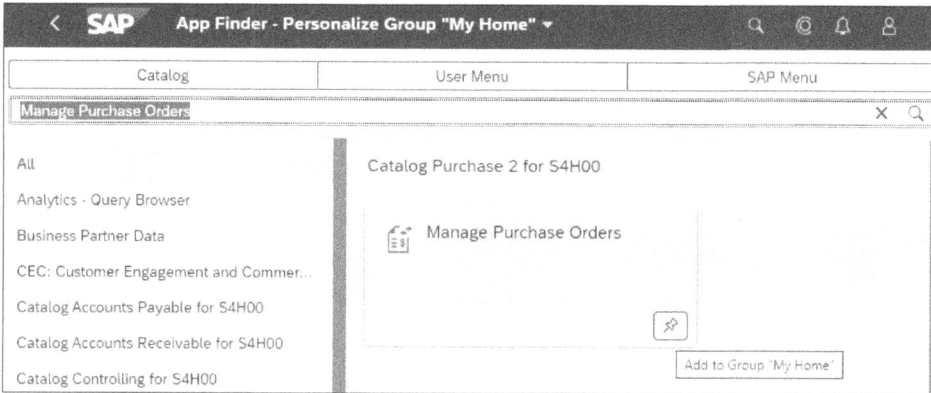

Figure 10.12 Add to My Home Group

6. Now click 📌 (**Add to Group "My Home"**). You'll next see the message '"Manage Purchase Orders was added to group 'My Home'". Voilà! You've now added a new tile to your SAP Fiori launchpad. You can tell because the add icon is now colored blue 📌, as shown in Figure 10.13.

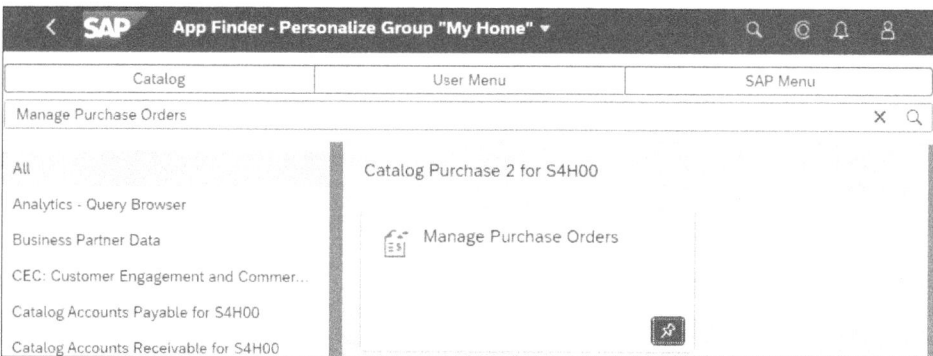

Figure 10.13 New Tile with Blue Add Icon

7. You can now add more apps to the previously selected group or return to edit mode by clicking ❮ (**Back**) several times from the shell bar.

8. You'll see the result of your work on the homepage after clicking the **Close** button.

10.2.2   Convert Tiles to Links

You use a lot of tiles, and your SAP Fiori launchpad homepage is too cluttered for you? You have to scroll up and down all the time? We have good news: apps can be visualized not only as tiles but also as links, so you can "shrink" rarely used tiles down to space-saving links. Such links show only the title and subtitle of a tile. Accordingly, no key performance indicators (KPIs), mini charts, or icons are rendered in links.

Follow these steps:

1. On the homepage, in the shell bar, click ⌸ (**User Menu**), and select the ✎ (**Edit Home Page**) menu item. This will put you in edit mode.
2. To turn any tile into a link, move the mouse cursor to that tile.
3. Press and hold the left mouse button and drag the tile with the left mouse button pressed into the area with the text **Drag apps here to create links**. This area exists at the end of each group.
4. Release the left mouse button, and you've now turned a tile into a link, as shown in Figure 10.14.

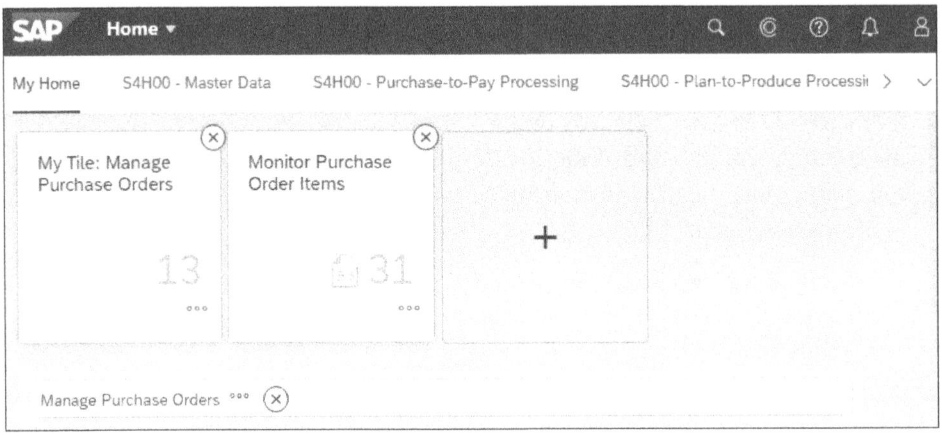

Figure 10.14 Manage Purchase Orders Link in Edit Mode

To turn the link back into a tile, simply drag the link back to the other tiles in the group.

10.2.3   Rename Tiles and Links

By renaming, you can also personalize a single tile or link! To do so, follow these steps:

1. On the homepage, in the shell bar, click ⌸ (**User Menu**), and select the ✎ (**Edit Home Page**) menu item. This will put you in edit mode.
2. Click ⋯ (**Tile Actions**) at the lower-right edge of a tile contained, as shown in Figure 10.15.

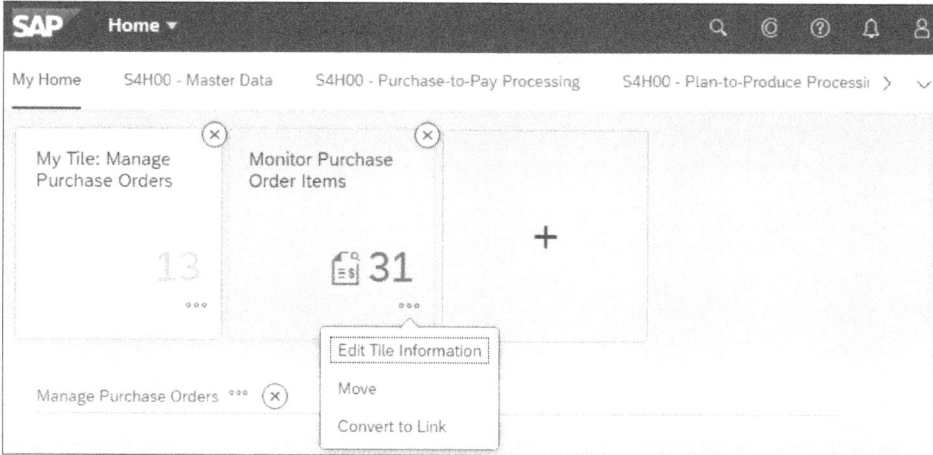

Figure 10.15 Clicking the Tile Actions Dropdown in Edit Mode

3. In the associated menu, click **Edit Tile Information**. The **Tile Information** popup window appears, as shown in Figure 10.16.

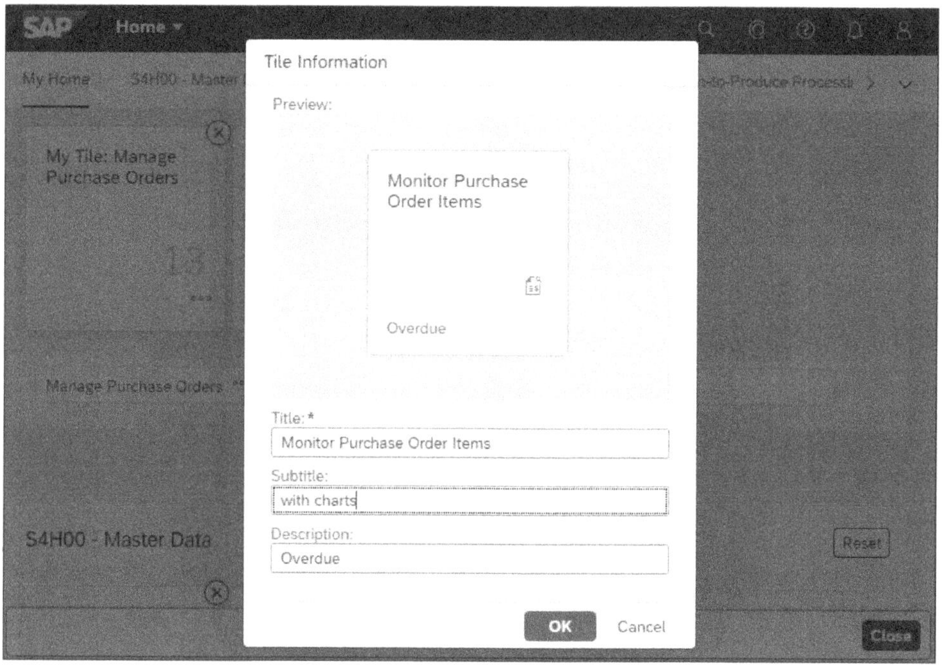

Figure 10.16 Tile Information in Edit Mode

4. As you can easily see, you can define the title, subtitle, and description yourself—there are no limits to your imagination. The preview shown in Figure 10.17 visualizes the finished tile.

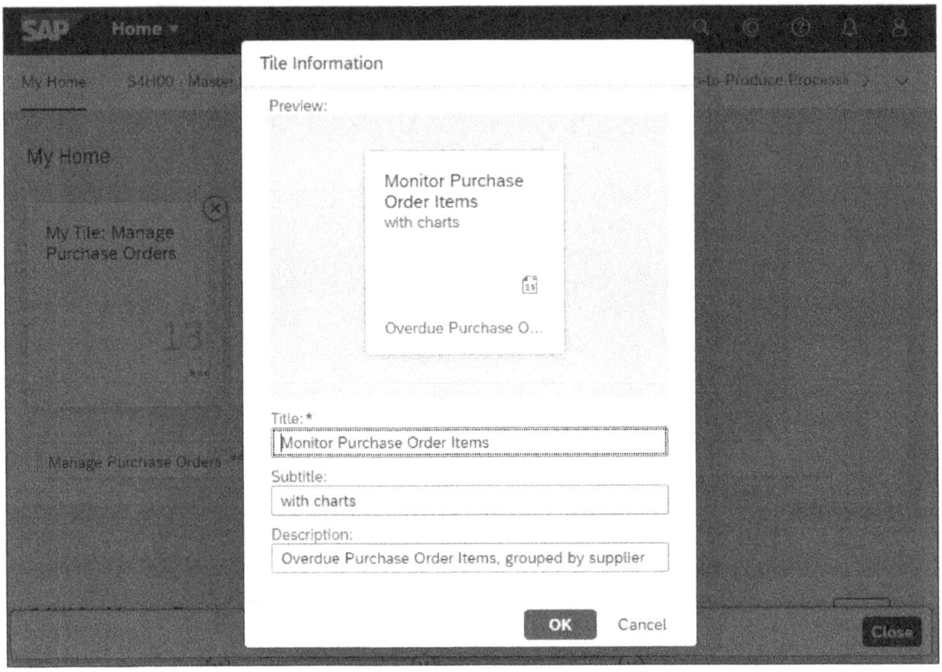

Figure 10.17 Preview of the Finished Tile in Edit Mode

5. Click the **OK** button to apply your changes. To view the result in "full bloom," exit edit mode by clicking the **Close** button at the bottom right.

Note that personalization of tile information isn't possible for all tile types.

When editing links, the procedure is the same: here, after clicking ∘∘∘, select the **Edit Link Information** option, and then you'll be allowed to change the title and subtitle of the link.

[+]   **Copy Operating Instructions to the Tile**

You can also copy larger amounts of text, such as an instruction manual for the app, into the description or subtitle.

Longer descriptions or subtitles of a tile are only displayed to a limited extent on the homepage. But if you position the mouse pointer a little longer on the tile, a separate window appears, as shown in Figure 10.18.

If you copy a larger text into the subtitle of a link, this text will be displayed completely on the homepage.

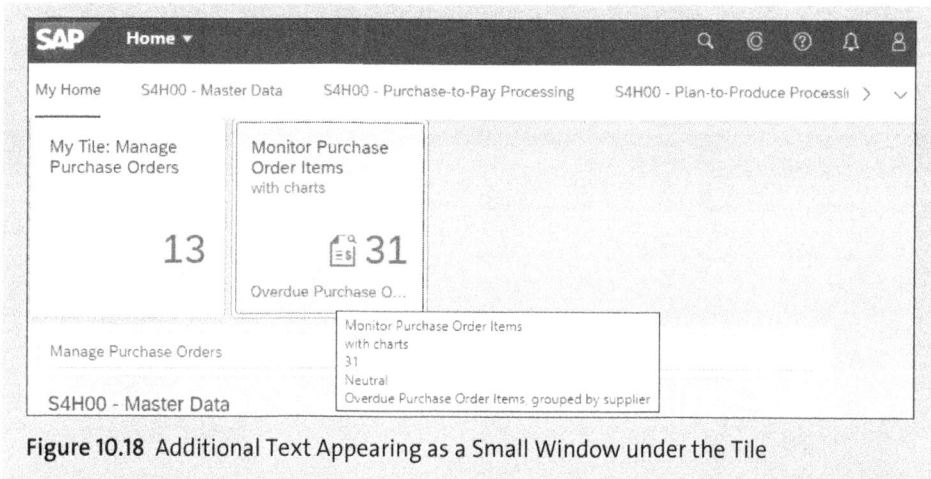

Figure 10.18 Additional Text Appearing as a Small Window under the Tile

10.2.4   Move Tiles and Links

You're allowed to drag and drop your tiles and links on the normal homepage as well as in edit mode. But imagine you have 50 different tiles or links sorted into 10 groups and want to move a tile or link from the first to the last group. That's not much fun with drag and drop, so it's faster and easier to follow these steps:

1. On the homepage, in the shell bar, click ⌂ (**User Menu**), and select the ✎ (**Edit Home Page**) menu item. This will put you in edit mode.

2. Now click ⚬⚬⚬ (**Tile Actions**) on the tile you want to move, as shown in Figure 10.19. This icon is always located in the lower-right corner of each tile or directly to the right of the title of a link. After the click, a menu will open.

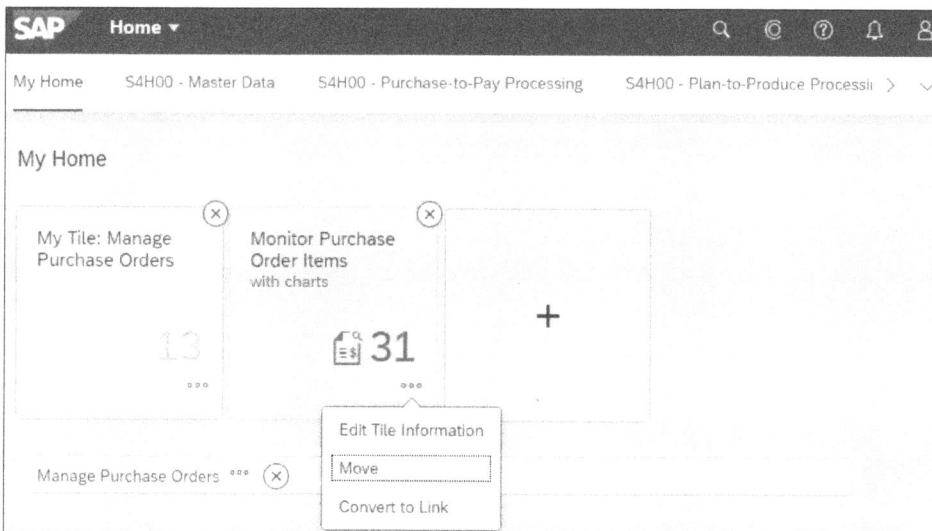

Figure 10.19 Open Tile Actions in Edit Mode

3. If you select **Move** from this menu, a separate popup window appears, as shown in Figure 10.20. In it, you can select the group as the destination of the move. Click on a group to move it. A message confirms that the move was successful.

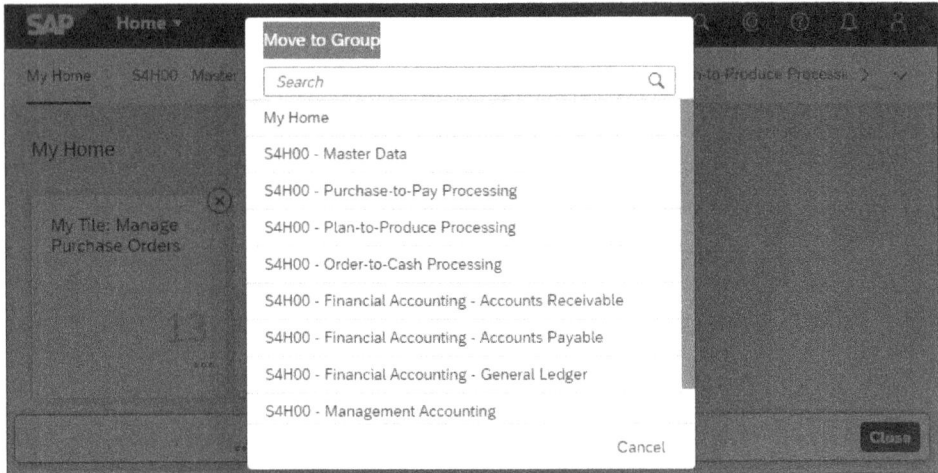

Figure 10.20 Move to Group Popup in Edit Mode

[»]

Move a Link in Front of the Tiles? Not Possible!

Links, by definition, can only be positioned at the end of the group after the tiles. If you drag a link on the normal homepage or in edit mode to a position above a tile, the link magically becomes a tile again.

10.2.5   Delete Tiles and Links

In this section, we'll discuss the same delete procedure for tiles and links.

Follow these steps:

1. As described previously, start edit mode in your homepage.

2. Now you're spoiled for choice and can designate one of the displayed tiles or links to disappear from the homepage.

   To do this, click ⊗ (**Remove**) in the upper-right corner of the respective tile, as shown in Figure 10.21.

3. If the tile has disappeared, you've proceeded correctly. SAP Fiori deletes the tile without any confirmation prompt.

4. You may now exit edit mode by clicking the **Close** button on the right at the bottom of the window.

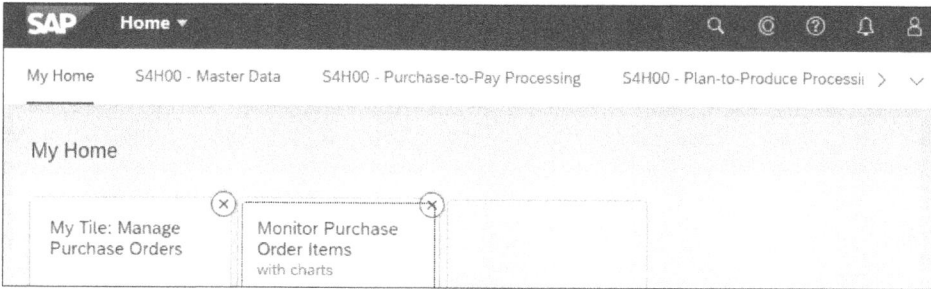

Figure 10.21 Removing a Tile in Edit Mode

Deleted Is Deleted!

As with deleting tile groups, what's gone is gone! Even closing the browser immediately after deletion won't save the deleted tile.

However, you can add an accidentally deleted tile back to your homepage using the App Finder, and in groups you didn't create yourself, there is a second option: the **Reset** function promotes the deleted tile back to the group (see Section 10.1.5).

10.3   Set Default Values for Real SAP Fiori Apps

In Chapter 5, Section 5.2, we showed you how to save a lot of typing in SAP GUI with user parameters. In SAP Fiori, however, there's good news and bad news.

First, the good news for all those who use legacy apps or "fiorized" transactions in SAP Fiori: the default values set in SAP GUI also apply to these apps in SAP Fiori.

And now the bad news: in "real" SAP Fiori apps, the default values from SAP GUI don't apply. Instead, you can predefine default values for a limited number of fields in SAP Fiori. These default values are then permanently and specifically valid for your user, just like the values from SAP GUI, but only in real SAP Fiori apps.

10.3.1   Define Default Values

In the following instructions, we'll show you the central place in SAP Fiori where you define all available default values.

Follow these steps:

1. Start SAP Fiori with the browser you trust, and click ⚊ (**User Menu**) to open the user menu. Then click the ⚙ (**Settings**) menu item, and the **Settings** popup window opens.
2. Display the possible fields in this popup window by clicking on **Default Values** in the lower-left corner. You'll now see all fields in plain text for which you can define default values, as shown in Figure 10.22. These are directly assigned to modules such

as controlling or financial accounting. Thus, in contrast to SAP GUI, you fortunately don't need cryptic parameter IDs. With the ⊡ icon, you can search for a value.

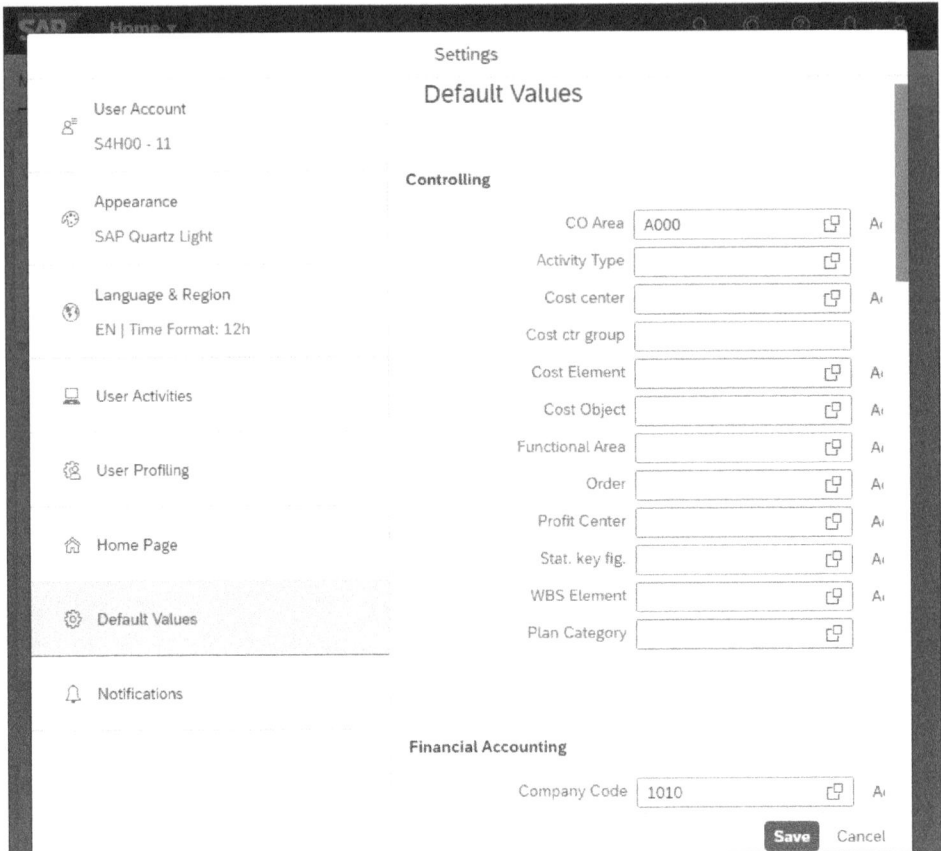

Figure 10.22 Defining Default Values

3. For example, enter the value "1010" in the **Company Code** field, as shown in Figure 10.22.

[»]

Define Additional Values for Filters

In contrast to SAP GUI, you can also define value ranges or intervals in SAP Fiori. To do so, click the **Additional Values** button, which is available only for selected fields.

But what is the use of this for you at all? It doesn't help you when entering new data, such as a new material or a new order, because you always need exactly one value. This option is only useful for filtering lists. For example, if you're specifically responsible for the customers from a number interval, you can specify the corresponding number interval here.

---

4. Save the company code default by clicking the **Save** button in the **Settings** popup window. SAP Fiori confirms the successful change in a small window.

10.3.2   Use Default Values

Seeing is believing, so let's test the new settings now. With this in mind, find and launch an app such as Display Supplier Balances that contains the **Company Code** field. As mentioned before, it should be a real SAP Fiori app because only here will the default values be applied. SAP Fiori has already filled in the **Company Code** field for you with the previously defined default value, as shown in Figure 10.23.

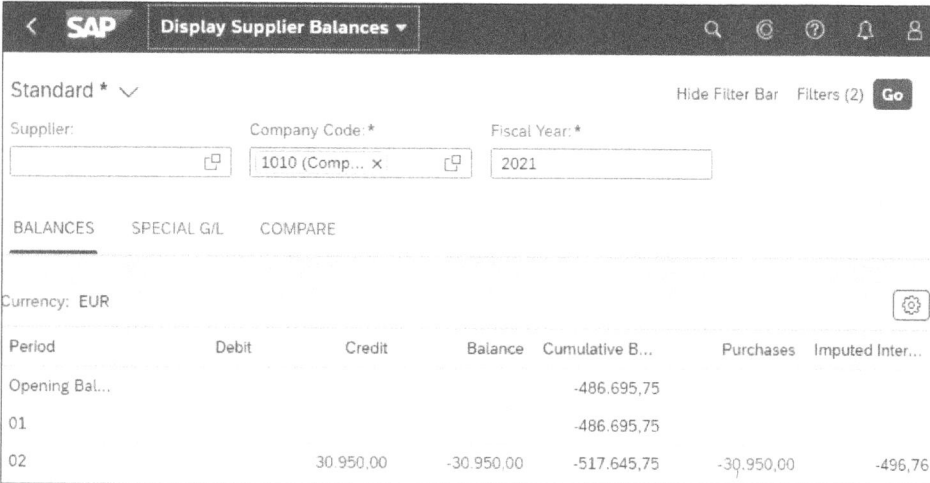

Figure 10.23 Display Supplier Balances App with the Default Value in the Company Code Field

The Default Value Is Missing?

As mentioned, "fiorized" transactions from SAP GUI ignore the default values defined in SAP Fiori. (You can tell whether it's a "fiorized" transaction by using the [i] (**About**) menu item in the user menu, as discussed in Chapter 7, Section 7.1.2.) In addition, there are unfortunately some real SAP Fiori apps that ignore the set default values.

10.4   Settings for Legacy Apps

Fortunately, most of the settings we showed you in Chapter 5, Section 5.2, are carried over to the corresponding legacy apps in SAP Fiori:

- Default values defined as parameter values in SAP GUI
- Personal value lists

There are some special settings for legacy apps that can only be made in SAP Fiori, and this includes the option for our favorite way to quickly invoke SAP GUI transactions: entering transaction codes in the command field. This allows you to launch legacy apps in SAP Fiori without using the corresponding tiles, but not SAP Fiori apps.

To activate the command field, you must first launch any legacy app. In our example, we start the Maintain Business Partner app and then follow these steps:

1. Choose **More • SAP GUI Actions and Settings • Settings**, which is offered only in legacy apps.
2. In the **SAP GUI for HTML - Settings** window, click on the **Interaction Design** tab, if necessary.
3. Here, as shown in Figure 10.24, turn on the **Show OK Code field** slider. It corresponds to the command field from SAP GUI and just has a different name here.

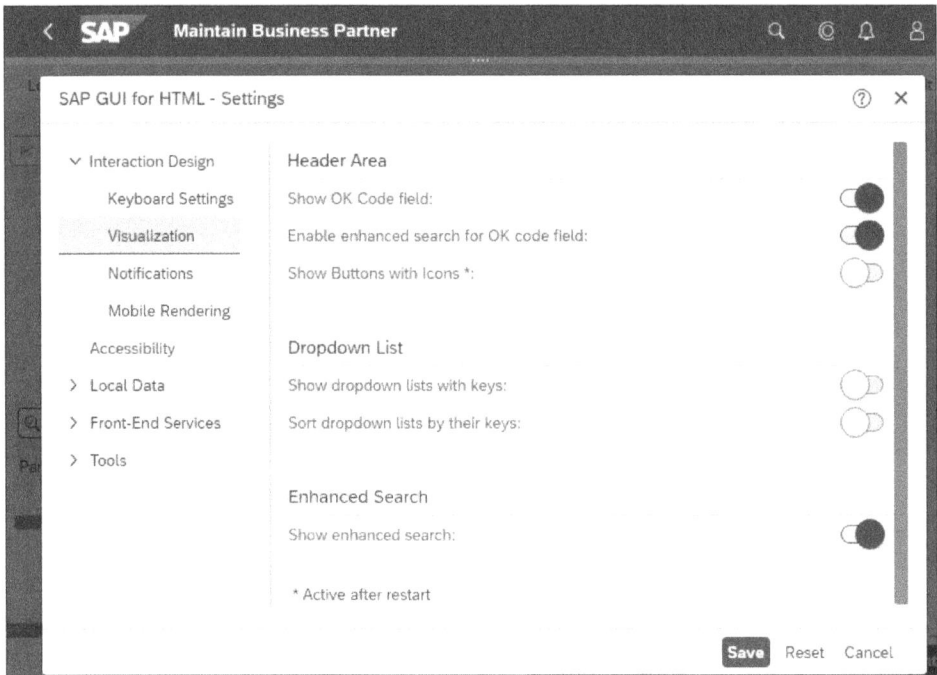

Figure 10.24 SAP GUI for HTML Settings

Confirm the setting with the **Save** button in the footer. Now you'll see the OK code field in the upper-left corner, waiting for you to enter the first transaction code.

4. Enter the command "/nMMBE" to start the Stock Overview transaction, as shown in Figure 10.25, and press Enter. You'll jump directly to the initial screen of the stock overview. Cool!

Figure 10.25 Starting Transaction MMBE

10.5 Summary

With this chapter, you've significantly expanded your knowledge in terms of personalizing your SAP Fiori environment. Now you should also implement this consistently. Don't forget: as soon as your requirements for the system change, for example, due to new tasks, you can constantly adapt and expand your settings in SAP Fiori—this is an ongoing process.

Congratulations, you've completed the second part of the book. We hope you've enjoyed our guide to SAP S/4HANA so far. If so, we recommend Part III of the book because the topics there are guaranteed to be even more fun!

10

PART III

Running Core Business Processes

Chapter 11

Materials Management: Coordinating Procurement

How are you feeling after the first two parts of the book? Hopefully well! Or is everything still a bit nebulous? Up to this point, you've only learned the most important rules and tools for *general* operation, which are, after all, only individual puzzle pieces of a big picture. Now you need an insight into the processes and practice with concrete procedures such as procurement, sales, and finance. In this third part of the book, we'll therefore show you the processes from practice and their handling in the system.

Working with the SAP system requires teamwork! Even if you don't carry out all these processes yourself in practice, you should look beyond your own nose and develop a basic understanding of what your colleagues do and where the data you enter goes. Some of the chapters build on each other, so it's a good idea to read them in their entirety and chronologically.

If you work in finance or controlling, you're one of the last "buyers" of the data in the processes described here. You'll often stumble upon problems resulting from incorrect or incomplete input from materials management or sales. Therefore, our recommendation is that you also take the information from this chapter and Chapter 12 with you.

We'll begin in this chapter with materials management. Procurement is an important part of the materials management module. Although we can't present here naturally the complete module, you'll gain an insight into this module and get a basic understanding of the processes in SAP S/4HANA. At the same time, you'll gain valuable practical experience in handling the apps.

So, what can you look forward to in this chapter? We'll cover the basic framework of the procurement process here, the *purchase-to-pay* process, which is often abbreviated to P2P. Because you can't run a business process without master data, you first review the vendor and material master data, which you need in the purchase-to-pay process. Then you enter a purchase order, the associated goods receipt, and the supplier invoice. The "pay" part of the process, that is, the payment of the invoice, will be covered in Chapter 13.

What You'll Learn

- Which subprocesses make up the purchase-to-pay process
- How to view and change the supplier data of a business partner

- How to display material data and material stocks
- How to create a purchase order and output it as a PDF
- How to enter a goods receipt with reference to a purchase order
- How to enter an incoming invoice

11.1   Purchase-to-Pay Process

Procurement is a process that consists of a set of activities. These activities are called partial processes. With the execution of these partial processes, documents develop. Figure 11.1 shows you the basic flow and the documents of the *purchase-to-pay* business process, in which a stock material is purchased.

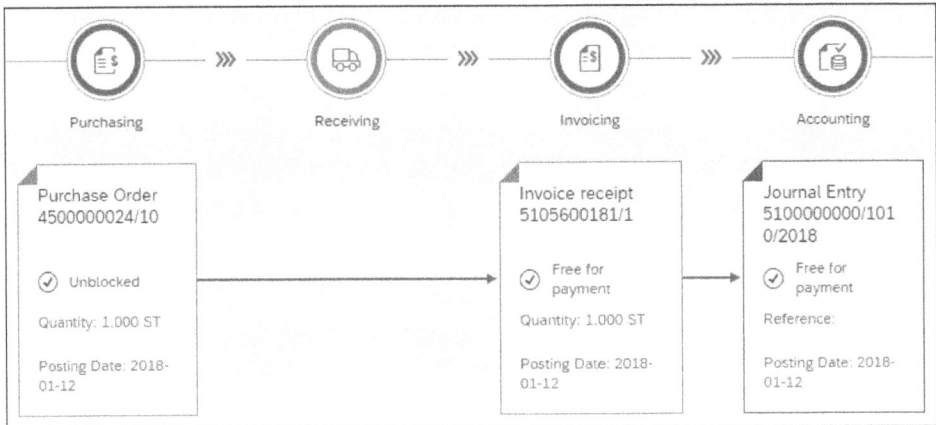

Figure 11.1 Subprocesses and Documents of the Purchase-to-Pay Business Process

We have a little story about our example business process and its documents. It starts with the fact that the stock of tires in your warehouse is too low and urgently needs to be reordered. Then, the following steps occur:

1. In purchasing, a *purchase order* is entered into SAP S/4HANA. The order then reaches the supplier, either automatically electronically, by email as a PDF file, or by fax.

   Data entered in the purchase order doesn't have to be entered again in *subsequent documents* such as goods receipts or incoming invoices. The advantage is that time-consuming and error-prone multiple data entry is eliminated when data entered once is used consistently in all departments.

2. The second step is the *goods receipt* recorded in the warehouse as a *material document*.

3. Now—with reference to the purchase order—the *incoming invoice is* entered and checked. If the goods receipt and invoice correspond to the order, and there are no complaints, the invoice is released for payment.

   Simultaneously with the invoice document for the materials management module, an *invoice document* for the financial accounting module is automatically created when the invoice is saved. This will be needed later for the payment of the invoice.

4. Now we're in *accounts payable* (other name: *vendor* accounting), which initiates the payment of the invoice and thus creates a *clearing document*. In practice, this is normally done largely automatically in a *payment run*. Further entries aren't necessary. What about bank details or other data? These don't have to be entered because they are in the supplier master data and in the preceding invoice document.

The components listed in Table 11.1 are affected by this process.

Subprocess	Component	Abbreviation of the Component
Order	Purchasing	MM-PUR
Goods receipt	Inventory management	MM-IM
Invoice receipt	Invoice verification	MM-IV
Payment of the invoice	Accounts payable	FI-AP

Table 11.1 Components in the Purchase-to-Pay Business Process

11.2   Organizational Units for Procurement

In the introduction to this book, you already learned about the basic organizational units in the "Company Structure in the SAP System" section. There we promised you that we would present them to you in more detail in this part of the book, so let's get to it.

The first thing you need is the *company code* because each purchase order must be assigned to a company. In addition, there are organizational units from logistics, as listed in the following sections.

11.2.1   Plant

A *plant* must be specified for each purchase order. If there is a head office, this is also defined as a plant in SAP S/4HANA. A plant doesn't necessarily have to be a production facility. For our order (see Figure 11.2), we use plant 1010.

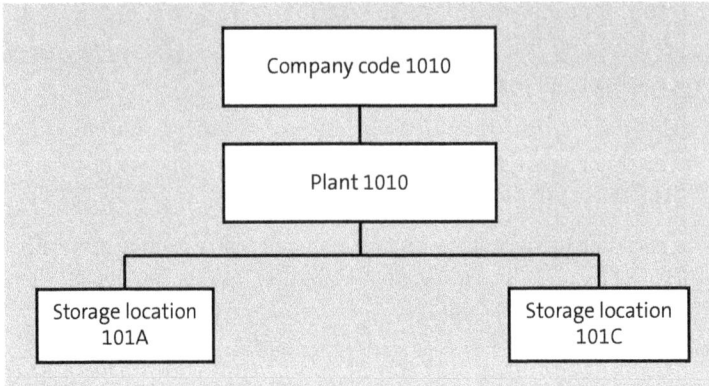

Figure 11.2 Company Code 1010 with Plant 1010 and Two Storage Locations

11.2.2   Purchasing Organization

There can be different *purchasing organizations* as purchasing departments, which are responsible either only for individual plants, only for individual company codes, or as central purchasing or group purchasing across all company codes. For our purchase order, we use purchasing organization 101. This is assigned to company code 1010.

11.2.3   Storage Location

Each plant has its own *storage locations*, for example, one for raw materials and another for finished products. If a plant consists of different halls, there can be one or more storage locations per hall. Storage locations are therefore used to manage and differentiate material stocks within a plant.

For example, in the SAP Live Access training system, there are the following storage locations:

- **101A (Std. storage 1)** for storage and shipping of finished products
- **101C (Raw mat.stoloc.)** for storage and receipt of raw materials

A storage location *must be* specified at the latest at the time of goods receipt of an inventory-managed material so that the stock is increased at the corresponding storage location. Our goods receipt is made to storage location 101C.

11.3   Presets and Training Data

Now, we'll discuss the default values and training data for the purchase-to-pay process. The trend, of course, is toward SAP Fiori, and that is why in this part of the book, we'll only show you examples of these processes in SAP Fiori.

11.3.1 Set Default Values for Real SAP Fiori Apps

We want you to be able to use the SAP Fiori apps comfortably. That's why—and you have to promise us this—you first must make some prerequisite settings that save you superfluous typing, in this case, for materials management.

Follow these steps:

1. In SAP Fiori, open the user menu by clicking 8 (**User Menu**), and then select the ⚙ (**Settings**) menu item. The **Settings** window opens.

2. Click **Default Values** in the lower-left corner.

3. You'll now see the fields on the right for which you can set default values, as shown in Figure 11.3. If you use the SAP Live Access training system, you can use the values that we've entered behind the field names. By clicking 🗗 (**Value Help**) in the respective field, you can search for values from a value list.

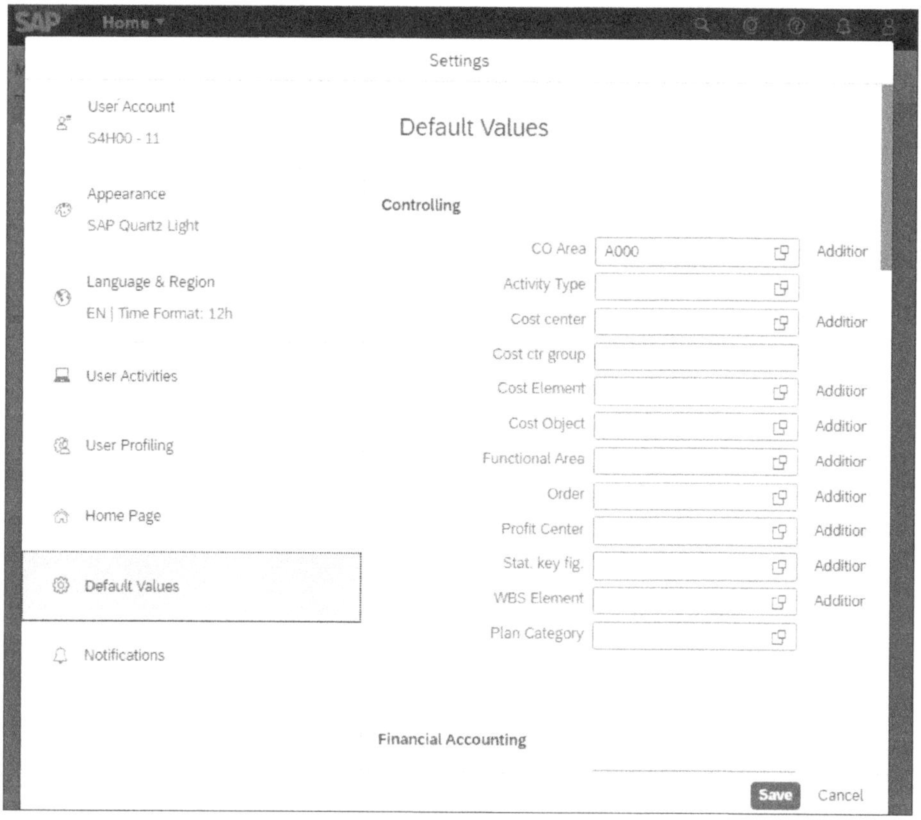

Figure 11.3 Entering Default Values

Continue scrolling down until you see the following fields and make these settings:

– **Company Code**: "1010"
– **Calculate Tax**: "Yes"

- **Plant**: "1010"
- **Stor. Loc.**: "101C"
- **Purchasing Org**: "1010"
- **Purch. Group**: "Z##" (replace "##" with your subscriber number)

4. Save your presets by clicking the **Save** button.

Legacy apps ignore the default values defined in SAP Fiori, so you define your default values with SAP GUI Transaction SU3 (Maintain User Own Data). In addition, there are unfortunately some "real" SAP Fiori apps that ignore the default values set.

11.3.2   Training Data

Now we need to let you in on some important information about the training data: For the processes in the system, you need master data and organizational units. For this part of the book, we used an SAP Live Access system that SAP was kind enough to make available to us. It's also used by SAP's training partners and provides great training data.

Here, master data is numbered consecutively for each participant. We've received the number 11. So, our numbers are correspondingly T-AV11 for the supplier or T-R511 for a material (you can see them in our screenshots). You then simply use your own number in the training system instead of 11.

If you don't have SAP Live Access but are in your company's training system, then look there for suitable master data such as business partners or materials. You'll usually find these with the value help or in lists—we'll support you.

We've also integrated some apps that aren't available on the SAP Fiori launchpad of the SAP Live Access training system. However, all apps we use according to SAP Training's S4H00 course, "SAP S/4HANA Overview" based on course version 17, can be accessed via the App Finder. Use the App Finder to launch these apps, as was described in detail in Chapter 6, Section 6.4.

11.3.3   When Things Get Stuck

If you run into a problem, check the footer line to see if there is a message or a message notice there. If so, click on it to get more information.

And one more tip: Because you'll not only gain a basic understanding of processes here but also get more practice in handling apps, we recommend that you click through the examples.

11.4   Supplier as a Business Partner

Now back to our story mentioned at the beginning. It begins with a colleague contacting you because there is far too little stock of the material "tire and tube." It has to be reordered urgently, so you want to call the supplier to see if the material can be delivered immediately. To do this, you need the phone number and the minimum order quantity from the supplier business partner data.

We'll walk through the business partner master data and see how to access it in the following sections.

11.4.1   Structure of a Business Partner Master Record

All departments involved, such as purchasing, warehouse, invoice verification, accounts payable, or controlling, use the same supplier master data because they all use the same supplier data from the business partner master record.

The same master record under the same number can also contain customer data that is used in *sales* and *distribution* and by *accounts receivable accounting* (other term: *customer accounting*) because a supplier can also be a customer. In other words, the same *business partner* can act as a supplier in a procurement process and as a customer in a sales process.

The many bits of data for a business partner are divided into *roles*. For the complete procurement process, you need a business partner with at least these two roles:

- **Supplier**
  This role includes information for purchasing such as incoterms, minimum order value, or payment terms. This supplier data is managed per purchasing organization. If there are several purchasing organizations in one client, each purchasing organization can create its own purchasing data. Before a purchase order can be created for a supplier, the data for the **Supplier** role must be entered.

- **FI Vendor**
  This role for accounts payable contains, for example, information on payment transactions such as a possible bank collection. This data is managed separately for each company code. If there are several company codes, the accounting department of each company code maintains its own data for a vendor. Before an incoming invoice from a vendor can be entered, its data must be entered with the **FI Vendor** role.

One Role for All

There is data that every employee needs and sees—whether in procurement or sales. This includes the company name, addresses, or bank details. This data belongs to the **Business Partner (Gen.)** role and is always created. The information is valid client-wide, that is, for all company codes as well as for all purchasing and sales organizations.

11

11.4.2   Three Ways to Business Partner Master Data

There are three methods to display information about business partners such as suppliers or customers:

- **Real SAP Fiori apps**
  With one of the three real SAP Fiori apps, Manage Business Partner Master Data, Manage Supplier Master Data, or Customer Master, you get a list of the corresponding business partners and switch from there to a detailed display of the corresponding master data. All these apps work almost the same way.

- **Enterprise search**
  Using enterprise search, you can very quickly display a few basic pieces of information, such as the postal address and the vendor or customer name (see Chapter 7, Section 7.8).

- **Transaction BP**
  Transaction BP (Business Partner) or the corresponding legacy app Maintain Business Partner can be used to maintain both supplier and customer master data.

In this chapter, we'll first show you only the first-mentioned way via the real Manage Business Partner Master Data app.

11.4.3   Display and Change Supplier Master Data

No sooner said than done! You now check the data of a master record and find out about the phone number of a supplier and about a possible minimum order value. To do this, you use the real Manage Business Partner Master Data app.

Follow these steps:

1. Start the Manage Business Partner Master Data app shown in Figure 11.4.

Figure 11.4 Manage Business Partner Master Data App

> **Do You Know a Supplier Number?** [+]
>
> If you know the supplier number, enter it in the **Business Partner** field right away.
>
> In the SAP Live Access training system, there is the prepared supplier with the number T-AV##. So, you could click **Go** right now on the far right. Then you can skip the next steps and start again at step 6. It doesn't hurt if you nevertheless follow the steps completely.

2. Now the "role play" begins! In the **Role** field, click ⊡. A long list of possible roles appears, as shown in Figure 11.5.

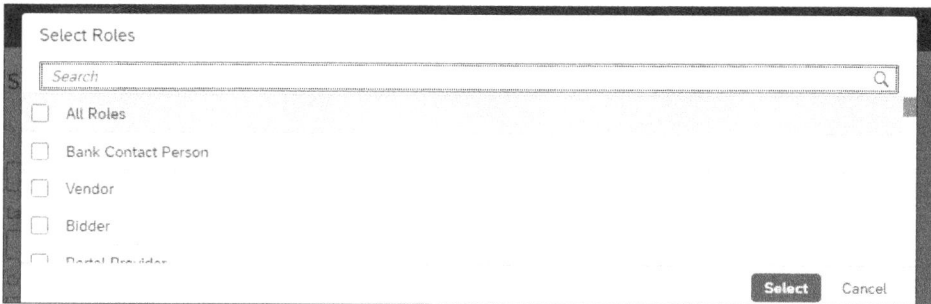

Figure 11.5 Select Role Popup

> **Why Do I Need to Select a Role?** [«]
>
> You select the role **Supplier** so that only business partners with supplier data are displayed. As you learned at the beginning of this section, customers are also business partners, but you won't find any supplier data in this **Customer** role.

3. In the upper **Search** field, enter the letter string "Sup" (for "supplier," not "stand-up paddle boarding"), and confirm the entry with the ⌴Enter⌴ key.
4. Select the **Supplier** entry, as shown in Figure 11.6. Then click the **Select** button.

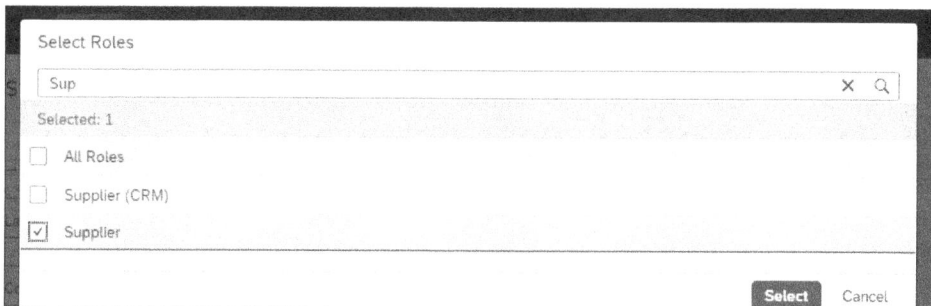

Figure 11.6 Searching for Supplier

5. The selected entry is now visible in the **Role** field, as shown in Figure 11.7. Click the **Go** button to get a list of the corresponding business partners.

Figure 11.7 Entry in the Role Field

6. In Figure 11.8, you see the **Role** column. A supplier usually has both the **FI Vendor** role with the accounting data and the role with the purchasing data. Our supplier needs both roles if you want to complete the procure-to-pay process with him.

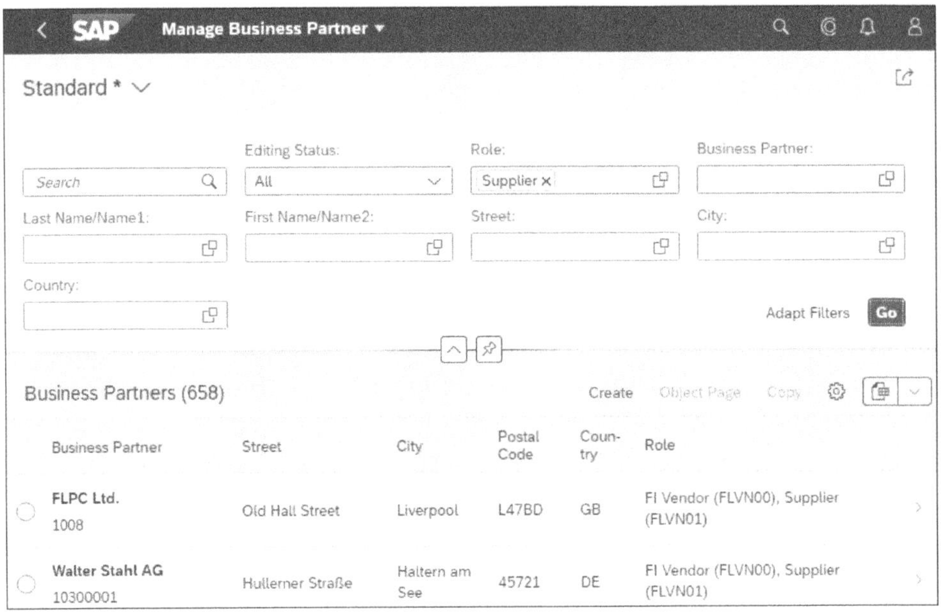

Figure 11.8 Displaying the Business Partner List with Roles

7. Do you want to see more supplier data? Click ⟩ on the far right in the row with the supplier. You're now in the master record of the business partner.

Click on the **Edit** button to switch to change mode in the master record to correct or add data. However, this only works in this main screen and not in the other screens.

8. Click on the **Roles** tab. You see the roles of the business partner, as shown in Figure 11.9.

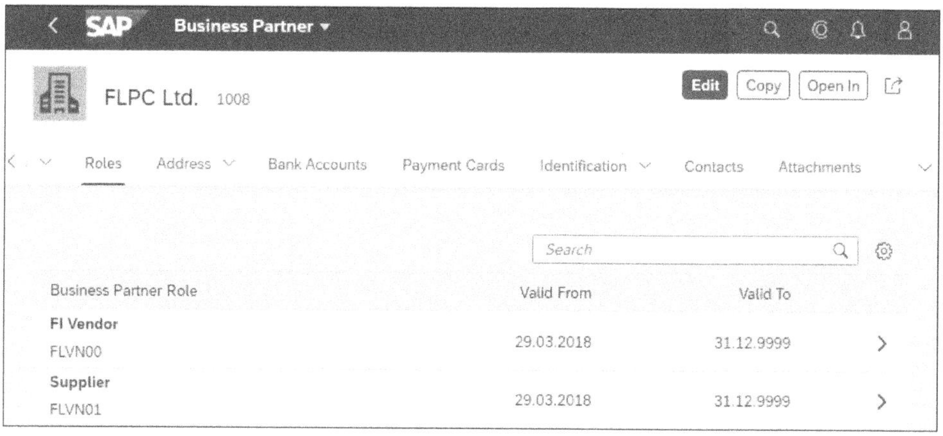

Figure 11.9 Displaying Business Partner Roles in the List

9. Click [>] on the far right in the row with the entry **Supplier FLV01**.

10. In the master record, there are several tabs where the data of the business partner is distributed. Click on the **Address** entry in the tab bar, as shown in Figure 11.10. You can also see the data by scrolling down.

   First you'll see the information about the postal address. If you scroll down a bit further, you'll find the fields for the phone numbers under **Standard Communication**, which aren't filled in our case.

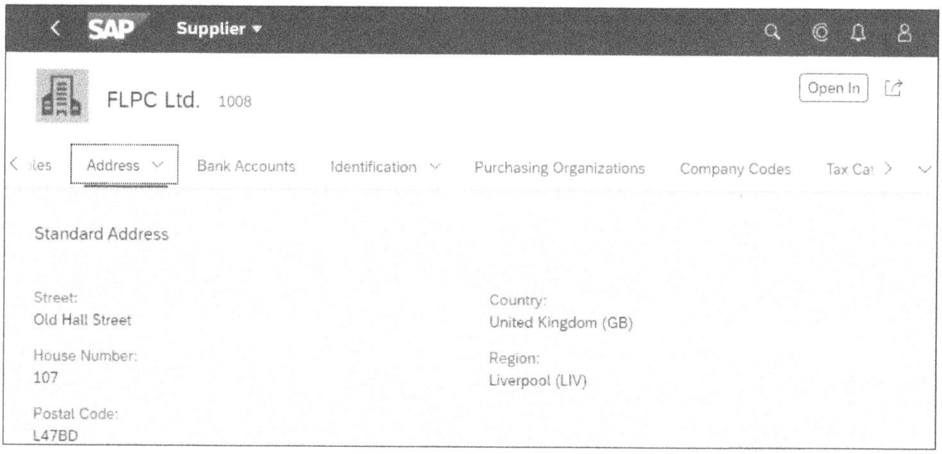

Figure 11.10 Address Tab

11. Click on the **Purchasing Organizations** entry in the tab bar, as shown in Figure 11.11. Now you can see the purchasing departments that work with this supplier and have created data for this. Behind the purchasing organization also hides special supplier data such as the minimum order value.

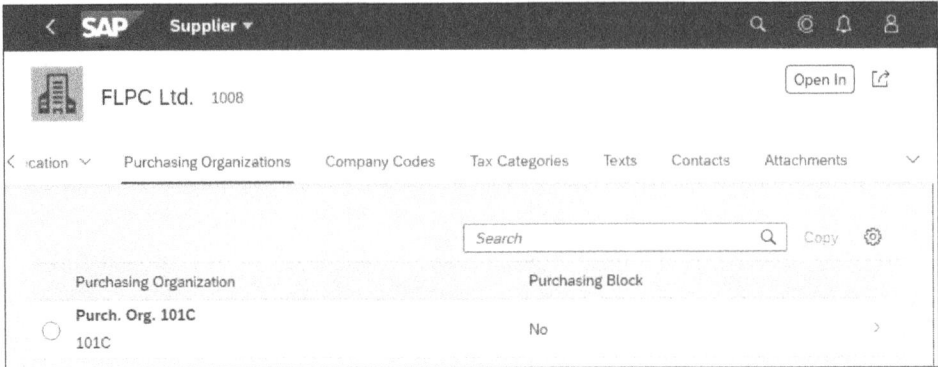

Figure 11.11 Purchasing Organizations Tab

[»]

Own Data for the Same Supplier

The data of the purchasing organizations may differ from each other. It may well be that, for example, a purchasing organization in another country has entered different incoterms or a different minimum order value for the same supplier than the purchasing organization to which you belong.

12. Click ⟩ to the right of the purchasing organization. Now you'll see the purchasing-specific data distributed on the **General Data** and **Purchasing Organizations** tabs, as shown in Figure 11.12.

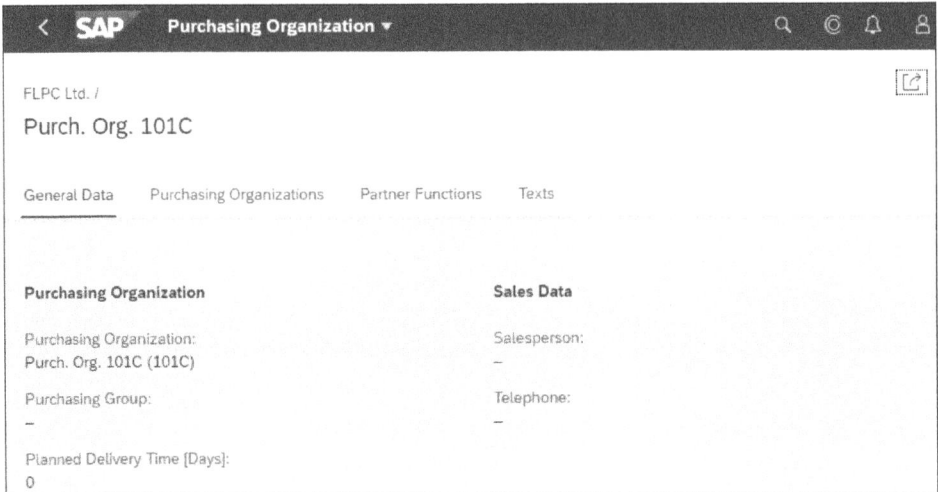

Figure 11.12 General Data Tab

13. Click on the **Purchasing Organizations** tab, or simply scroll further down. Now you'll see the special purchase data, as shown in Figure 11.13. You'll be pleased to see that no specific amount is specified for the minimum order value.

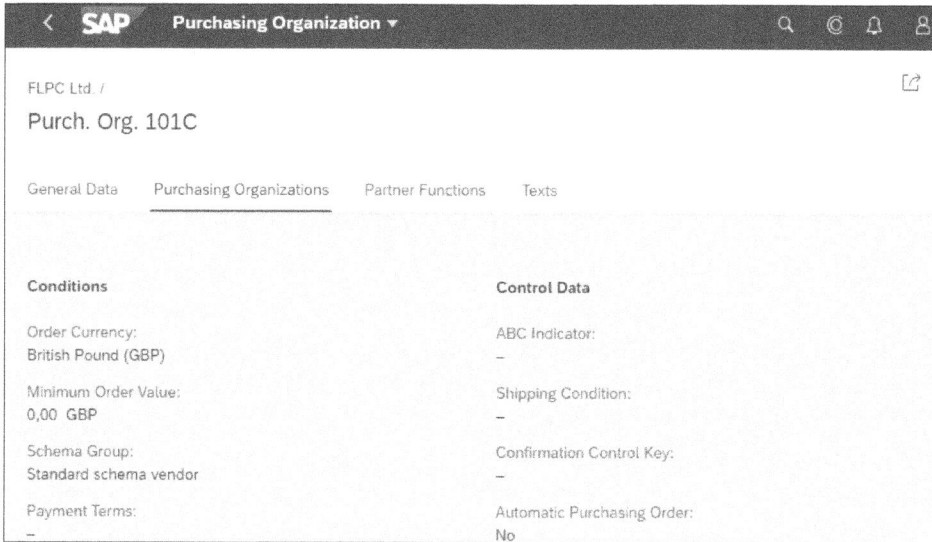

Figure 11.13 Purchasing Data

14. Click ❮ (**Back**) in the header line to go back to the main page of the supplier master data, as shown in Figure 11.14.

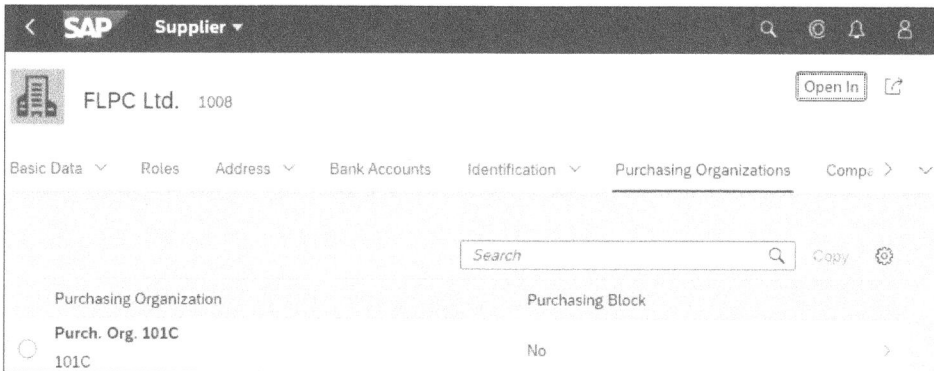

Figure 11.14 Return to Purchasing Organizations Tab

Get More Information

You may have noticed the ⚙ (**Settings**) icon on the right side of many tabs. As a future professional, you'll use this icon to show many more frequently needed fields (e.g., **Incoterms, Minimum Order Value**, or **Planned Delivery Time**) for the purchasing organizations table, if needed.

15. Click [SAP] from the shell bar to return to the homepage.

[»]

Who Maintains the Supplier Data?

Supplier data is shared between accounting and purchasing. Often the purchasing data is created first with the **Supplier** role, and later the accounting data is added with the **FI Vendor** role. Sometimes, there are departments for complete master data maintenance, which are responsible for all data. The latter is called *central data maintenance*.

[»]

Switching from SAP ERP to SAP S/4HANA

Everything is new for the supplier master! In SAP GUI, there are still older transaction codes for vendor master data, which start with FK0, XK0, or MK0. However, when calling these transactions, you'll usually be sent directly to Transaction BP (Business Partner) after confirming the initial screen.

Other basic innovations are that multiple addresses can be entered and also maintained on a time basis with the specification of a validity period. We're particularly pleased about the modern and convenient real SAP Fiori apps, such as Manage Business Partner Master Data, with which—in contrast to SAP GUI—flexible lists of master data can be created. In addition, we like to use the extensive setting options for tables in these apps via [⚙] (**Settings**), which aren't available in SAP GUI.

When creating a new business partner, it must be assigned one of three *business partner categories*:

- Person: Private person
- Group: Several people (shared)
- Organization: Company (legal entity)

Depending on the selected business partner, special data can be entered. For example, a legal form or a foundation date can only be entered only with the **Organization** business partner type. Your advantage: you only have the fields on the screen that you need.

11.5   Display Purchasing Data in the Material Master

In the company, the *material master* is the central source for retrieving material-specific information such as size, weight, or unit of measure. The material master defines how a product is purchased, stored, manufactured, distributed, and costed.

The basic philosophy is the same as for business partners: all material data from all departments is stored and used together in a single master record under one material number each. We'll walk through how to access, navigate, and use the material master in the following sections.

11.5.1   Three Ways to Material Data

As with the business partner master, there is a wide range of methods for getting to the information in the material master:

- **Enterprise search**
  Using enterprise search, you can very quickly view a few basic pieces of information, such as the weight or manufacturer. Using links, you can switch from there to the Display Material legacy app.
- **Display Material app**
  With this legacy app, you directly display all information of the material master.
- **Manage Product Master Data app**
  With this real SAP Fiori app, you get a list of materials from which you can switch to a detailed, but not complete, display of the material master data.

11.5.2   Display Material Data with Enterprise Search

Enterprise search lets you view basic material data in a snap:

1. Open the enterprise search by clicking Q (**Search**) in the shell bar.
2. In the shell bar, click ✓ to the right of **All**. This will open the object list, as shown in Figure 11.15.
3. Click the **Materials** entry in the object list.

Figure 11.15 Opening the Object List

4. Enter the number of a material, as shown in Figure 11.16, and press the Enter key.

Figure 11.16 Entering the Material Number

5. Now some basic material data such as the material description or the weight are displayed, as shown in Figure 11.17.

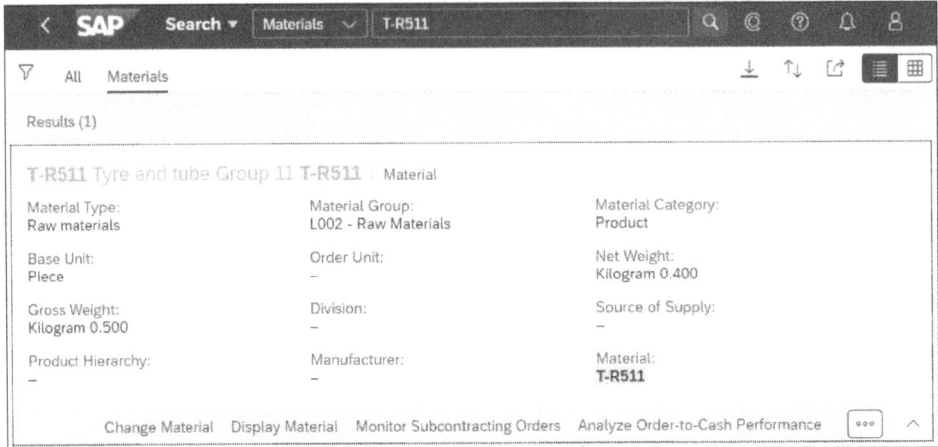

Figure 11.17 Displaying Material Data

6. Is the desired information missing here? With a click on the **Display Material** button in the bottom line of the screen, you get to the initial screen of the **Display Material** legacy app. There you can see (almost) everything, but more about that later.

11.5.3 Structure of the Material Master

As already mentioned, the material master record is used by many departments. The components of the material master record are called *views*. Just as with the supplier master data with its roles, there are also views for different departments with the material data, for example:

- Purchasing
- Warehouse
- Production
- Sales
- Accounting
- Costing

SAP S/4HANA provides each department with views in which the department-specific data can be maintained and retrieved. This data can be different per plant or per storage location.

11.5.4 Display Material Data with a Legacy App

In our example, a raw material is procured as a stock material. You now check the corresponding material master record with the Display Material legacy app. In SAP GUI, you use Transaction MM03 (Display Material), which works according to the same pattern.

Follow these steps:

1. Launch the Display Material app.

2. In the initial screen, enter the **Material** number, as shown in Figure 11.18. Then click on the **Select View(s)** button at the top left, or press the Enter key.

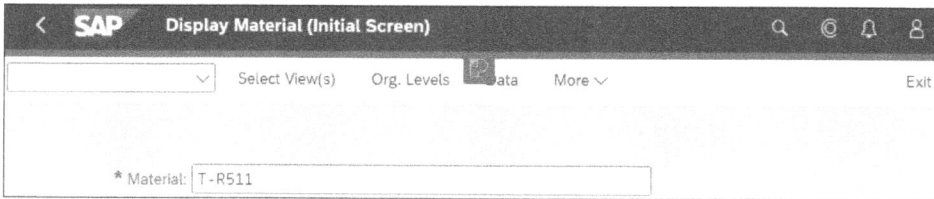

Figure 11.18 Entering the Material Number

3. SAP S/4HANA now shows you the views that have been created for this material. In our example, you belong to the purchasing team. Therefore, in the **Select View(s)** window (see Figure 11.19), select the **Basic Data 1** and **Purchasing** views. Then confirm your selection with the Enter key.

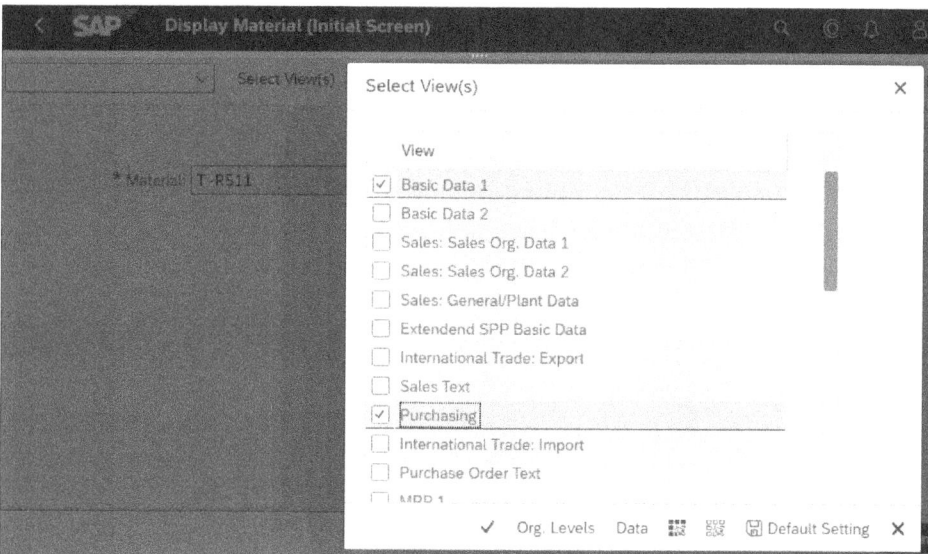

Figure 11.19 Selecting Views

Two Views for All

The first two views, **Basic Data 1** and **Basic Data 2**, are the same for all departments. They contain cross-departmental information such as the material number and the description as well as other details, such as the base unit of measure (piece, liter, etc.) or the weight. No matter whether in a plant in the United States, Canada, India, South

Africa, or Great Britain, this information is the same all over the world and thus applies to all organizational units in the entire client!

4. Purchasing data can vary from plant to plant and be stored separately. Therefore, SAP S/4HANA now requires you to enter your plant number. Enter it in the **Plant** field, as shown in Figure 11.20, and confirm with Enter .

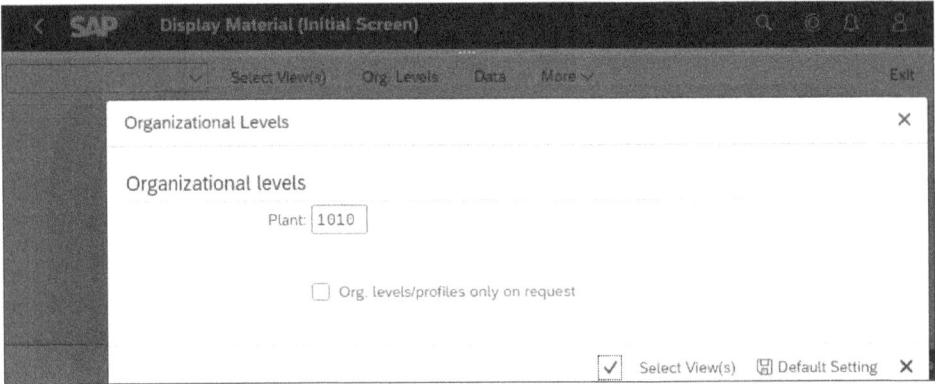

Figure 11.20 Entering the Plant Number

5. All views that exist for the selected material will then appear as tabs. The tabs for the previously selected views are marked with a circular icon .

The **Basic Data 1** tab is displayed first, as shown in Figure 11.21. This data, such as the base unit of measure or, further down, the weight, applies to the entire client, that is, to all company codes, purchasing organizations, and plants.

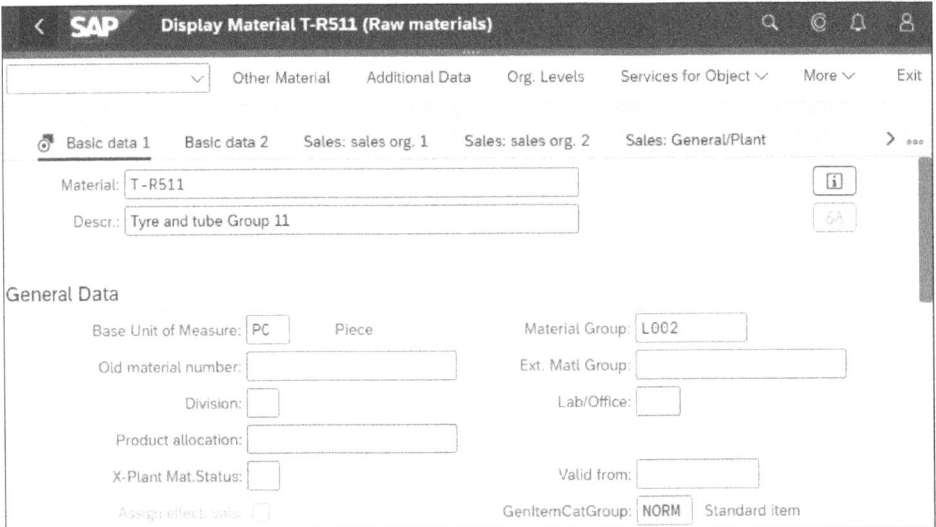

Figure 11.21 Basic Data 1 Tab

Are You Interested in Individual Fields?

[+]

View the field information! In legacy apps like our example app, you simply click in the field you're interested in and press the [F1] key to see background information about what that field contains.

If a field contains the 🗗 icon, such as the **Material Group** field, click on it. From the list of values, you can often see for what purpose the particular field was invented.

6. Unlike the real SAP Fiori apps, you can't scroll down to the other data in legacy apps; you have to use the tabs. So, press the [Enter] key to go to the next tab, or click on the **Purchasing** tab further to the right (see Figure 11.22). Don't see that? Then click ⟩ in the tab bar on the far right.

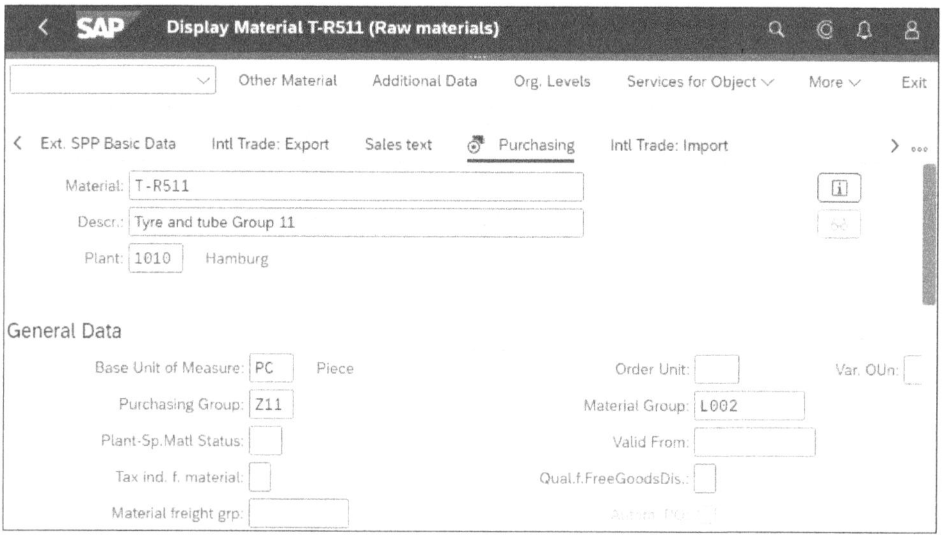

Figure 11.22 Purchasing Tab

Fields in the Purchasing Tab

[«]

Some fields of the **Purchasing** tab are exciting for our purchase-to-pay business process, for example:

- The **Order Unit** field defines a unit of measure that differs from the **Base Unit of Measure** field, specifically for purchase orders, for example, pallet instead of piece. SAP S/4HANA then requires a conversion ratio, such as 1 pallet = 120 pieces. If the field remains empty, the base unit of measure also applies as the order unit.
- The **Purchasing Group** field shows which department or which person is responsible for material procurement. In practice, this field is often used for the name of the buyers.

The corresponding values are transferred to the order when it's created.

7. Click ![SAP] from the shell bar. You're now back on the start page.

11.5.5   Call Up Material Data with the Manage Product Master Data App

Do you need an overview list for your materials? This function doesn't exist in the Display Material legacy app. Instead, the real Manage Product Master Data app helps you here.

Follow these steps:

1. Launch the Manage Product Master Data app.
2. Enter a part of the material name in the **Search** field at the top left, for example, "Tyre" (see Figure 11.23), and then click the **Go** button.

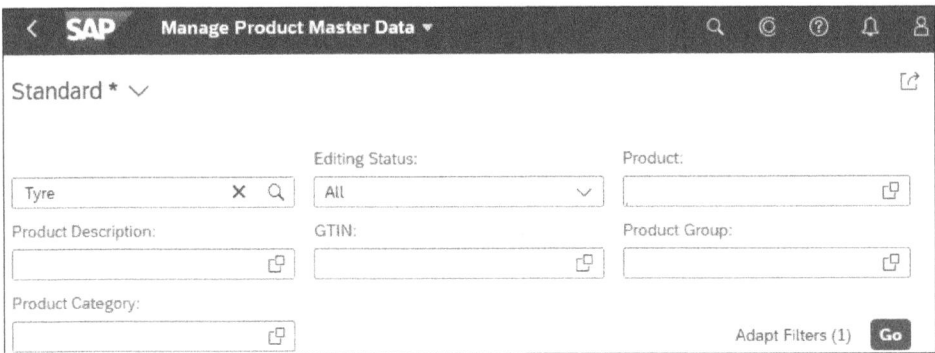

Figure 11.23 Entering a Part of the Material Name

3. Now the list of the corresponding materials appears, as shown in Figure 11.24. To display further material information, click $>$ at the far right of the list, or click on the material number.

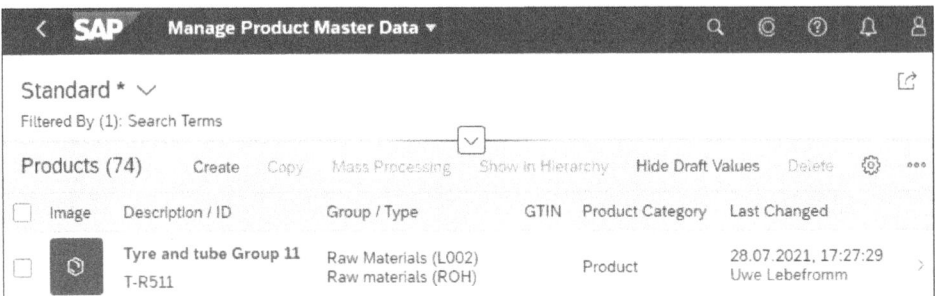

Figure 11.24 Displaying the Material

4. Now you see the detailed information, as shown in Figure 11.25. You may click on the different tabs or simply scroll down again here because you're in a real SAP Fiori app. The grouping of material data differs from the views of the Display Material legacy app.

5. Click **SAP** to return to the SAP Fiori launchpad homepage.

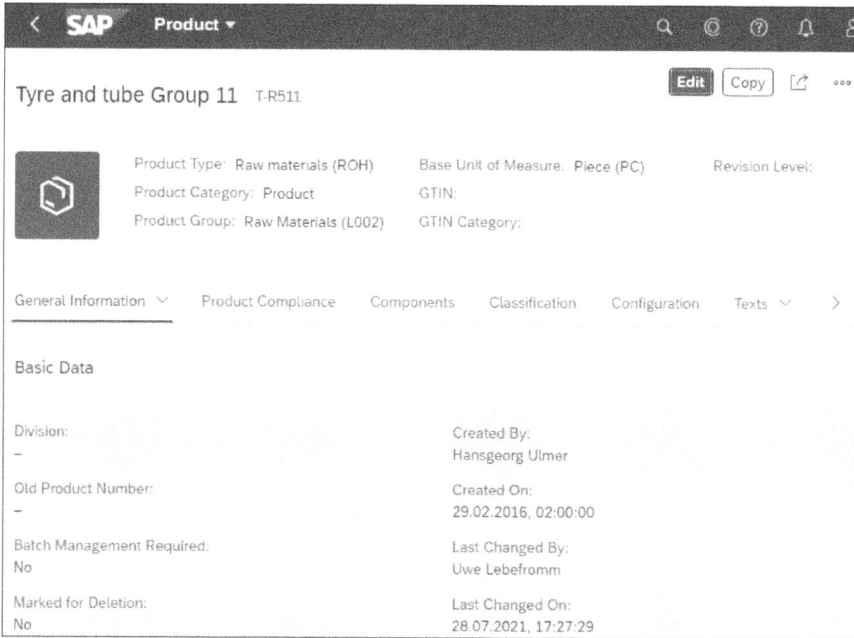

Figure 11.25 Display Material Details

11.6   Display the Material Stock

Trust is good, but control is better! Before you order, check the material stock. To do this, you can use Transaction MMBE (Stock Overview) in SAP GUI, which is available in SAP Fiori as the Stock – Single Material legacy app. However, we'll show you the process here with the real app, Stock – Single Material.

Follow these steps:

1. Start the Stock – Single Material app using the App Finder (see Chapter 6, Section 6.4).

2. Enter the **Material** number, as shown in Figure 11.26, and confirm it with the [Enter] key.

Figure 11.26 Entering the Material Number

3. Now you see the *stock overview*, as shown in Figure 11.27. It shows all stocks in the client, that is, for all company codes, plants, and storage locations. The display is separated according to the individual stock types.

4. Make a note of the current stock of the material in the **Raw mat. stoloc.** storage location so that you can check the stock receipt after the goods receipt. As shown in Figure 11.27, we have no stock in plant 1010.

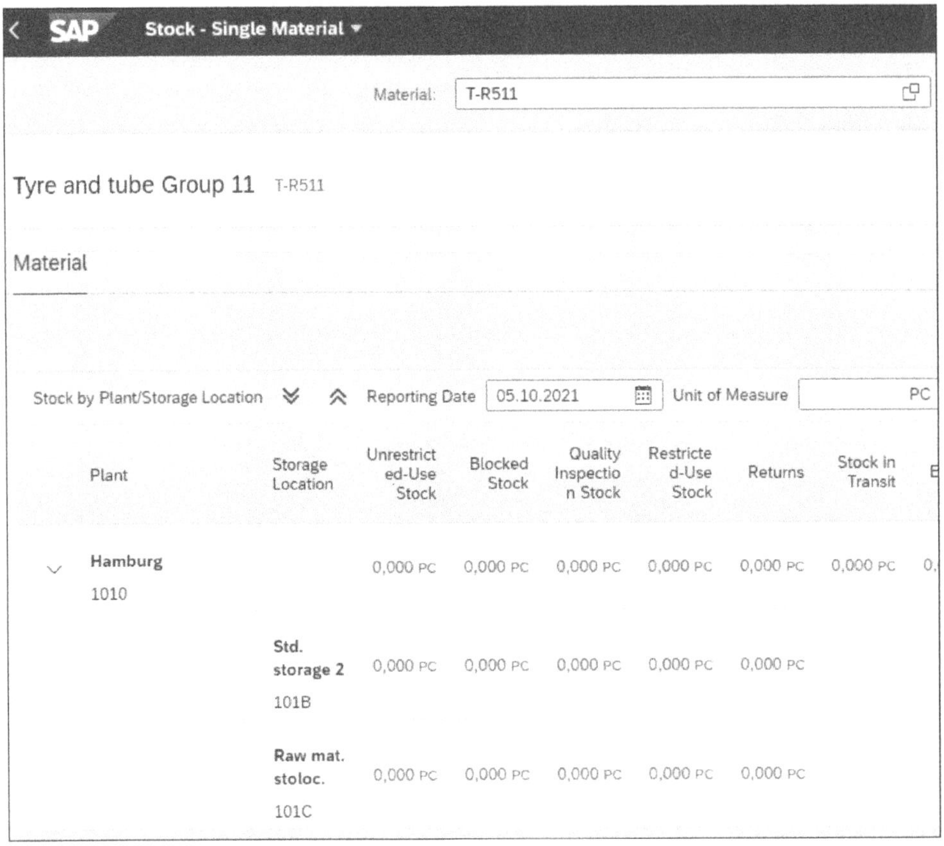

Plant	Storage Location	Unrestricted-Use Stock	Blocked Stock	Quality Inspection Stock	Restricted-Use Stock	Returns	Stock in Transit
Hamburg 1010		0,000 PC	0,000 PC	0,000 PC	0,000 PC	0,000 PC	0,000 PC
	Std. storage 2 101B	0,000 PC	0,000 PC	0,000 PC	0,000 PC	0,000 PC	
	Raw mat. stoloc. 101C	0,000 PC	0,000 PC	0,000 PC	0,000 PC	0,000 PC	

SAP Stock - Single Material ▾

Material: T-R511

Tyre and tube Group 11   T-R511

Material

Stock by Plant/Storage Location   Reporting Date 05.10.2021   Unit of Measure   PC

Figure 11.27 Stock Overview

What Are the Stock Types?

In SAP S/4HANA, there are, among others, the following *stock types*:

- **Unrestricted-use stock**
  Materials that are in the warehouse and can be withdrawn immediately.
- **Blocked stock**
  Materials that are in the warehouse but are blocked due to a quality defect, for example, and therefore can't be withdrawn.
- **Quality inspection stock**
  Materials that are in quality inspection and therefore can't be withdrawn.

5. An exciting feature is the stock check (see Figure 11.28): Click ![chart icon] (**Chart View**) right above the table. You see stock types for each plant and storage location, although we don't yet have any stocks in our example.

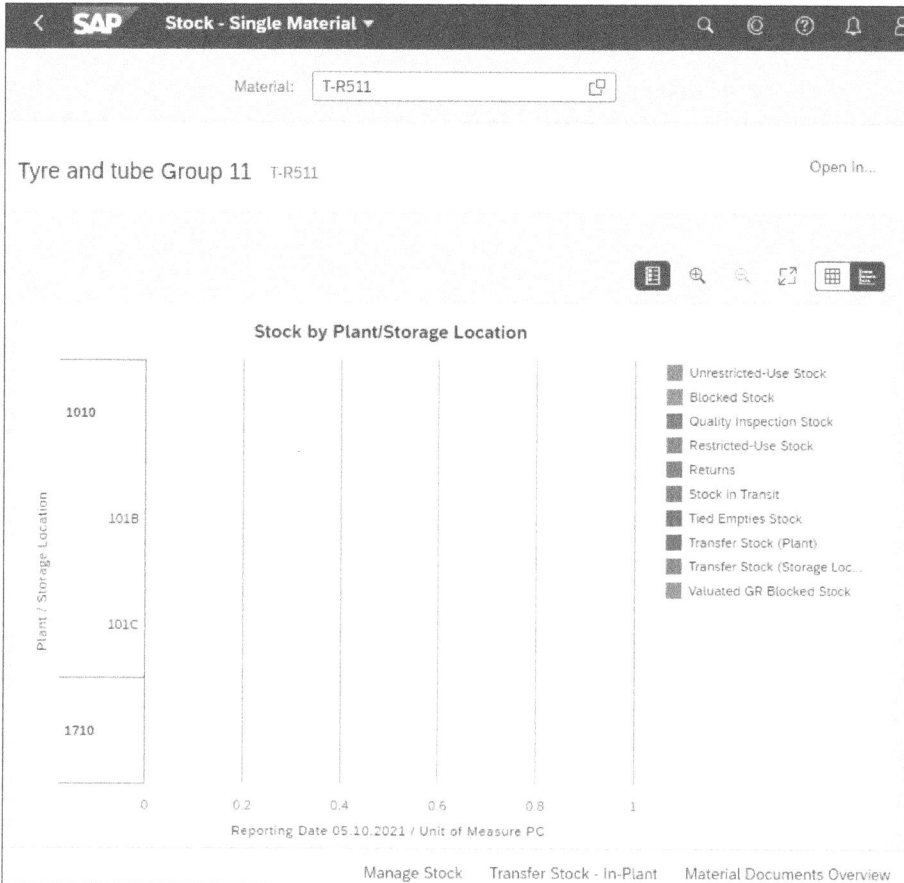

Figure 11.28 Chart of Stock Developments

Here ends the prelude to our purchase-to-pay process. In the next section, we finally get started with your first order.

Switching from SAP ERP to SAP S/4HANA

There are only a few basic innovations in the material master. The length of the material number can be extended from 18 to 40 characters in Customizing. In addition, there is a new material type called SERV for service master data. When using this material type, registers and fields that are not required for services are hidden.

11.7   Create an Order

In our purchase order, we manually enter data such as customer and material number. This wouldn't be necessary if you had another document, such as a purchase requisition, from which you could copy this data. Copying is sometimes referred to as *creating with reference* in SAP S/4HANA. In practice, purchase orders can be created with reference to the following other documents:

- Purchase requisitions
- Request for quotation
- Purchase orders
- Contracts

The advantage is that if a purchase order is created with reference to one of the other documents mentioned, its data (e.g., material data, supplier data, and prices) is transferred to the new purchase order. In other words, you save yourself a lot of typing. Unfortunately, in our example, you don't have another document. But everyone starts small.

[+]   **What Data Do I Use in the Next Flow?**

Don't have an SAP Live Access training system? Then it's smart to use data from orders that have already been successfully completed. But how do you find them?

Start the Monitor Purchase Order Items app, set the **Fully Delivered & Invoiced** filter there to **Yes**, and select an entry for the **Display Currency** field.

Note the following fields: **Plant, Purchasing Organization, Purchasing Group, Company Code, Supplier,** and **Material**.

You now enter a purchase order with the real Manage Purchase Orders app. In SAP GUI, you use Transaction ME21N (Create Purchase Order), which is available in SAP Fiori as the Create Purchase Order – Advanced legacy app.

Follow these steps:

1. Start the Manage Purchase Orders app. You'll now see the filters and an empty table, as shown in Figure 11.29.

2. Click the **Create** button directly below the header.

   And now the typing begins? Not necessarily. You only enter the supplier number, and when you then move to the next field, many other fields are filled automatically, for example, the **Currency**, **Purchasing Organization**, and **Company Code** fields, as shown in Figure 11.30.

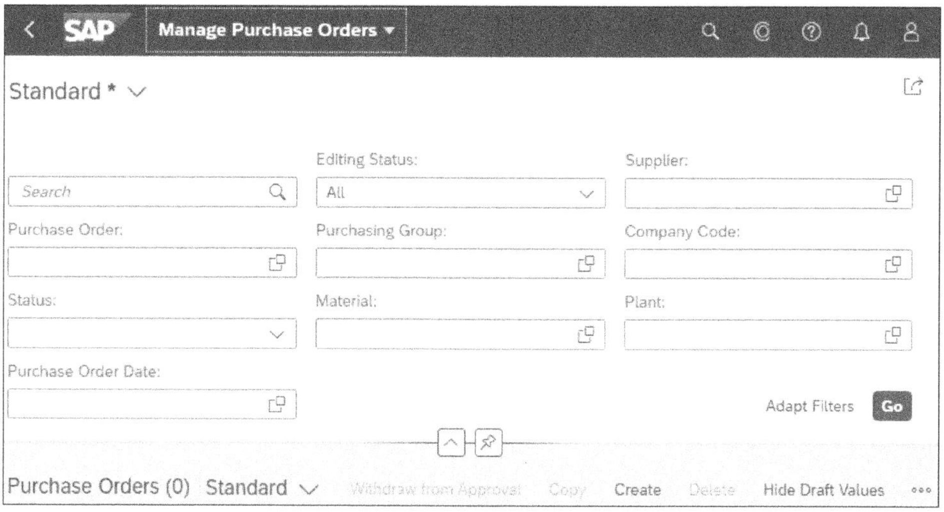

Figure 11.29 Manage Purchase Orders App

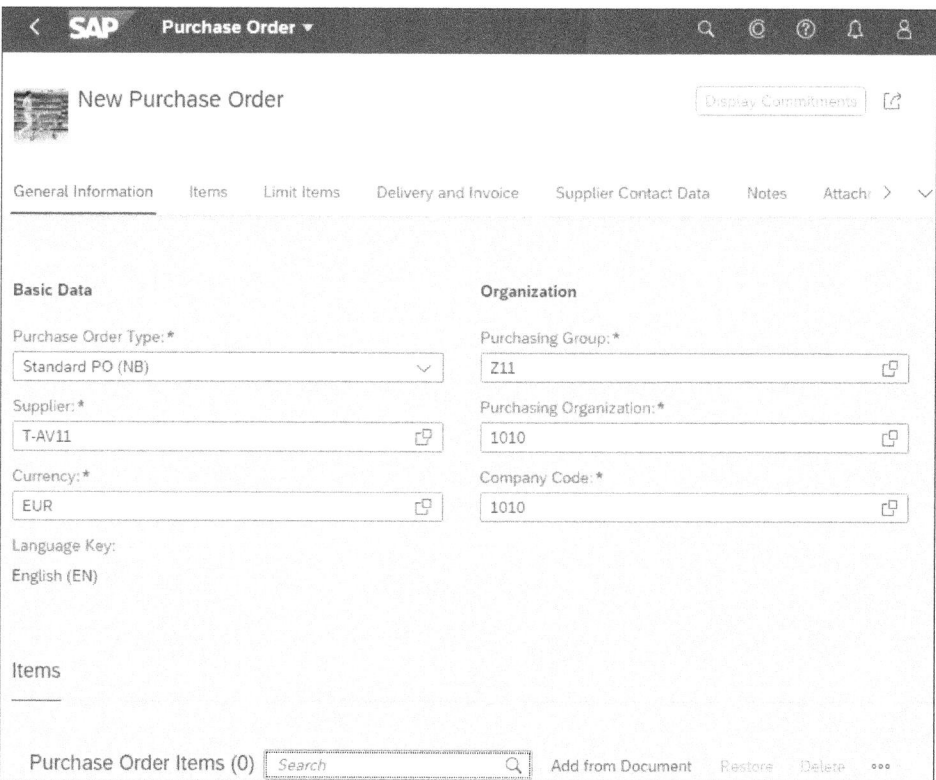

Figure 11.30 New Order Entry

3. Fill in all the fields in the **Basic Data** and **Organization** sections as shown in Figure 11.30; usually you have to enter the **Purchasing Group** manually. If you see the **Purchase Order Type** field at the top left, select the **Standard PO (NB)** entry for this. (NB is the abbreviation for the German word *Normalbestellung*, which means "Standard Purchase Order".)

4. Still missing are material number, price, and quantity. This information is entered as a *purchase order item*. To do this, click the **Create** button in the **Items** area. If you don't see the **Create** button, click ⊙⊙⊙, and then click on the **Create** menu item, as shown in Figure 11.31.

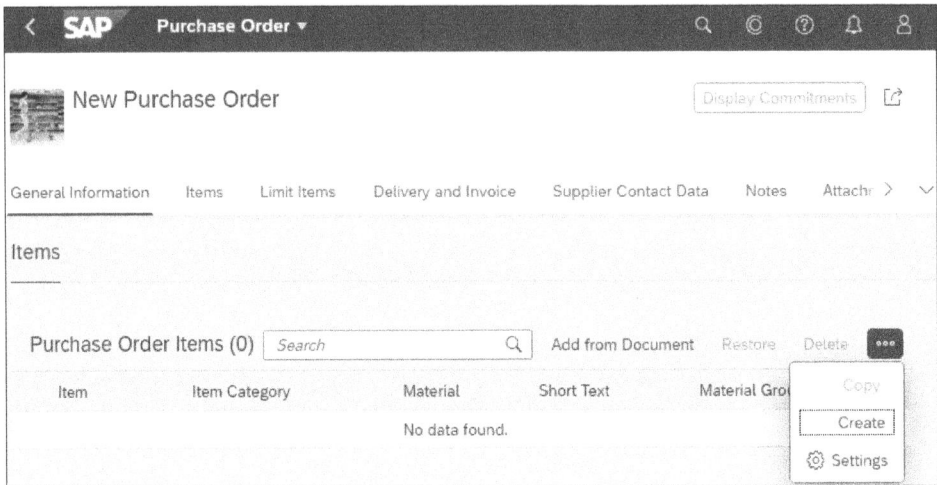

Figure 11.31 Create Order Item

5. This opens the most important item fields. On the screen shown in Figure 11.32, you make entries in the **Material** and **Order Quantity** fields. The material master data must of course be maintained for the corresponding plant.

> **Other Items**
>
> A purchase order can, of course, contain further items with further materials. In this case, you create additional items with the **Create** button. However, we'll content ourselves here with one item.

6. Click the **Order** button at the bottom right. This saves the purchase order. You'll then see a display of the order and the document number of the order at the top left, as shown in Figure 11.33. Make a note of this document number and the net value visible on the right side. You'll need the document number in the next step at goods receipt and the value at invoice entry.

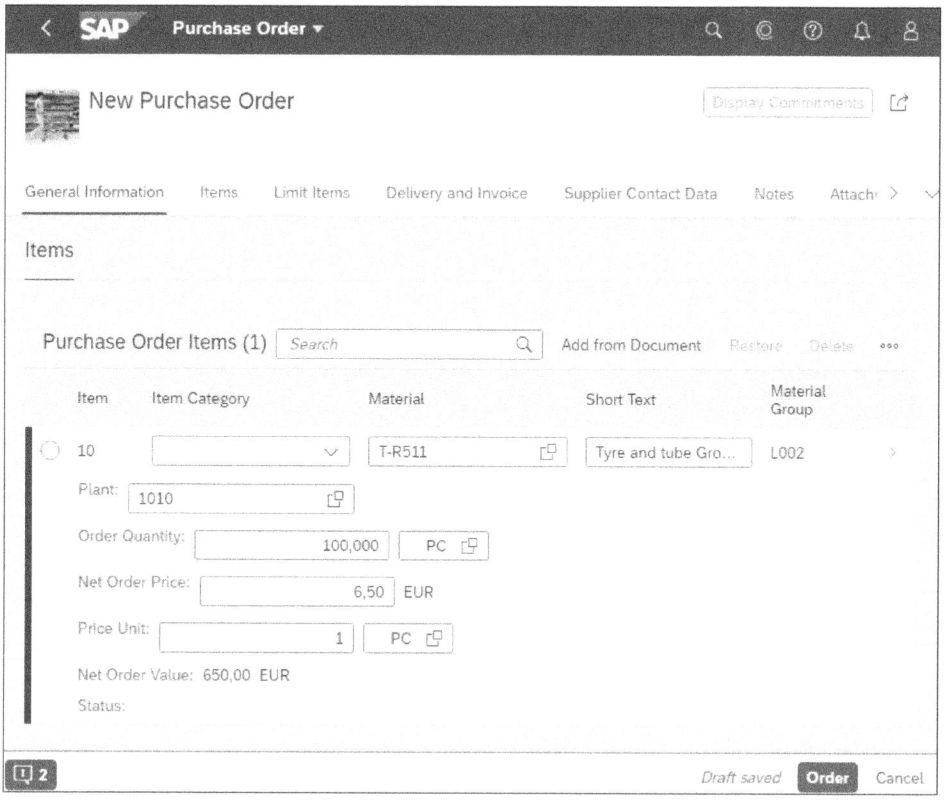

Figure 11.32 Order Item Details

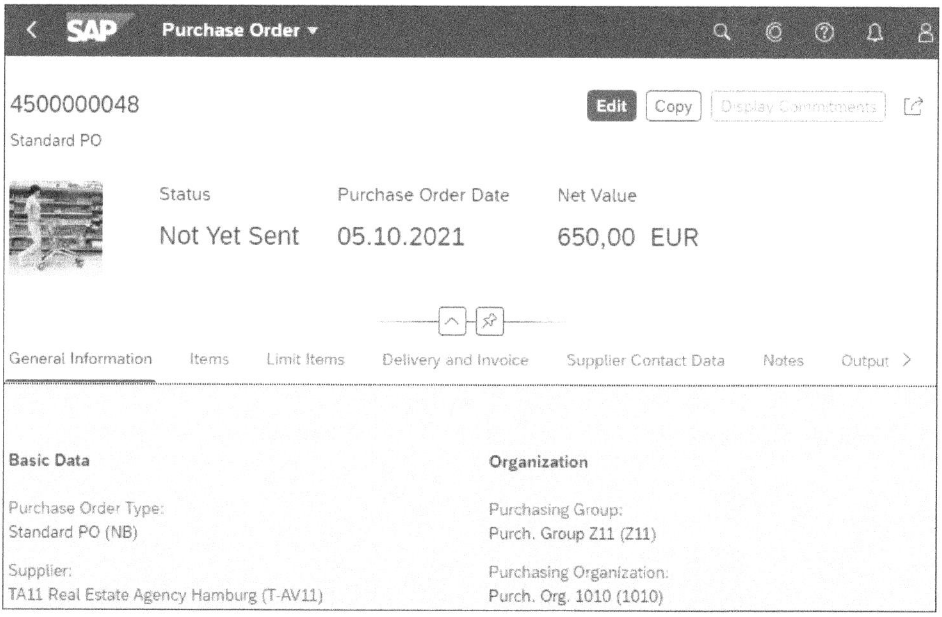

Figure 11.33 Display of the Order and Document Number

7. Click **SAP** to get back to the SAP Fiori launchpad start page.

You're done! After saving the order, it's transmitted to the supplier. With a corresponding default setting in SAP S/4HANA, this is done automatically via electronic data interchange (EDI). Alternatively, of course, a PDF file can also be created and sent to the supplier by email.

[+] **Similar Order Available? Just Copy It!**

Do you see a similar order in the list of the Manage Purchase Orders app before you enter it? Select it on the far left and click the **Copy** button under the header. This will switch you to the entry form, and you'll see the data of the copied order, which you can adjust there. This is especially worthwhile for larger orders with multiple items.

11.8   Display the Order and Output It as a PDF

With the real Manage Purchase Orders app, you can display a purchase order first in the list, and by clicking on the corresponding purchase order number, you can see the details. In SAP GUI, you use Transaction ME23N (Display Purchase Order) for this purpose.

Follow these steps:

1. Launch the Manage Purchase Orders app.

2. Enter the **Purchasing Group** in the header and click on the **Go** button on the far right, as shown in Figure 11.34.

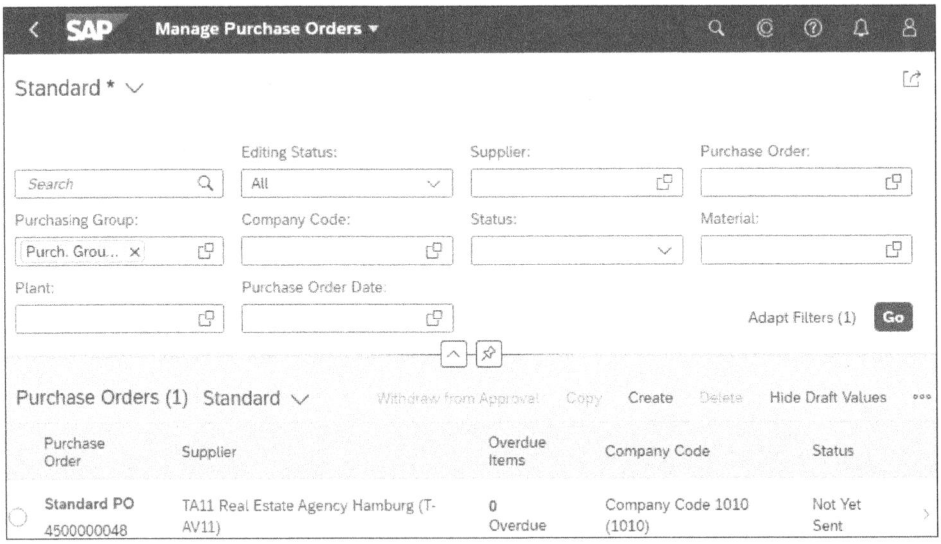

Figure 11.34 Manage Purchase Orders App

3. By clicking on the order number on the far left of the list, which in our case starts with "45", SAP S/4HANA displays the order.

4. The order is to be sent to the supplier by email as a PDF. The PDF is created automatically in the SAP Live Access training system. To view the PDF, select the **Output Management** tab. If the tab isn't displayed, open the tab overview on the far right, and click the **Output Management** menu option, as shown in Figure 11.35.

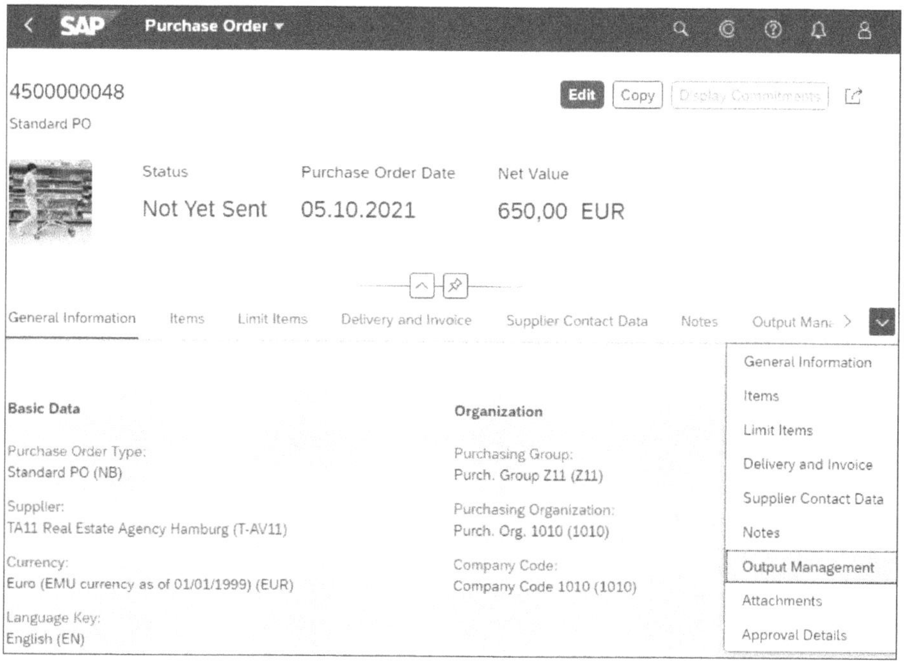

Figure 11.35 Output Management

5. In the **Display** column, click 🔲 (**PDF**), as shown in Figure 11.36.

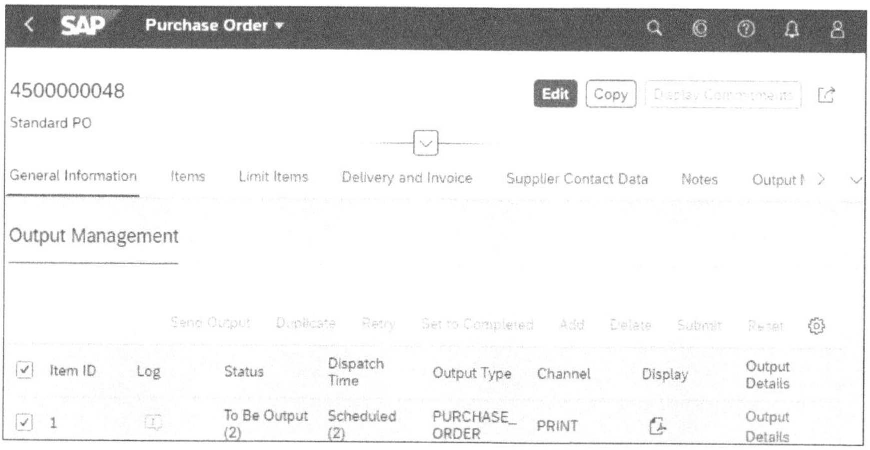

Figure 11.36 Clicking the PDF Icon

6. Now SAP S/4HANA shows you the PDF in an additional tab of your browser, as shown in Figure 11.37. From the PDF display, you can right-click to download or print it.

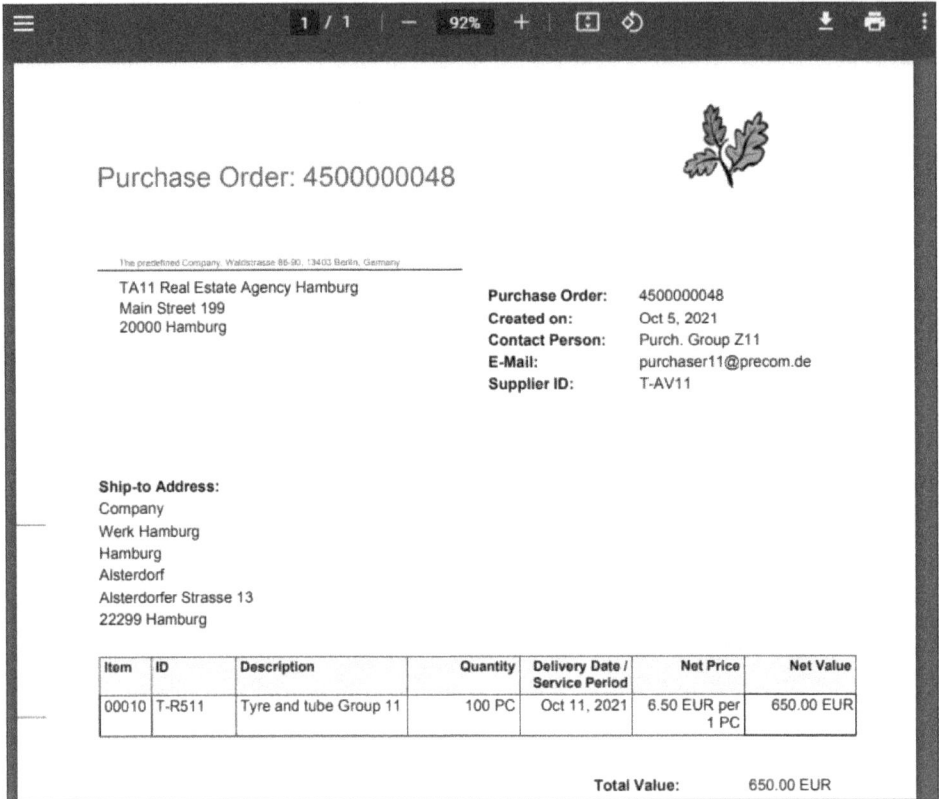

Purchase Order: 4500000048

The predefined Company, Waldstrasse 86-90, 13403 Berlin, Germany

TA11 Real Estate Agency Hamburg
Main Street 199
20000 Hamburg

Purchase Order:	4500000048
Created on:	Oct 5, 2021
Contact Person:	Purch. Group Z11
E-Mail:	purchaser11@precom.de
Supplier ID:	T-AV11

Ship-to Address:
Company
Werk Hamburg
Hamburg
Alsterdorf
Alsterdorfer Strasse 13
22299 Hamburg

Item	ID	Description	Quantity	Delivery Date / Service Period	Net Price	Net Value
00010	T-R511	Tyre and tube Group 11	100 PC	Oct 11, 2021	6.50 EUR per 1 PC	650.00 EUR

Total Value: 650.00 EUR

Figure 11.37 Displaying the PDF

7. Close the new tab, and click the **SAP** icon to return to the SAP Fiori launchpad homepage.

11.9   Enter a Goods Receipt

Job rotation! You've left the purchasing department and moved to the warehouse. There, the package sent by the supplier is on your desk, and your colleagues from production would like to take it right away because the stock has dropped to zero in the meantime. However, this goods issue can only be processed after the goods receipt has been entered. Otherwise, you would have a minus stock, and a minus stock can exist in our bank accounts, but not in SAP inventory management for a material.

> **Goods Movements**
>
> We now welcome you to the material management–inventory management component. This is where all goods movements are recorded and documented. In addition to goods issues or stock transfers, this also includes what we need now: a goods receipt for our purchase order.

Goods receipts are goods movements with which the receipt of materials from an external supplier or from production is posted. A goods receipt leads to an increase in stock and the creation of a material document, as shown in Figure 11.38.

Figure 11.38 Material Document for the Goods Receipt is the Second Document of the Purchase-to-Pay Business Process

For the following instructions, you'll use the real Post Goods Receipt for Purchasing Document app. In SAP GUI, you'll instead use Transaction MIGO (Goods Movement), which in turn is also available in SAP Fiori as the Post Goods Movement legacy app.

Here's one more piece of good news before you start: there is very little typing required here. Usually, there are only two entries: the respective purchase order number and the delivery note number. SAP S/4HANA transfers the remaining data, such as vendor, material number, and order quantity, from the purchase order to the goods receipt document.

Follow these steps:

1. Start the Post Goods Receipt for Purchasing Document app.
2. In the **Purchasing Document** field, as shown in Figure 11.39, enter the number of the purchase order you created in the previous sections. (You've made a note of the number).

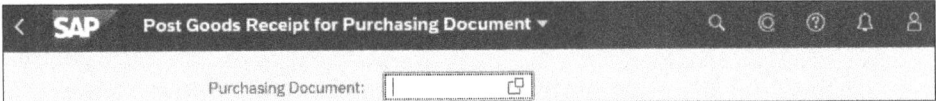

Figure 11.39 Purchasing Document Field

[+]

How Do I Find My Order Number?

You didn't write down the order number? No problem, you have several ways to find it:

- Enter part of the supplier's name in the **Purchasing Document** field. This will list all the corresponding purchase orders. Unfortunately, this doesn't work with the name of the material.
- If that doesn't help, in the **Purchasing Document** field, click ⌗ to find the purchase order using a search text (we explained the entry and search helps in Chapter 7).
- If that doesn't help either, use the Manage Purchase Orders app to search. Here you can search for all the orders you've created and their numbers in the **Search** field at the top left by specifying your username.
- And the last method (you may also remember it from Chapter 7) is to use enterprise search in the shell bar if you know, for example, the material number, the material description, or the supplier names.

3. After confirming the order number, the screen for goods receipt appears, as shown in Figure 11.40. Scroll down to the items.

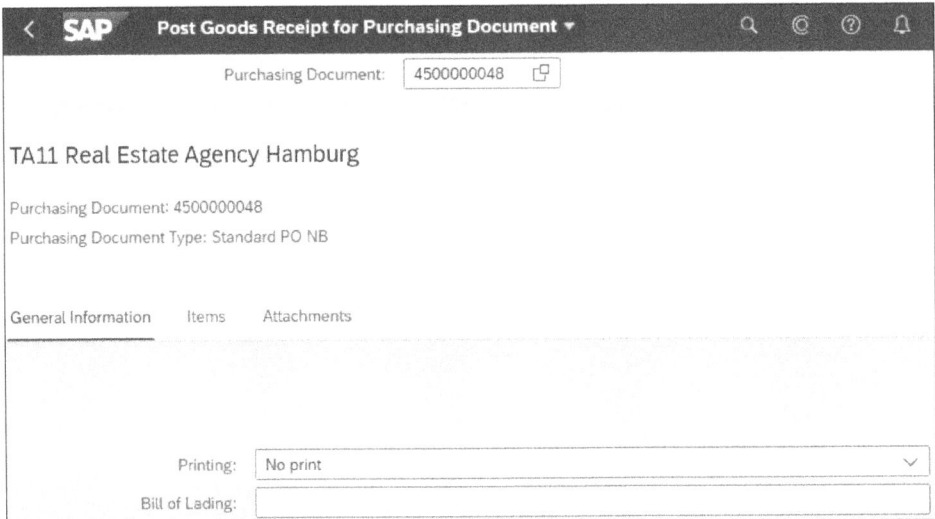

Figure 11.40 Goods Receipt Screen

4. Click on the **Items** tab, and place a checkmark on the far left for the item(s) delivered, as shown in Figure 11.41. The ordered quantity is already entered in the **Delivered** field.

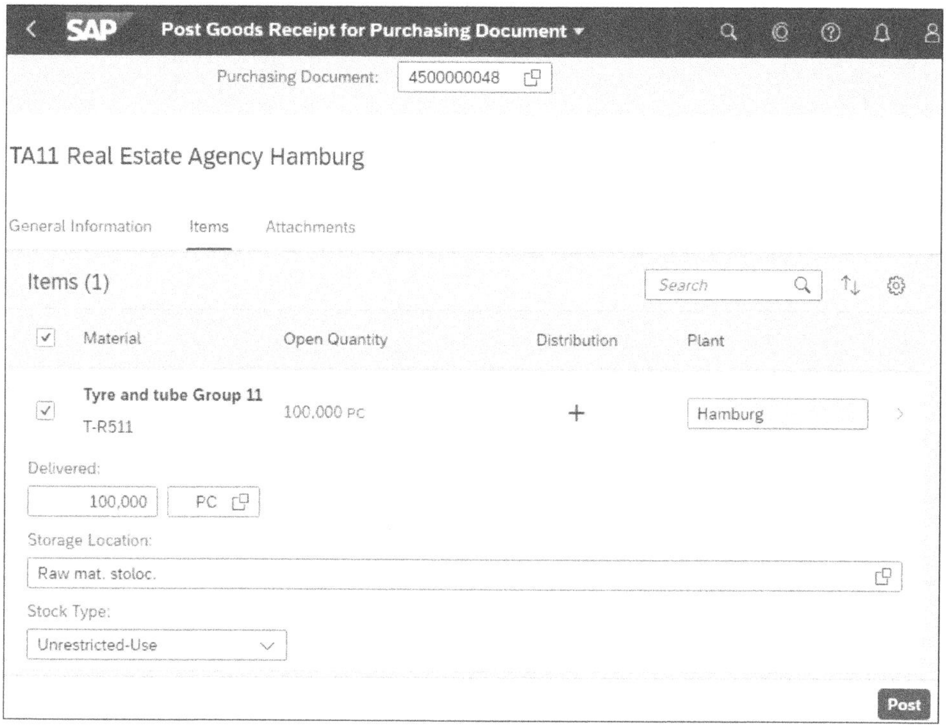

Figure 11.41 Goods Receipt Item

More Fields

In practice, there is often a requirement to enter the **Delivery Note** field, in which the delivery note number is transferred.

For a partial delivery, change the quantity in the **Delivered** field.

In the **Storage Location** field, you can adjust the receiving storage location from the purchase order.

If the material isn't to be withdrawn because, for example, its quality must be checked before it's withdrawn, change the stock type in the **Stock Type** field from **Unrestricted-Use** to **Blocked** or **Quality Inspection**.

In addition, you'll find further fields in the detail screen for each item, which you can access by clicking on the $\boxed{>}$ icon on the far right.

---

5. Confirm your entries with the **Post** button at the bottom right. After saving, a message appears with the information on the material document number, as shown in Figure 11.42.

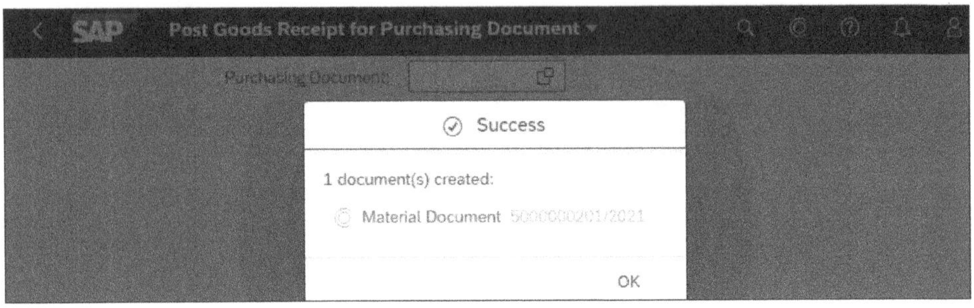

Figure 11.42 Message with Material Document Number

6. Confirm this friendly message with ⌐Enter⌐ or by clicking the **OK** button.

7. Click **SAP** to close the app for entering the goods receipt.

The goods receipt has increased the stock. If you want to be on the safe side and check the success of the goods receipt, follow the instructions from Section 11.6 where you (hopefully) noted the stock of the material before the goods receipt.

With the real Material Documents Overview app, you display the entered goods receipts in the list and also in detail by clicking on the corresponding material document number. In SAP GUI, you use Transaction MB51 (Display Material Documents) for this purpose.

> [»]
>
> **Switching from SAP ERP to SAP S/4HANA**
>
> The following transactions from SAP ERP no longer exist in inventory management with SAP S/4HANA: MB01, MB02, MB03, MB04, MB05, MB0A, MB11, MB1A, MB1B, MB1C, MB31, MBNL, MBRL, MBSF, MBSL, MBST, and MBSU. They have been replaced by the universal goods movement Transaction MIGO or by corresponding apps.

11.10   Enter a Supplier Invoice

Attention, now it's about money: the supplier invoice (other term: *incoming invoice*) has arrived, been checked, and has to be entered. SAP S/4HANA supports you during the check because it takes into account the purchase order or goods receipt that has already been saved. This reference ensures that only what was ordered or delivered is paid.

As a result of the invoice posting, SAP S/4HANA generates two documents, as shown in Figure 11.43:

- A document for the incoming invoice for materials management
- An accounting document for accounts payable and general ledger in financial accounting that is used in financial accounting to pay the incoming invoice

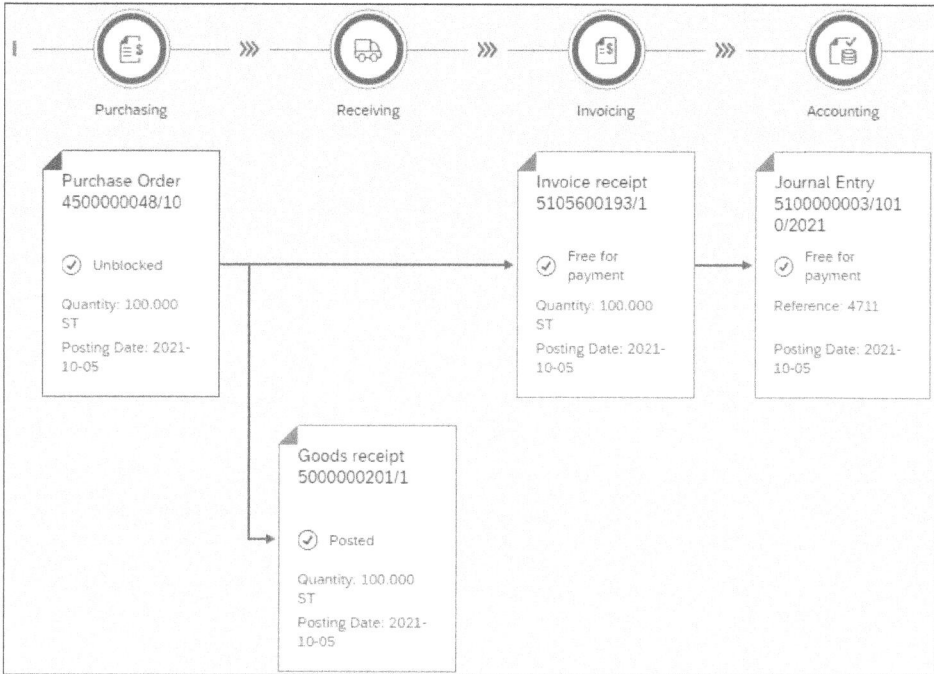

Figure 11.43 Two Documents after Entering the Incoming Invoice: Invoice Receipt and Journal Entry

What to Do in Case of Deviations?

SAP S/4HANA alerts you to quantity and price variances. In case of discrepancies between purchase order or goods receipt and invoice, SAP S/4HANA shows a warning, and, depending on the system configuration, the posting is denied or the invoice is blocked for payment.

You now enter the incoming invoice with the real Create Supplier Invoice app. In SAP GUI, Transaction MIRO (Enter Incoming Invoice) is used for this purpose, which in turn is available in SAP Fiori as the Create Supplier Invoice – Advanced legacy app.

Follow these steps:

1. Launch the Create Supplier Invoice app.

2. Make entries in the following fields in the topmost **Basic Data** area, as shown in Figure 11.44:

   – **Invoice Date**: Enter the current date here.

   – **Company Code**: If the field is not filled by a default value, enter the company code that was specified in the purchase order.

- **Gross Invoice Amount:** Enter the gross amount of the invoice. To compute this, multiply the net amount from the purchase order (**650.00 EUR**, per our example in Section 11.8, Figure 11.37) by 1.19.
- **Reference:** Enter the invoice number of the supplier.

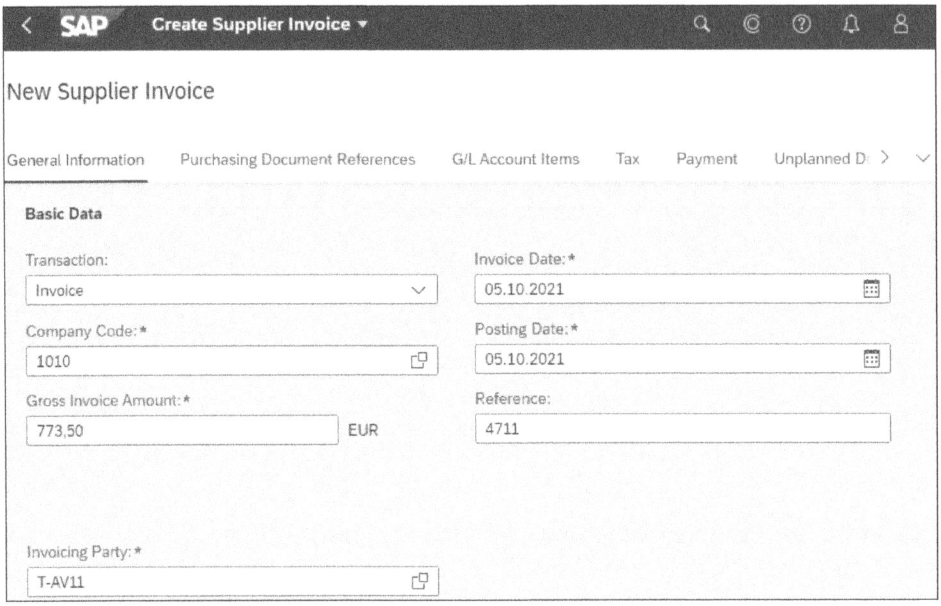

Figure 11.44 Basic Data Entries

[+]

Don't You Remember the Invoice Amount?

Of course, this procedure only applies to the training and not to the practice! If you don't know the invoice amount, leave the **Gross Invoice Amount** field at the default value "0.00". Carry out step 3 by entering the order number. After confirming with `Enter`, you'll see the deviation from the order amount in the **Balance** field in the upper-left corner. This value, of course positive and not a minus amount, is the gross amount from the order.

---

3. Scroll down and make entries in the following fields under the **Purchasing Document References** heading, as shown in Figure 11.45:
   - **Reference Document Category:** Select **Purchase Order/Scheduling Agreement** from the list.
   - **Purchase Order/Scheduling Agreement:** Enter the number of the order.

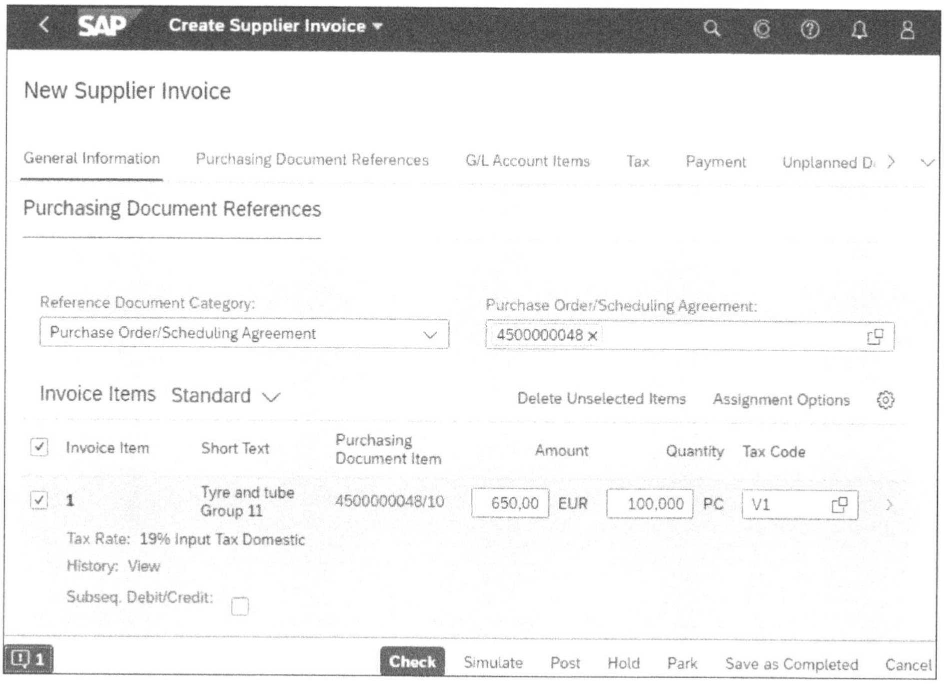

Figure 11.45 Purchasing Document References Entries

4. Press [Enter], and other fields such as **Invoicing Party** (corresponds to the business partner) are automatically filled in, and the items with details of material, price, and quantity are displayed.

5. Click the **Check** button in the footer.

For Accountants — Where Is the Debit?

Accountants will now be suspicious because an entry should consist of debit *and* credit. The information about the invoice amount and the invoicing party corresponds to the credit entry. The debit posting to the corresponding general ledger accounts isn't visible here. It's done automatically in the background, as you'll see later in Chapter 13. But if you're curious, before posting you can click on the **Simulate** button in the footer of the entry screen.

---

6. All good? Then the message **Invoice Was Successfully Checked** appears, and you can click the **Post** button.

7. After completing the posting, you'll be asked if you want to create another invoice. Click **No** here.

11.11   Display a Supplier Invoice

Use the Supplier Invoices List app to display the materials management invoice documents (you'll encounter the accounting document in Chapter 13).

Follow these steps:

1. Launch the Supplier Invoices List app.
2. After starting the app, filter by the posting date, for example. If you've posted the document today, enter today's date in the **Posting Date** field.

[+]

How to Save Typing in Date Values

In a date field, it's sufficient to enter "T". The letter T automatically shows a list with the words "today" or "tomorrow". After that, you immediately make further entries or click on the **Go** button. This trick works just as well with the input of "Y for "yesterday" or of "A" for "April".

Do you need the invoices from a certain period, for example, from the past seven days? Then simply type in a number, for example "7", and SAP S/4HANA will suggest entries such as **Last 7 days** or **Next 7 days**.

---

3. Click the **Go** button to see the list of corresponding invoices, as shown in Figure 11.46.

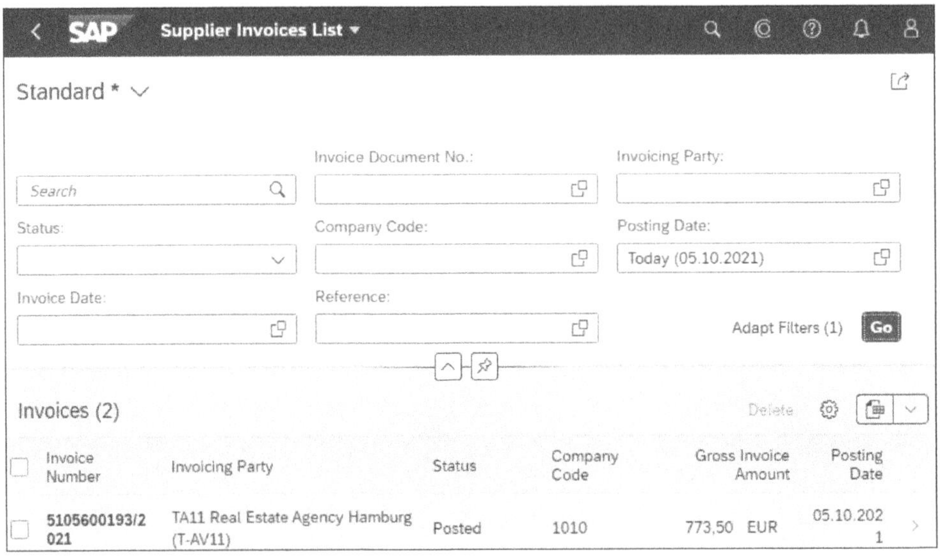

Figure 11.46 List of Supplier Invoices

4. If you click on an invoice number in the list, you'll then see the detailed information about that invoice, as shown in Figure 11.47.

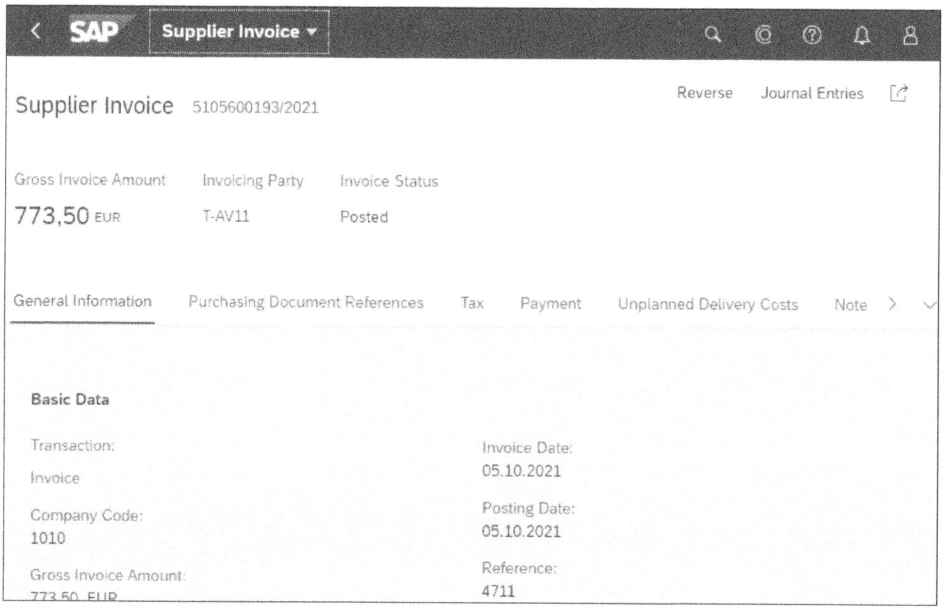

Figure 11.47 Invoice Details

5. Click on the **Journal Entries** button at the top right. A new tab opens in your browser showing the process flow with its documents (see Figure 11.48), but the material document for the goods receipt is suppressed.

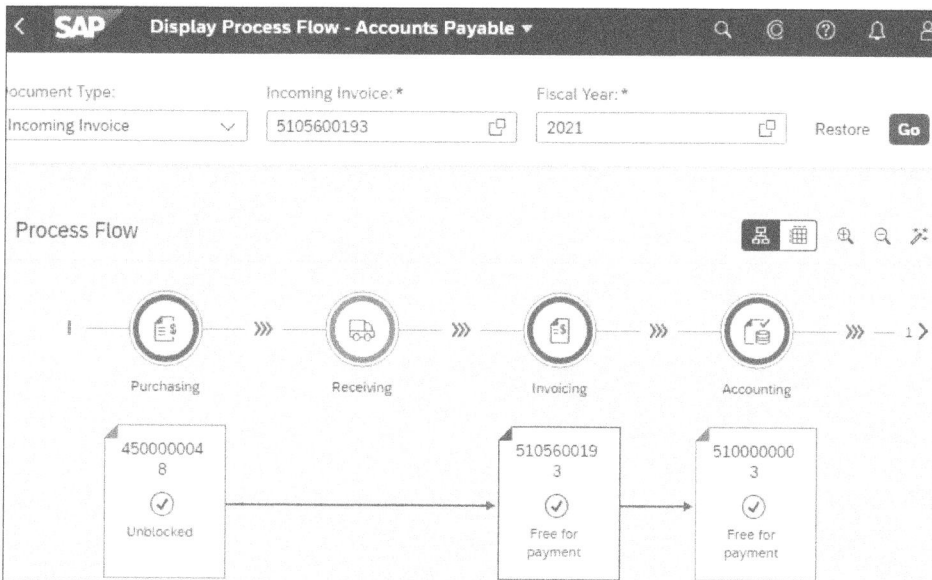

Figure 11.48 Process Flow

[+]

How to Get into the Process Flow Faster

There are also apps that can be called directly for the process flows. The Display Process Flow – Accounts Payable app shows documents from the purchase-to-pay process, and the Display Process Flow – Accounts Receivable app shows documents from the order-to-cash process.

After starting these apps, all you need to do is enter any document number, such as an order number or a purchase order number; confirm it with the **Go** button; and you'll see the respective process diagram.

6. Click on the number of the document under the **Accounting** item (on the right).

7. Now more information about this accounting document will be displayed, for example, the due date of the invoice, as shown in Figure 11.49.

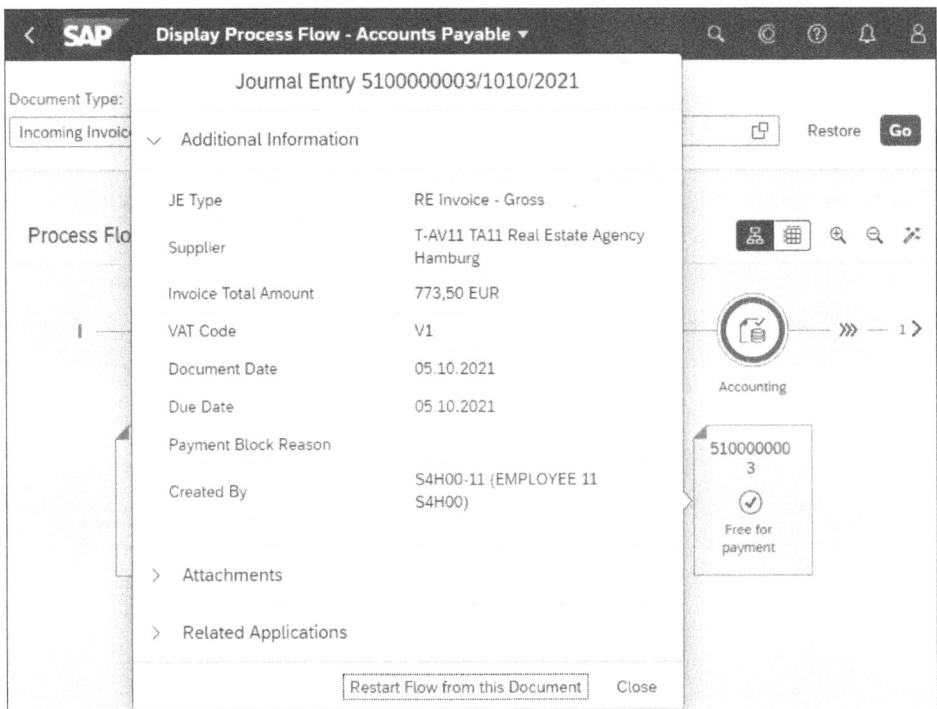

Figure 11.49 Accounting Document

8. Click the **Close** button and then the ![SAP] icon to go back to the SAP Fiori launchpad homepage.

11.12 Summary

Bravo, now you've gone through a complete process! However, the procurement process exists in many variants, for example, as an order for services or with a delivery schedule for a scheduling agreement. We would like to show you all of this, but it would go beyond the scope of a basic book. Our goal was to give you a basic understanding of processes in SAP S/4HANA using a common process as an example, and at the same time give you the opportunity to gain practical experience in handling the apps.

In the next chapter, we'll move onto a second core business process: order-to-cash.

11

Chapter 12

Sales and Distribution: From the Order to the Invoice

If you've gone through the previous chapter with the purchase-to-pay process, then you'll have an easier time with the *order-to-cash* process (also abbreviated as O2C) in this chapter. It's similar, only "mirror-inverted"! Instead of a purchase order, you enter a *sales order*, instead of a goods receipt you enter a *goods issue*, and instead of an incoming invoice you create an *outgoing invoice*.

As a sales clerk, you should know that your order is the basis for the delivery and the subsequent invoice, and, vice versa, accountants should understand that the data for the customer invoice to be posted comes from the original order. Everyone should think outside the box, as this helps to avoid mistakes and clarify discrepancies. Therefore, click along with the instructions, even if some topics aren't in your area of expertise. In addition, to avoid repetition, we assume in this chapter that you understand what was covered in the preceding chapter.

What You'll Learn

- Which subprocesses the order-to-cash process consists of
- How to view and change the customer data of a business partner
- How to display the sales data of a material
- How to create an order
- How to make a delivery
- How to create and display an outgoing invoice

12.1 The Order-to-Cash Process

Figure 12.1 shows you the basic flow and documents of the *order-to-cash* business process in which a stock material is sold.

Again, we have a little story that will take you through the following sections. Our company has been awarded a contract in response to a bid, and the customer's written order has been received.

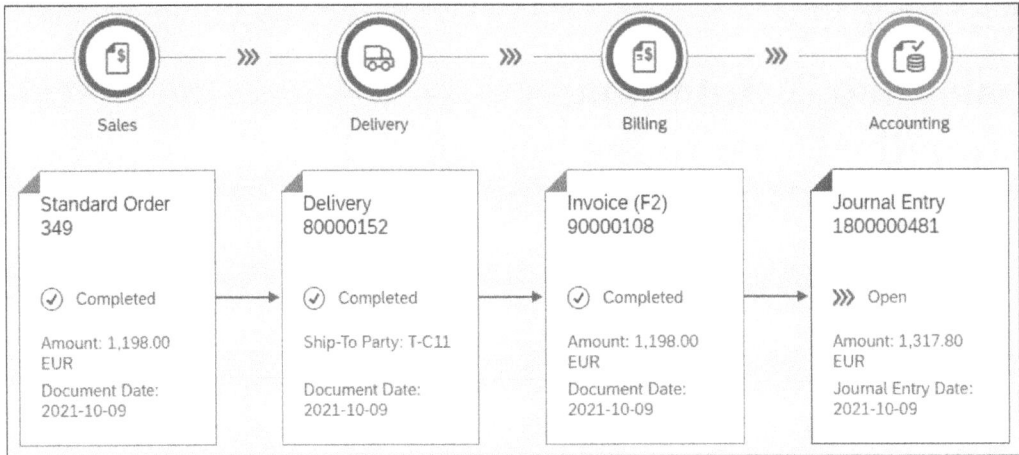

Figure 12.1 Subprocesses and Documents of the Order-to-Cash Business Process

Customer inquiries and quotations as presales documents could also be created and managed with the sales and distribution module. In practice, however, special *customer relationship management* (CRM) software is more often used for this purpose. Let's walk through our example:

1. Our example starts in *sales* with order entry. A lot of data doesn't have to be entered, usually only the customer number, the material number, and the quantity. Based on the customer and material master data, SAP S/4HANA adds much more data, such as the material description and the delivery address.

2. So that you can ship the material, you create an *outbound delivery*.

    – After staging for shipping, the material is picked in the outbound delivery document. *Picking* means that it has been removed from the storage bin and is ready for shipment.

    – The *goods issue* takes place on the basis of the delivery document and, as with the goods receipt, also creates an accounting document.

3. Now the *billing* takes place. In the billing process, the *outgoing invoice* (customer invoice) is created for the customer. Of course, you don't need to reenter the data entered in the order. SAP S/4HANA retrieves the data required for this from the customer master data, the material master data, and the order data.

4. Simultaneously with the billing document for the sales and distribution module, a second document for the financial accounting module is automatically created when the invoice is saved. This will be needed later for the payment of the invoice.

5. Now it's the turn of accounts receivable accounting (other name: customer accounting): It monitors the payment, creates reminders if necessary, and posts the payment of your outgoing invoice. As long as no payment has been made, the document hasn't yet been cleared.

The components listed in Table 12.1 are affected by this process.

Subprocess	Component	Abbreviation of the Component
Order entry	Sales	SD-SLS
Delivery and picking	Shipping	SD-SHP
Goods issue	Inventory management	MM-IM
Outgoing invoice	Billing	SD-BIL
Reminder and payment of the invoice	Accounts receivable	FI-AR

Table 12.1 Components in the Order-to-Cash Business Process

12.2   Organizational Units

From the "Company Structure in the SAP System" section in the Introduction, you already know the *company code*, which corresponds to a company, and the *sales organization*. If there are several sales regions, these can be subdivided according to sales organizations. Each sales organization can in turn consist of several departments, which are differentiated according to the *distribution channel* and *division* organizational units, as shown in Figure 12.2. We'll walk through them in the following sections.

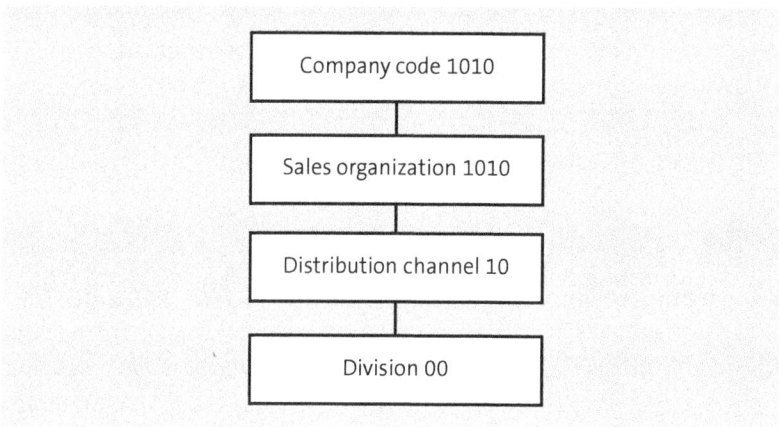

Figure 12.2 Company Code 1010 with the Sales Organization, Distribution Channel, and Division

12.2.1   Distribution Channel

The *distribution channel* organizational unit identifies the route by which products reach customers, for example, via direct sales or wholesalers.

12.2.2   Division

The *division* is formed with regard to the sales responsibility or profit responsibility of product groups. Examples of divisions are motorcycles, services, or food.

12.2.3   Sales Area

A *sales area* results from a complete combination of sales organization, distribution channel, and division. It corresponds to a single sales department. Think of each sales area as a separate sales department that can maintain different data for the customer in SAP S/4HANA, such as different payment terms or even different prices.

[»]

> **Your Organizational Units in Sales**
>
> We would like to give you an example that corresponds to the organizational units in the SAP Live Access training system. Sales area 1010/10/00 is made up of the following organizational units:
>
> - Sales organization 1010 (Germany)
> - Distribution channel 10 (direct sales)
> - Division 00 (cross-divisional)

You may wonder what all these subdivisions in the organizational units are for. One important reason is that the company can make accurate evaluations based on these subdivisions: sales analyses and sales statistics in any combination, separated by sales organizations, divisions, and distribution channels. To make these evaluations possible, SAP S/4HANA forces you to specify these organizational units when you enter orders.

12.3   Preferences

As in Chapter 11, you now make some default settings that reduce your typing effort. To enter orders, you use a "fiorized" transaction (legacy app), but unfortunately the default values from SAP Fiori aren't taken into account there. Therefore, this time, the default settings are made in the user defaults of SAP GUI Transaction SU3 (Maintain User Own Data). However, there is no app in the standard system that you can call for this transaction. So, do you have to go to SAP GUI to do this? No, because with the following trick, you can start SAP GUI transactions directly in SAP Fiori even without tiles!

12.3.1   Show the Command Field

You may remember Chapter 10, Section 10.4, where we've already shown you how to enable the legacy app command field. To recap, follow these steps:

1. Launch any legacy app in SAP Fiori. In our example, we use the Create Sales Orders app.
2. The **Order Type**, **Sales Organization**, **Distribution Channel**, and **Division** fields are to be preassigned in the initial screen, as shown in Figure 12.3. To do this, you need SAP GUI Transaction SU3 (Maintain User Own Data). So that you can start this in SAP Fiori, show the command field by choosing **More • SAP GUI Actions and Settings • Settings**.

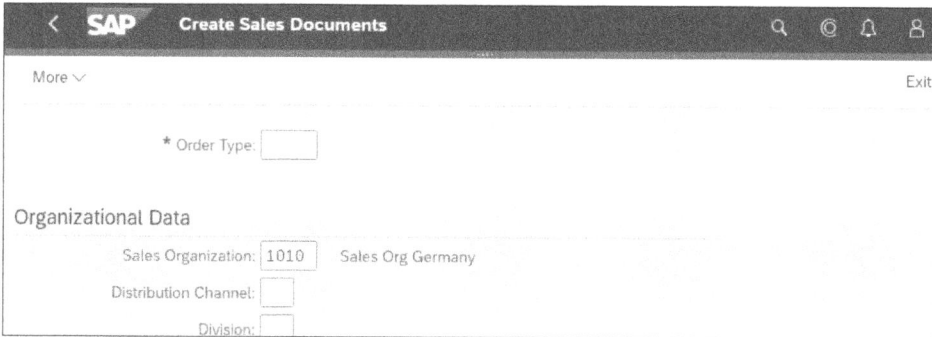

Figure 12.3 Fields in the Initial Screen of the Create Sales Orders App

3. In the **SAP GUI for HTML – Settings** window, select, if necessary, **Interaction Design • Visualization**. Here, select the slider for the **Show OK Code Field** checkbox, as shown in Figure 12.4.

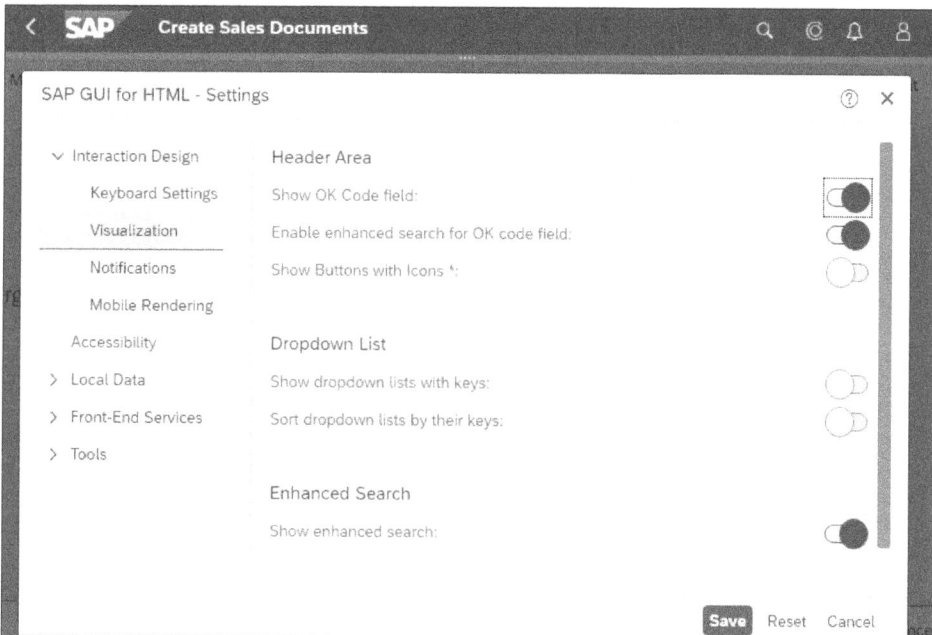

Figure 12.4 Show OK Code Field Slider

4. Confirm the setting in the footer with the **Save** button.

12.3.2   Set Default Values for Legacy Apps

Now you'll see the OK code field (empty field with down arrow) in the upper-left corner, as shown in Figure 12.5, waiting for you to enter the first transaction code.

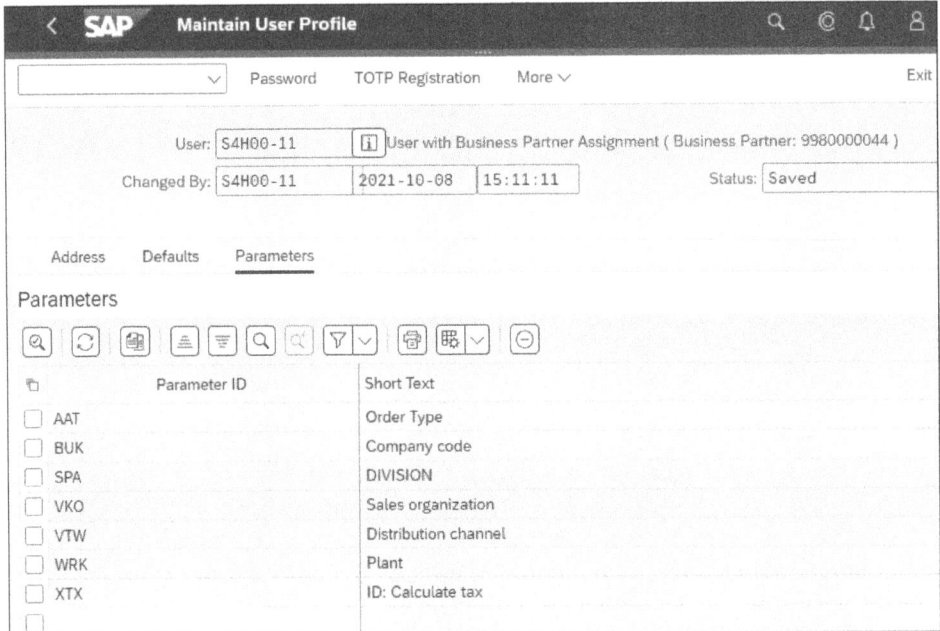

Figure 12.5 Parameters Tab

Follow these steps:

1. Enter "/nSU3" in the **OK Code** field at the top left and confirm with $\boxed{\text{Enter}}$.

2. Click the **Parameters** tab, as shown in Figure 12.5.

3. On the **Parameters** tab, click in the first free field in the **Parameter ID** column on the left. Enter the following parameter IDs and parameter values here:

   - **Order Type**: "AAT" parameter ID, "OR" parameter value.
   - **Company Code**: Parameter ID "BUK", parameter value "1010".
   - **Division**: Parameter ID "SPA", parameter value "00".
   - **Sales Organization**: Parameter ID "VKO", parameter value "1010".
   - **Distribution Channel**: Parameter ID "VTW", parameter value "10".
   - **Plant**: Parameter ID "WRK", parameter value "1010".
   - **ID: Calculate Tax**: Parameter ID "XTX", parameter value "X".

4. Press the $\boxed{\text{Enter}}$ key. Now SAP S/4HANA shows the meaning of the parameter IDs in the third column **Short Text**, as shown in Figure 12.5.

5. Save your entries by clicking the **Save** button at the bottom right of the footer line.

Our notes on training data from Chapter 11, Section 11.3, also apply to this chapter.

12.4   Customer as a Business Partner

You've already become familiar with the business partner master record in Chapter 11. Therefore, we concentrate here particularly on the business partner data.

As you know, the many bits of business partner data are kept apart with the help of *roles*. This time, you use the **Customer** role.

You've already become acquainted with the "real" SAP Fiori Manage Business Partner Master Data app, which provides you with a list of the corresponding business partners. The counterpart to this is the Customer Master app, which works according to the same rules (and so it's not covered here). This time, we use the following methods to display the customer master data:

- **Enterprise search**
  This function allows you to quickly display a few basic pieces of information, such as the postal address and the customer name, and from there branch further into detailed data of an object page.

- **Maintain Business Partner**
  This legacy app corresponds to SAP GUI Transaction BP (Maintain Business Partner).

12.4.1   Show Customer Data in the Object Page

Of course, to display customer data, you first use the quick method: displaying a special *object page* for customers using the enterprise search function. Follow these steps:

1. Open the enterprise search by clicking $\boxed{Q}$ (**Search**) in the shell bar.
2. In the shell bar, click $\boxed{\vee}$ to open the object list.
3. In the list, scroll down, as shown in Figure 12.6, and click the **Customers** entry.

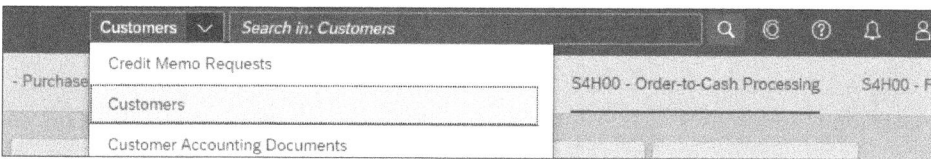

Figure 12.6 Object List after Scrolling Down

4. To the right of the field that now shows **Customers**, enter the number of the customer in the next field, as shown in Figure 12.7, and press the ⌈Enter⌋ key.

Figure 12.7 Entering the Customer Number

5. Now some basic customer data will be displayed, as shown in Figure 12.8. Click on the name of the customer or on its number, to go to the details, which show you a lot of further information.

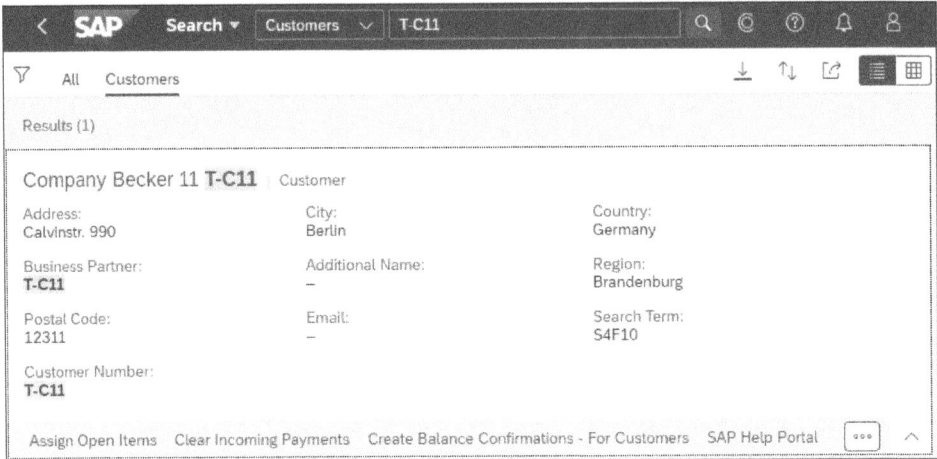

Figure 12.8 Displaying Customer Data

6. Use the tab bar or scroll to navigate to the desired information:

   – In the **General Information** tab, as shown in Figure 12.9, you can see, for example, the address or telephone numbers.

   – In the **Sales Area** tab, you'll find the sales departments that serve the customer with the details of the sales organization, distribution channel, and division.

   – The **Company Code** tab displays information for the colleagues in the accounting department.

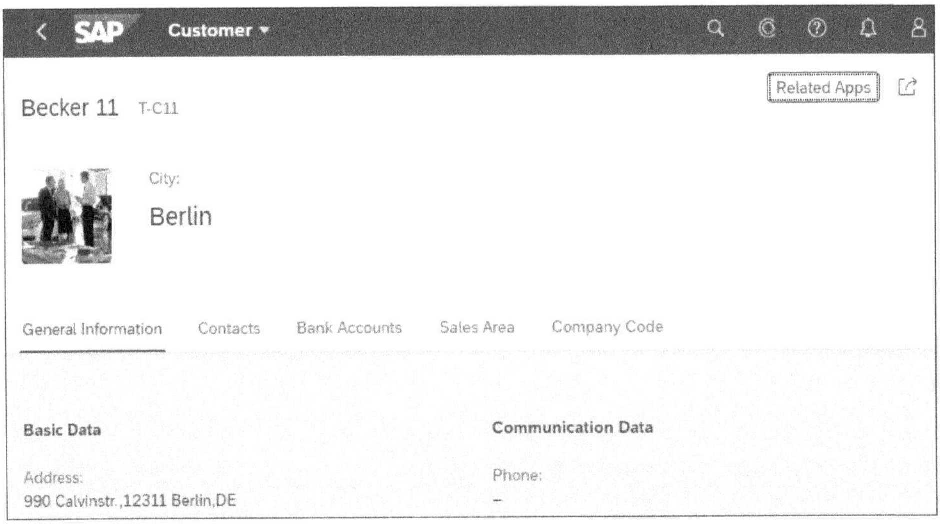

Figure 12.9 General Information Tab

7. Click **SAP** to leave the object page and return to the SAP Fiori launchpad start page.

However, you can't change data in the object page; you can only do that in the business partner master record. Displaying customer data in the business partner master record instead of via the object page is a bit more complex but sometimes necessary because the object page doesn't show all the information of the business partner master record; among other things, you don't see payment terms here.

12.4.2   View and Modify Customer Data with a Legacy App

Let's use the Maintain Business Partner legacy app this time, which corresponds to Transaction BP (Maintain Business Partner) in SAP GUI.

Follow these steps:

1. Start the Maintain Business Partner app.
2. When you see the **Worklist** and **Find** tabs on the left, as shown in Figure 12.10, click the **Locator On/Off** button at the top left. This way, you turn off the left part of the screen labeled **Locator** and have the same state as in our book.

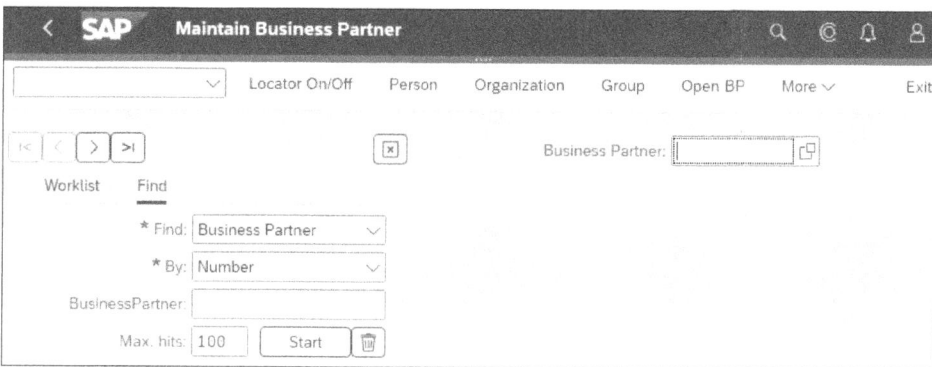

Figure 12.10 Worklist and Find Tabs

3. Enter the customer number in the **Business Partner** field, as shown in Figure 12.11, or use the ⧉ (**Value Help**) icon on the right side of the field. If there is already an entry, use the **Open BP** button.

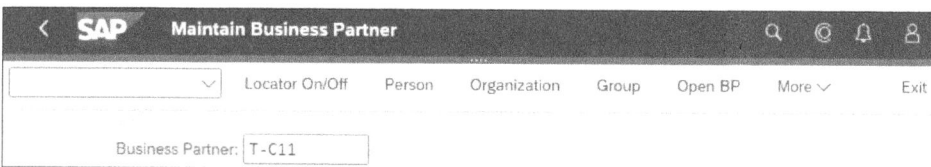

Figure 12.11 Entering Customer Number

4. Confirm the number with the ⌐Enter⌐ key. You're now in the screen with the detailed data. In Figure 12.12, you see the general data, as you can recognize by the entry in the **Display in BP Role** field.

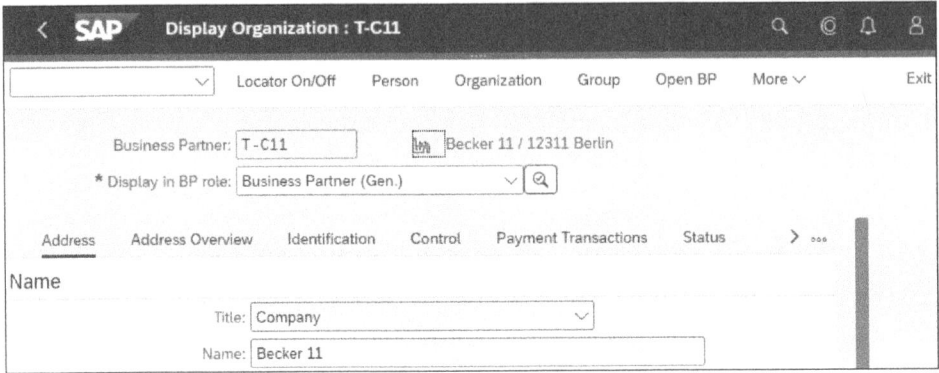

Figure 12.12 Displaying General Data

> **What Was General Data Again?**
>
> You may still be familiar with the term *general data* from Chapter 11, Section 11.4. This is the data that is the same for all departments, such as the main address. In contrast, the *sales data* can be different for each sales area.

5. To view the sales data, click on the ☑ icon to the right of the **Display in BP Role** field, ☑ and select the **Customer** entry from the list, as shown in Figure 12.13.

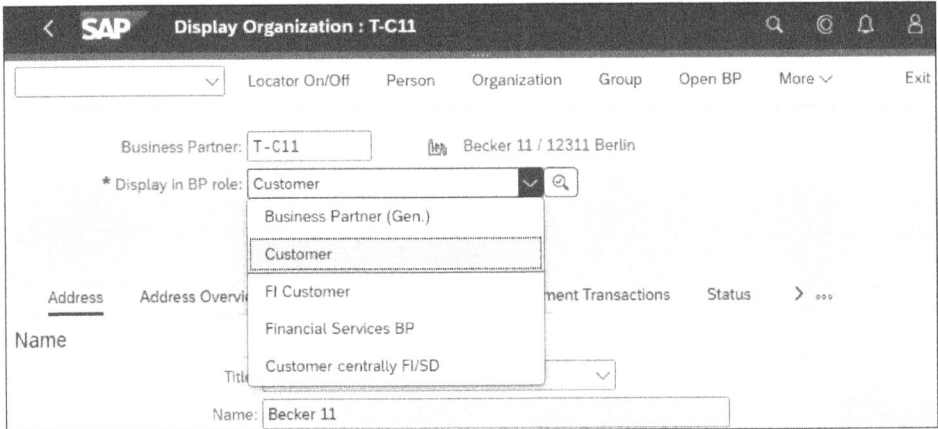

Figure 12.13 Selecting the Customer Role to View Sales Data

6. Our "role play" isn't over yet because you want to see the sales data of the **Customer** role. Only when you click on the **Sales and Distribution** button in the header (or select this in the **More** menu) does SAP S/4HANA shows you the corresponding data.

   If the **Sales and Distribution** button isn't displayed as in Figure 12.14, the window is too narrow. In this case, open the **More** menu, and select the **Sales and Distribution** command there.

In both cases, you'll now see several sales-specific tabs at the bottom, such as **Orders**, **Shipping**, **Billing**, and **Partner Functions**, which apply to the preset combination of sales organization, distribution channel, and division. This combination is already filled in thanks to your default settings, and you can see it under the **Sales Area** heading.

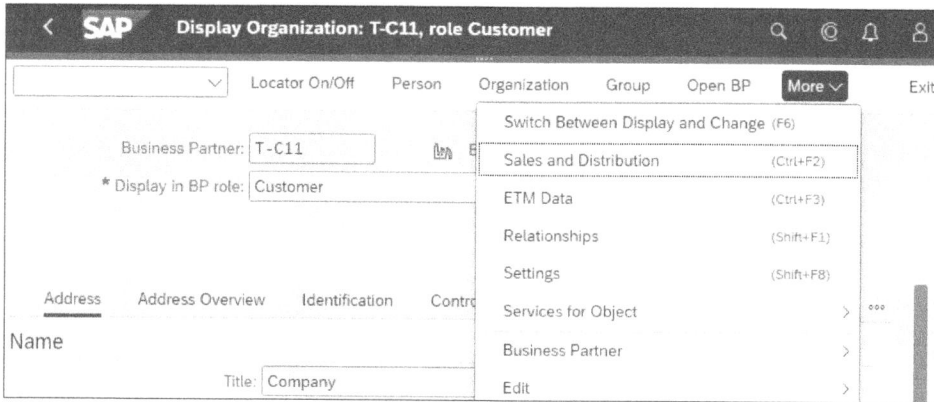

Figure 12.14 Sales and Distribution Command in the More Menu

7. Click on the **Billing** tab, as shown in Figure 12.15. This tab shows the *Incoterms* and further down the *payment terms*. This information as well as other fields will be transferred to the order later.

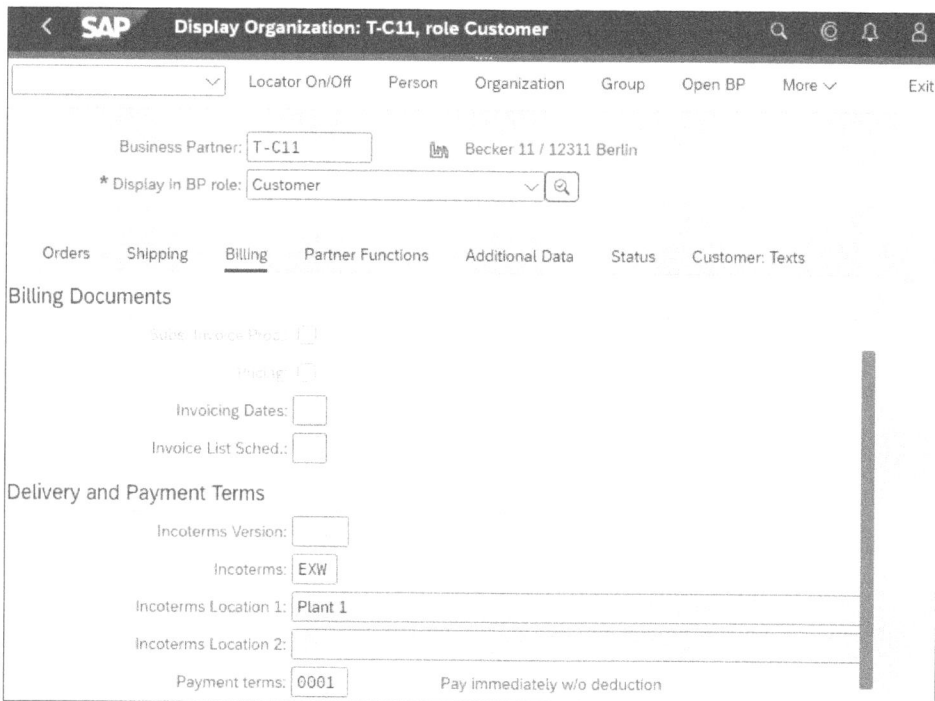

Figure 12.15 Billing Tab

8. Click on the **Partner Functions** tab, as shown in Figure 12.16. Here you can see in our example that the sold-to party is also the ship-to party, the bill-to party, and the payer.

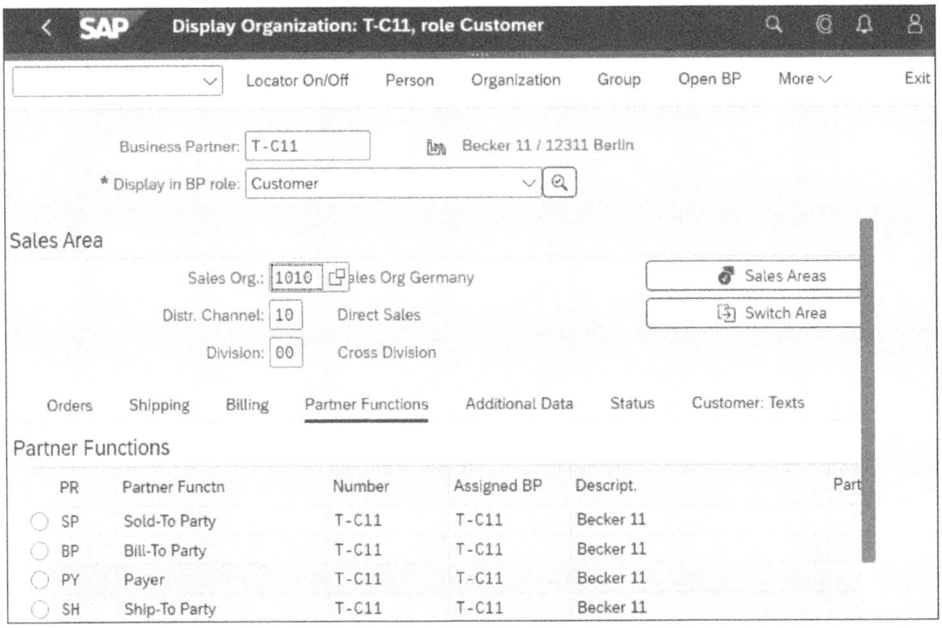

| < | **SAP** | **Display Organization: T-C11, role Customer** | | | | Q | ⊚ | ⏷ | 8 |

| | | Locator On/Off | Person | Organization | Group | Open BP | More ⌄ | | Exit |

Business Partner: `T-C11`   〔ᴍ〕 Becker 11 / 12311 Berlin

* Display in BP role: `Customer`   ⌄ Q

Sales Area

Sales Org.: `1010` 〔〕ᴀles Org Germany   🧭 Sales Areas

Distr. Channel: `10`   Direct Sales   〔⇅〕 Switch Area

Division: `00`   Cross Division

| Orders | Shipping | Billing | Partner Functions | Additional Data | Status | Customer: Texts |

Partner Functions

PR	Partner Functn	Number	Assigned BP	Descript.		Part
◯ SP	Sold-To Party	T-C11	T-C11	Becker 11		
◯ BP	Bill-To Party	T-C11	T-C11	Becker 11		
◯ PY	Payer	T-C11	T-C11	Becker 11		
◯ SH	Ship-To Party	T-C11	T-C11	Becker 11		

Figure 12.16 Partner Functions Tab

〔≫〕

Partner Functions

Here is an example: You send your mother-in-law a bouquet of flowers you ordered from an online florist. Your partner roles are **Sold-To Party** and **Payer**. Your mother-in-law has the **Ship-To Party** role because a separate address is required for this. A separate customer master record is created for this role, and its master record number is entered here in the **Partner Functions** tab.

9. You can use the Maintain Business Partner app and the corresponding SAP GUI Transaction BP (Maintain Business Partner) to display, change, and create new data.

   By default, you're in display mode. You can switch to change mode by choosing **More • Business Partner • Display <-> Change**. To do this, you must already be displaying business partner master data. Once in change mode, check out the lower-right corner of the window where you'll see the two additional buttons **Save** and **Cancel**, with which you can complete or reset the change.

10. To finish, click 〔SAP〕 to go back to the SAP Fiori launchpad homepage.

Customer data is shared between accounting and sales. As a rule, the sales data with the **Customer** role is created first, and the accounting data with the **FI Customer** role is added later. Often, there are also separate departments for master data maintenance, which are responsible for all data. The latter is called *central data maintenance*.

Switching from SAP ERP to SAP S/4HANA   [«]

In the previous SAP ERP system, customer and supplier master data had to be maintained separately and with different transactions. If there was a supplier who was also a customer, two different master records were entered, each with its own customer and supplier number. If, for example, the address changed, this had to be maintained in both places. Cumbersome!

In SAP S/4HANA, customer and supplier master data for the same business partner can be managed centrally under one business partner number and can be maintained in one transaction. A business partner can be used in processes under the same number as both a supplier and a customer.

In SAP GUI, the transaction codes for customer master data that start with FD0, XD0, or VD0 still exist, but when you call these transactions, you're taken directly to Transaction BP (Business Partner).

12.4.3   View and Modify Customer Data via the List

In Chapter 11, Section 11.4, you used the real Manage Business Partner Master Data app. In the same way, there is the Customer Master app. Follow these steps to display customer data:

1. Immediately after launching the app, you filter the business partners using the **Customer** role.

2. In the details, use the **Sales Areas** tab to switch to the sales-specific data by clicking on the $>$ icon.

12.5   Display Sales Data in the Material Master

Before you create your first order, you want to find out about the material you're selling in that order. As you already know from Chapter 11, Section 11.5.1, three paths lead to the material master:

- **Manage Product Master Data app**
  This app gives you a list of materials.

- **Enterprise search**
  This allows you to view a few basic pieces of information, such as weight or manufacturer.

- **Display Material legacy app**
  This app displays all the information for a material master.

The individual components of the material master record are called views. In the order-to-cash process, for example, the following views are of interest, which you call up for display immediately afterward:

- **Sales: Sales Org. Data 1** with information on division, delivery plant, and output tax
- **Sales: General/Plant Data** with information on weight, shipping dates, and availability check

In SAP Fiori, we use the Display Material legacy app to show you how to display sales-relevant data. In SAP GUI, you use Transaction MM03 (Display Material), which works in the same way.

Follow these steps:

1. Launch the Display Material app.
2. In the initial screen, as shown in Figure 12.17, enter the material number, and confirm it with the Enter key.

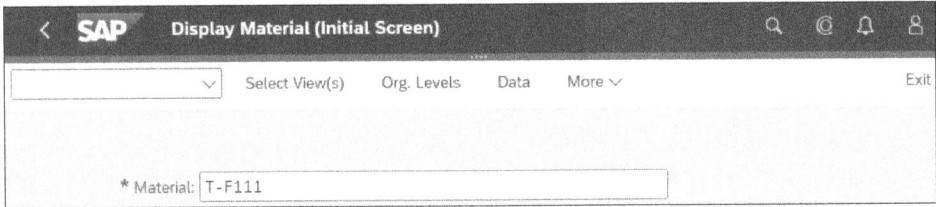

Figure 12.17 Entering the Material Number

3. SAP S/4HANA shows you the views that have been created for this material. In the **Select View(s)** window, as shown in Figure 12.18, select the **Sales: Sales Org. Data 1** and **Sales: General/Plant Data** views. Then confirm your selection by pressing the Enter key.

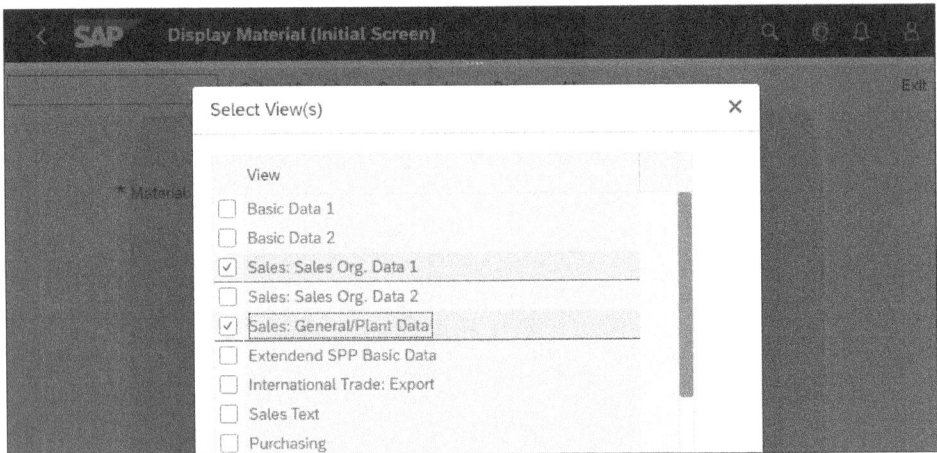

Figure 12.18 Selection View(s) Window

4. SAP S/4HANA now requires you to enter the *organizational levels* (another term for organizational units), as shown in Figure 12.19. Enter "1010" in the **Plant** field, "1010" in the **Sales Org.** field, and "10" in the **Distr. Channel** field. Confirm these entries with Enter as well.

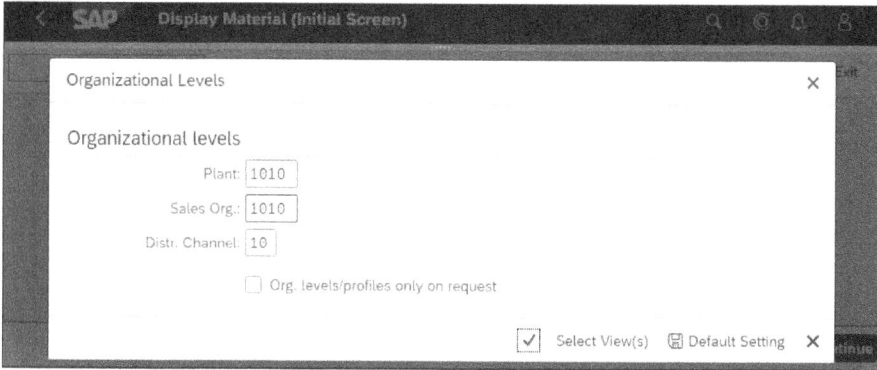

Figure 12.19 Entering Organizational Levels

> **How to Find the Organizational Units with Maintained Material Master Data**
>
> Material master data isn't always maintained for all organizational units. If, for example, you want to know through which sales organizations a material can be sold, simply click in this field and then on ☐ (**Value Help**).

5. SAP S/4HANA displays the **Sales: Sales Org. 1** tab, as shown in Figure 12.20, with information on division, units of measure, delivering plant, and output tax further on down. Press the Enter key.

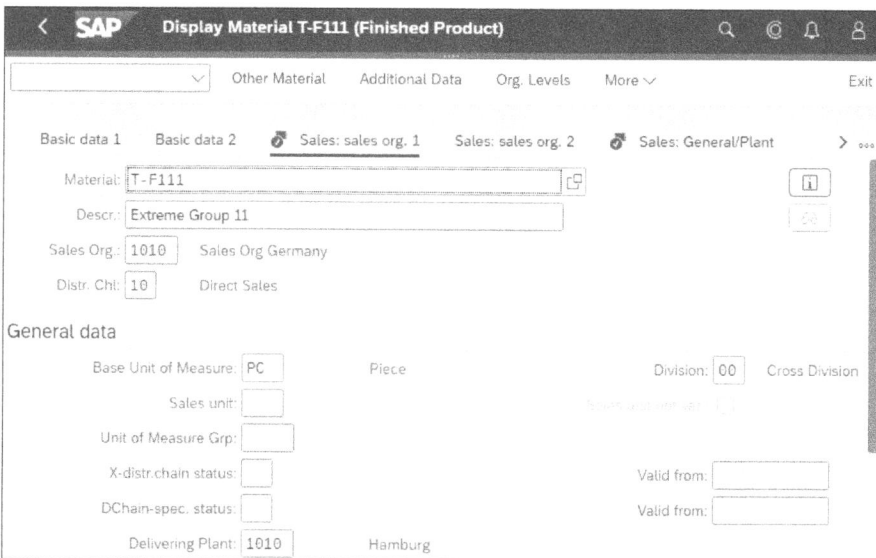

Figure 12.20 Sales: Sales Org. 1 Tab

351

6. You'll see the **Sales: General/Plant** tab, as shown in Figure 12.21, with information on weight and availability check, as well as shipping and packaging material data further on down.

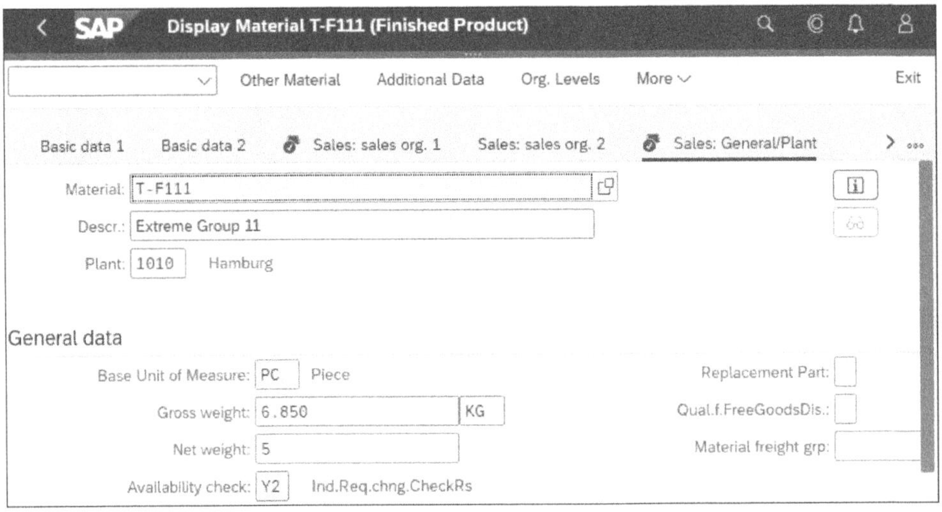

Figure 12.21 Sales: General/Plant Tab

7. Click on the **SAP** logo from the shell bar, and you'll be back on the SAP Fiori launchpad homepage.

[»]

Material Master Data for Sales and Distribution in the Manage Product Master Data App

The real Manage Product Master Data app also shows the previously mentioned material master data for sales. However, the data there is grouped differently than in the Display Material app. After switching from the list to an individual material, you'll find the corresponding information there on the **Sales** and **Distribution Chains** tabs, for example. For these tabs, click on the > icon to get to the desired data.

12.6 Create a Sales Order

Entering the sales order doesn't require much typing if you have well-maintained master data and default values. You need the following data:

- The order type, which determines whether the order is, for example, a forward order with a delivery, a cash sale without a delivery, or a repair order
- The organizational unit's sales organization, distribution channel, and division, which are mostly the same in practice, which is why you set default values for them

- The customer number, the material number, and the quantity of the material, which are used to fill many other order fields with data from the master records
- The number and date of the customer purchase order

Now we start entering the order with the Create Sales Orders legacy app , which corresponds to Transaction VA01 (Create Sales Orders) of the same name in SAP GUI.

Follow these steps:

1. Start the **Create Sales Orders** app.

2. On the initial screen as shown in Figure 12.22, in our example, you need to enter "OR" for standard order in the **Order Type** field. Below this, in the **Sales Organization**, **Distribution Channel**, and **Division** fields, you determine the organizational units to which the order belongs.

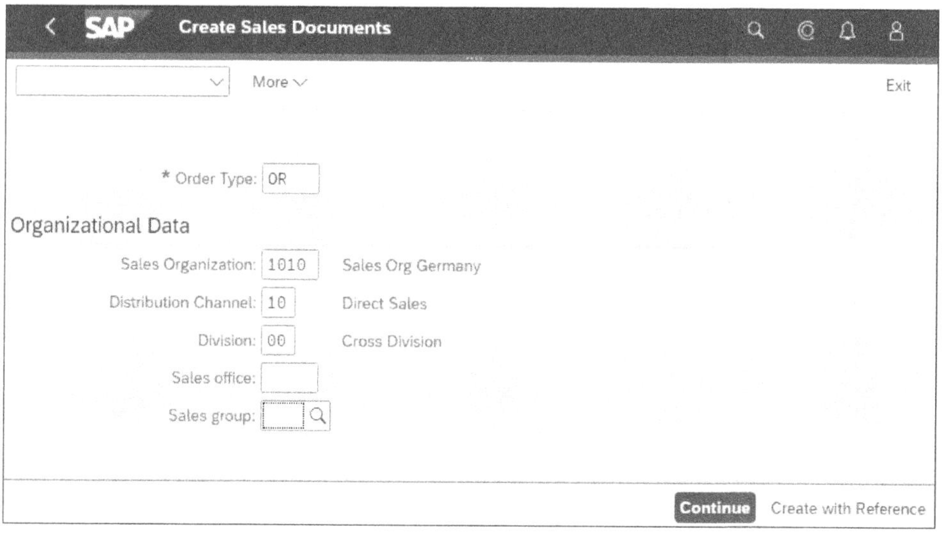

Figure 12.22 Create Sales Order App: Initial Screen

Create with Reference

Just like purchase orders, orders can be entered with reference to other documents from which the information is copied to the order. These include the following documents:

- Inquiries and quotations
- Other orders or billing documents (outgoing invoices); this is particularly worthwhile for orders with many items
- Sales contracts or sales scheduling agreements from which deliveries can be called off

In our example, however, you don't yet have a document to which you can relate.

3. Confirm the entries with the **Continue** button. This takes you to the **Create Standard Order** screen, as shown in Figure 12.23.

4. First enter the customer number in the **Sold-To Party** field at the top. You can leave the **Ship-To Party** field empty if, as in our case, the **Ship-To Party** is the same as the customer.

5. In the **Cust. Reference** field, enter any number as the customer's order number, and in the **Cust. Ref. Date** field, enter the current date as the order date. By the way, do you know the trick for legacy apps and SAP GUI transactions for how to determine the current date in an empty field in no time? You simply press the (F4) and (F2) keys in succession!

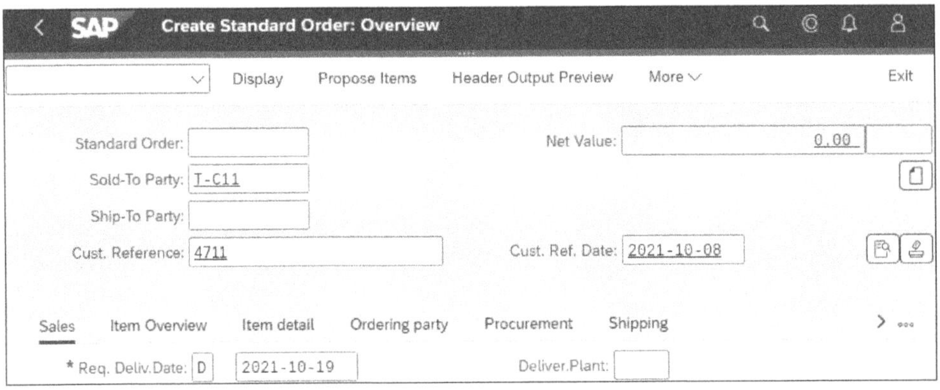

Figure 12.23 Creating the Sales Order

6. In the **Req. Deliv.Date** field, change the date to today's date. Confirm your previous entries with the Enter key.

What Happens after Confirmation?

When you confirm with Enter, SAP S/4HANA automatically fills in many other fields based on the customer master data, such as **Ship-To Party**, **Pyt Terms** (payment terms), and **Incoterms**, which are all fields that you don't need to fill in yourself if you have well-maintained master data.

7. Now scroll all the way down, if necessary. There you'll see a table with the heading **All Items**, as shown in Figure 12.24. Ignore the many columns. Here you only need to enter the material number in the **Material** column and the quantity in the **Order Quantity** column.

8. Click on the **Save** button in the footer to save the sales order. That's it! After saving, the success message **Standard order ### has been saved** appears. Make a note of the order number that is displayed in the footer.

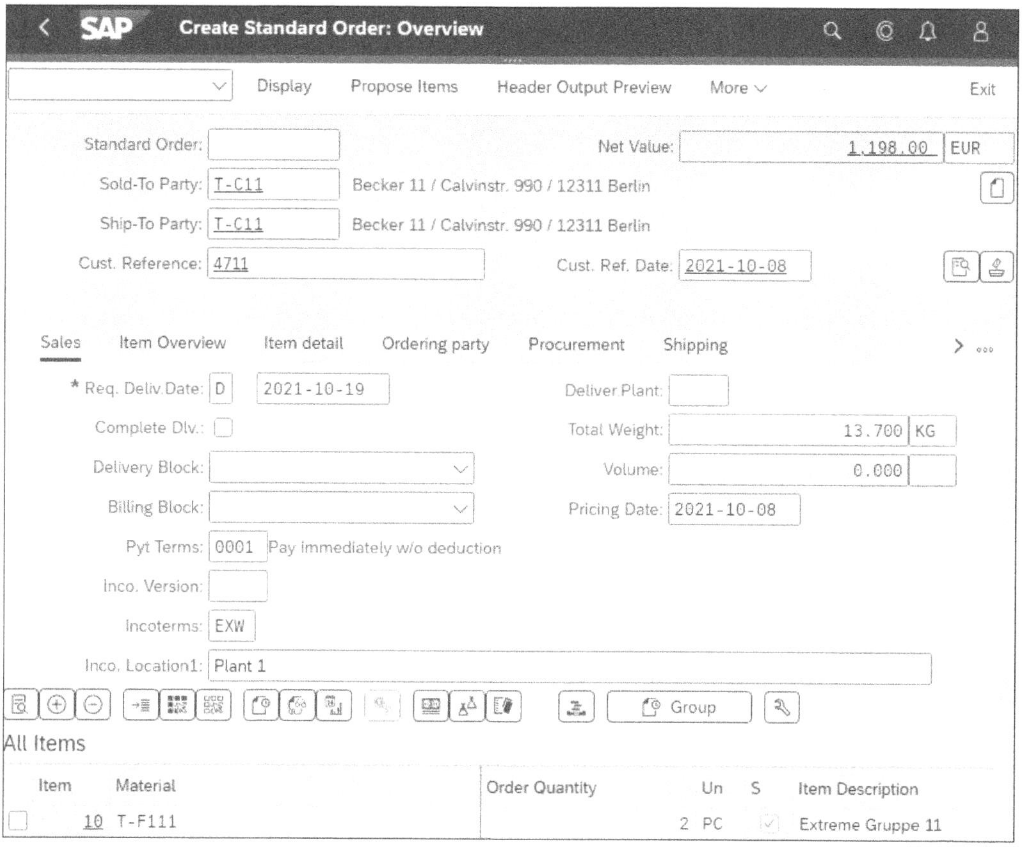

Figure 12.24 All Items Table

9. Click SAP to return to your SAP Fiori launchpad homepage.

12.7   Show Order List and Order Confirmation

Your customer is waiting for your order confirmation. You'll find the corresponding PDF file in the order display. This is created automatically in the SAP Live Access training system based on a corresponding preset.

In SAP Fiori, you use the real Manage Sales Orders app to display an order first in the list and click on the corresponding order number in the detail view. In SAP GUI, you use Transaction VA05 (List of Sales Orders) to display the list and Transaction VA03 (Display Order) to display an individual order.

In the following steps, you'll see both representations: first a real SAP Fiori display with a process diagram, which you can look forward to now, and a classic representation based on Transaction VA03 (Display Order).

Follow these steps:

1. Launch the Manage Sales Orders app.

2. Enter the customer number in the **Sold-To Party** field or, alternatively, the order number in the **Sales Order** field (see Figure 12.25).

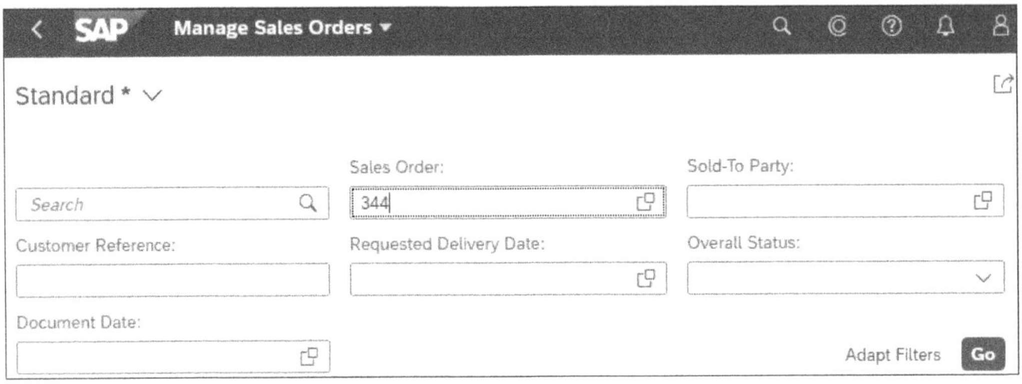

Figure 12.25 Entering the Order Number

3. Click on the **Go** button on the far right. Now the **Sales Orders** list appears, as shown in Figure 12.26.

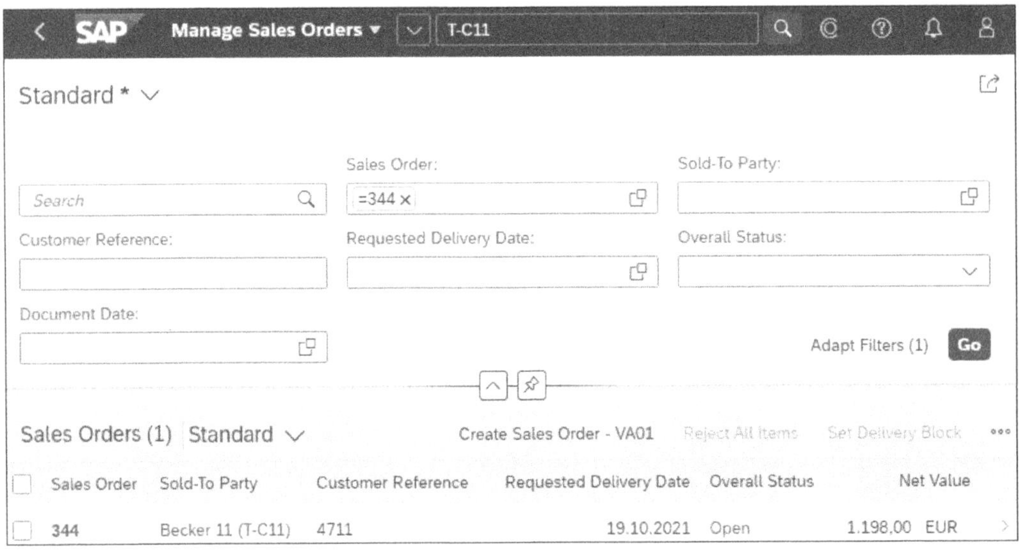

Figure 12.26 Displaying the Sales Orders List

4. Click on the number of your sales order on the left and on the same order number again in a small additional window that appears. SAP S/4HANA shows you information about the order in the SAP Fiori design, as shown in Figure 12.27.

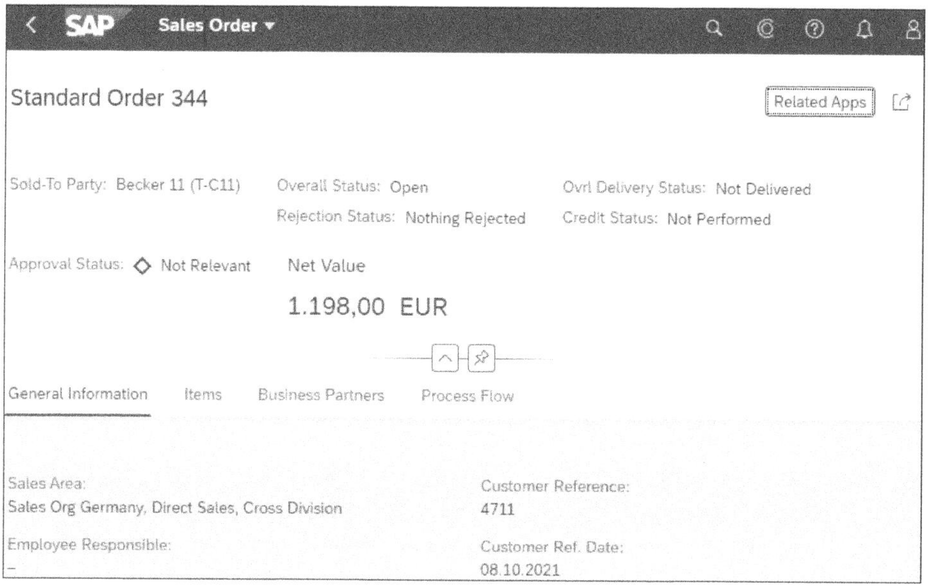

Figure 12.27 Order Information

5. The process status is exciting to see, so click on the **Process Flow** tab, as shown in Figure 12.28, or scroll all the way down.

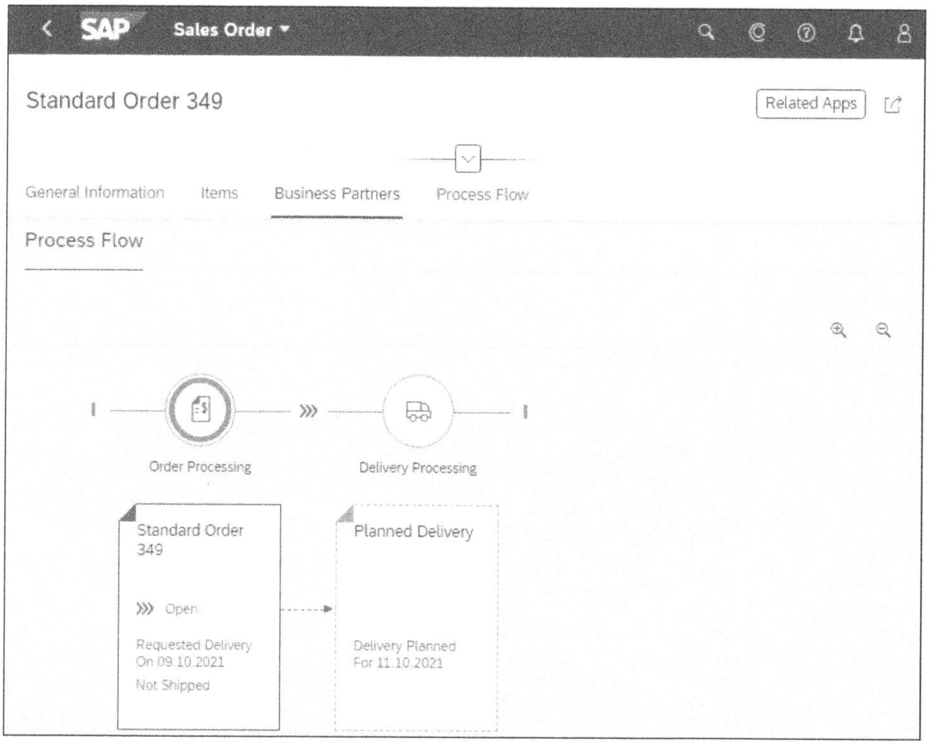

Figure 12.28 Process Flow

[Ex]

What Do I Use the Process Flow For?

Here's a practical example: your customer is urgently waiting for a delivery, and an employee calls you and wants to know if the package has already been sent. Thank you, dear SAP developers! In this very nice and clear process diagram, you can see when the delivery is planned for or if the delivery has already been picked or shipped.

And here's another example: a customer has received the invoice, but has queries. By clicking on the invoice document displayed here later, you can see the details of the invoice. In addition, you'll find out whether the invoice has already been paid.

6. Click on ◁ (**Back**) in the upper-left corner. You're back in the **Sales Orders** list, as shown in Figure 12.29.

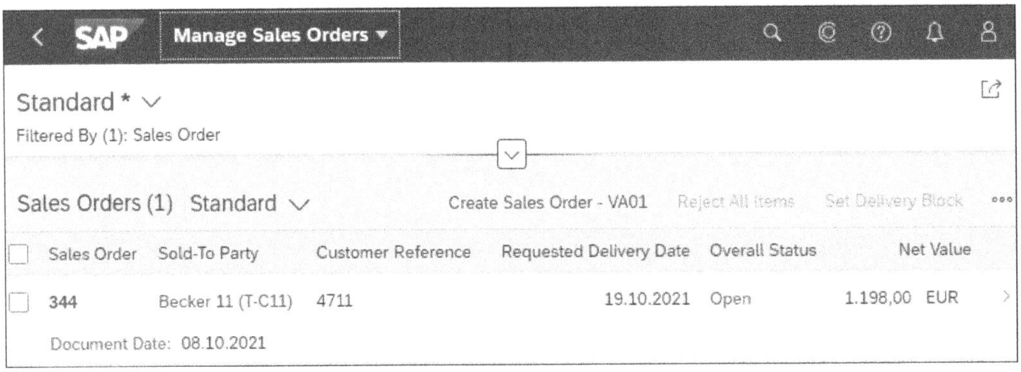

Figure 12.29 Returning to the Sales Orders List

We now switch from the order list to the display of the legacy app, which corresponds to SAP GUI Transaction VA03 (Display Order). To do this, follow these steps:

1. In the order list, click ▷ on the far right of the order line. Now you'll see the order in a form that corresponds to the form from the order entry (see Figure 12.30).

[»]

How Do I Change an Order?

Life is dynamic. Shortly after you've entered the order, the next email from your customer arrives: he wants to double the quantity. You can make this order change at this point after clicking the **Change** button.

2. To see the order confirmation, click ⚸ (**Display Output Request**) on the far right-middle area of the screen.

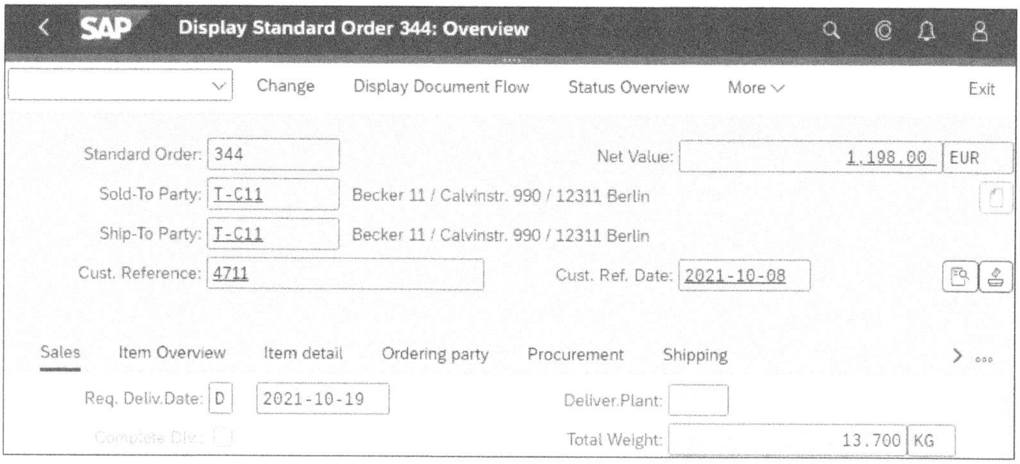

Figure 12.30 Sales Order with the Display Output Request Icon

3. In the output management, click on the **Display Document** button, and now you see the order confirmation as a PDF (see Figure 12.31). From this PDF display, you can also download the document.

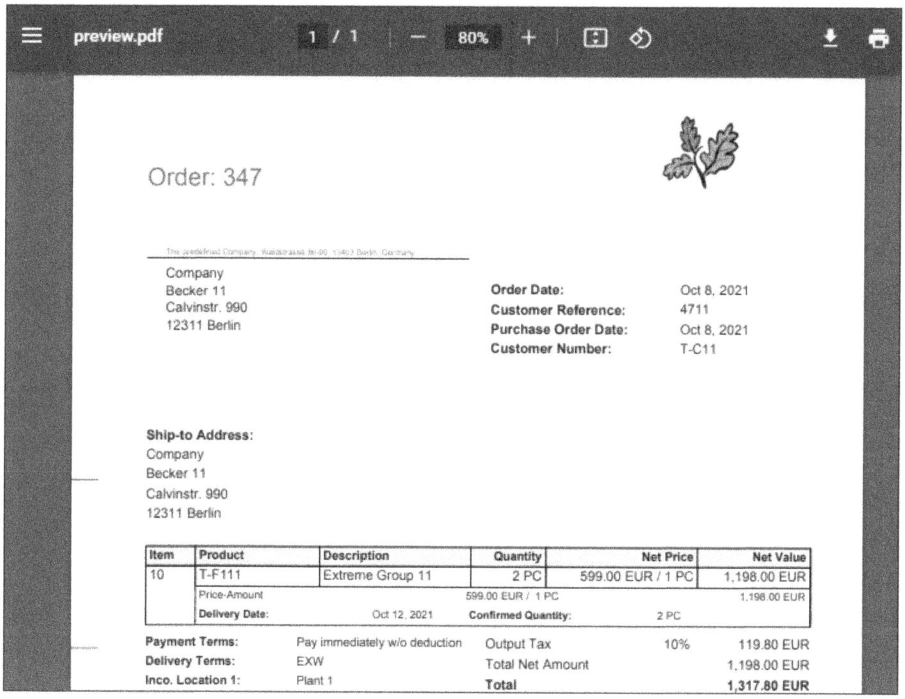

Figure 12.31 Viewing the Order Confirmation as a PDF

4. Close the PDF window, and click **SAP** to return to the SAP Fiori launchpad home-page.

12.8   Create an Outbound Delivery

So that the materials can be prepared for delivery in the warehouse, you now create an *outbound delivery* for your order. Note that Immediately after the creation of the outbound delivery, there is an outbound delivery document, but not yet a goods issue. An outbound delivery is only the "message" to the warehouse staff to make the material ready for shipment.

With the appropriate system setting, outbound delivery documents can also be created automatically. But we don't want to miss out on this fun so we'll create the outbound delivery manually. You almost don't need to make any entries (e.g., material numbers or quantities) because this data is copied from the sales order into the outbound delivery document, as shown in the process diagram in Figure 12.32.

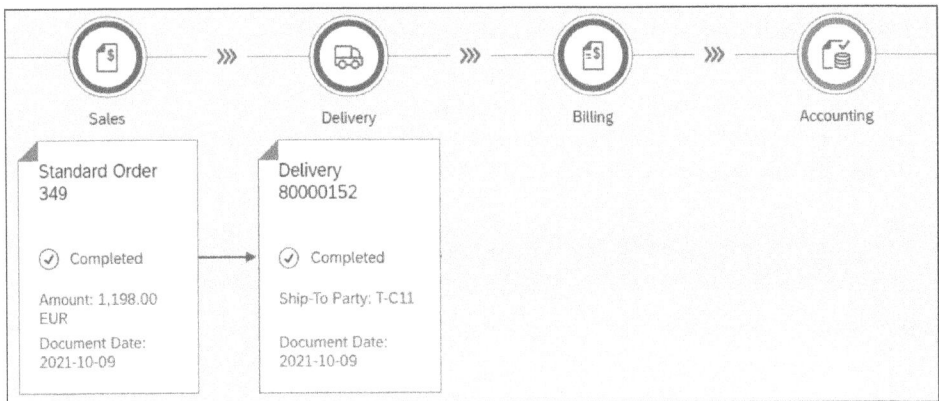

Figure 12.32 Process Diagram with Order and Delivery

To do this, you use the real Create Outbound Deliveries – From Sales Orders app in SAP Fiori. The app for creating a single outbound delivery corresponds to Transaction VL01N (Create Outbound Delivery with Order Reference) in SAP GUI.

Follow these steps:

1. Start the Create Outbound Deliveries – From Sales Orders app.
2. If necessary open the header by clicking on ☑.
3. In the **Ship-To Party** field, enter the customer number, and delete the date restriction in the **Planned Creation Date** field, as shown in Figure 12.33. In addition, you can enter the order number from the previous section in the **Sales Document** field. When you're finished, click the **Go** button.

> [»]
>
> **What Exactly Is a Shipping Point?**
>
> In practice, the **Shipping Point** filter is often used. A *shipping point* is responsible within a plant for shipping the goods to the customer. Shipping points can be, for example, a

loading dock for trucks, a freight station, or a mailroom. The shipping point in the SAP Live Access training system has the number 1010.

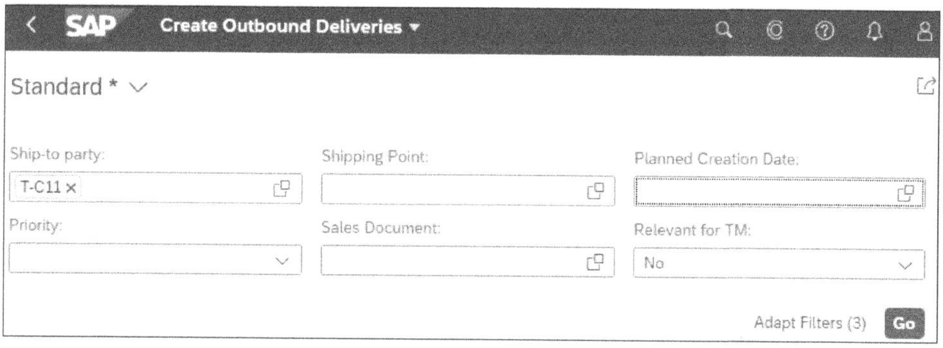

Figure 12.33 Header of the Create Outbound Deliveries App

4. Now you can see your order in a list, as shown in Figure 12.34. The list corresponds to the worklist for deliveries. Select your order, and then click the **Create Deliveries** button.

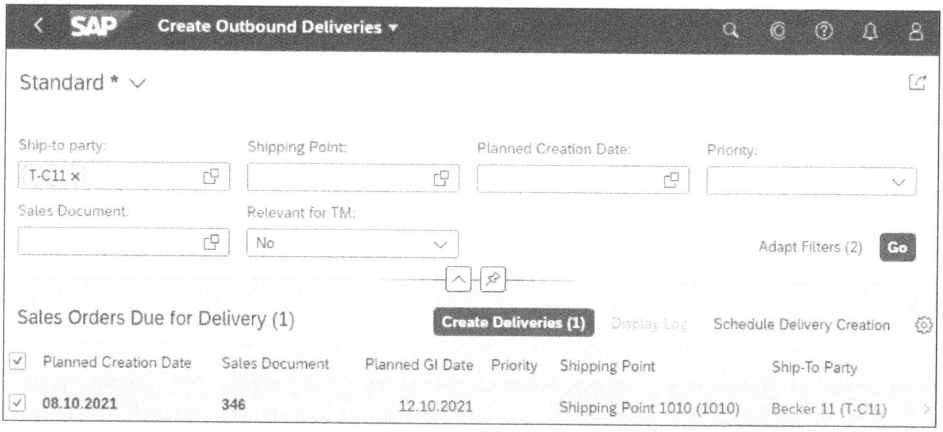

Figure 12.34 Order Shown in a List

Must Exactly One Delivery Always Be Created for Each Order?

There doesn't always have to be exactly one delivery for every order. You have all the options:

- Deliveries may also be partial deliveries that don't include the complete order quantity.
- Several individual orders can be combined into one delivery.

5. The order disappears from the list, and the delivery document is saved. To see the number of the delivery document, click the **Display Log** button.

In the delivery log, as shown in Figure 12.35, you'll see the number of the delivery document at the bottom left under the **Delivery** column. Make a note of this number because you'll need it in the next section.

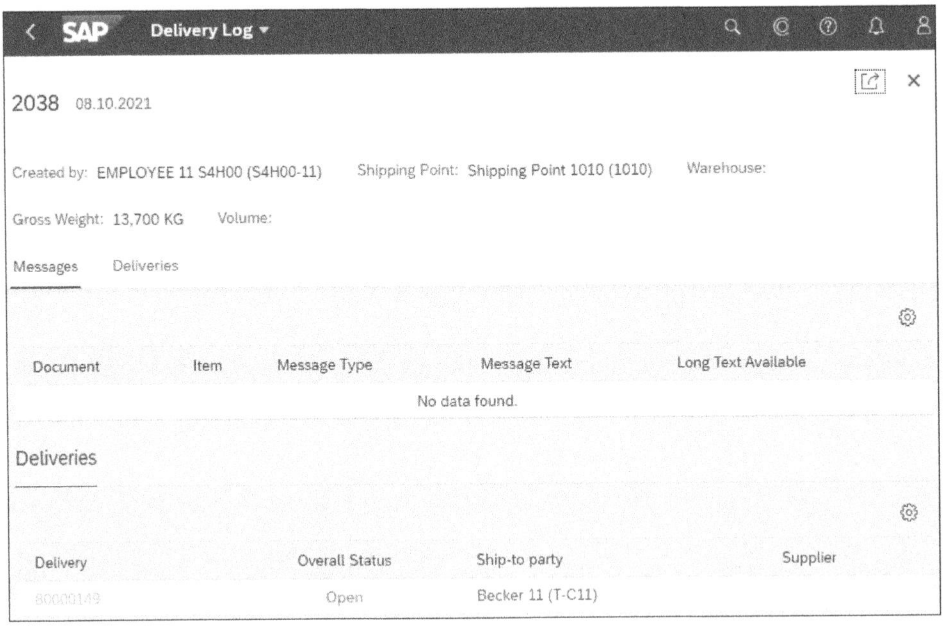

Figure 12.35 Delivery Log

6. Click ![SAP] to switch back to the SAP Fiori launchpad start page.

The outbound delivery document that you've just created is the basis for all further shipping activities. This includes picking, packing, transport, and goods issue.

In the next section, we'll discuss the mandatory activities of shipping: picking and goods issue.

12.9   Confirm Picking and Post Goods Issue

For the next job rotation, we leave the sales department and move to the warehouse because after creating the delivery document, the work begins there. The requested materials are taken from the warehouse shelf and prepared for delivery.

Picking takes place in the warehouse management system. This is used to plan and execute a transport of the material to the picking area. In this area, the materials are packed and prepared for shipping, for example. As soon as this process is completed, it's documented in the outbound delivery document. And this is exactly the process you now perform with the real Pick Outbound Delivery app. In SAP GUI, you confirm the picking in Transaction VL02 (Change Outbound Delivery).

After picking, the *goods issue* takes place, whereby the data for the goods issue is transferred completely from the outbound delivery document.

Follow these steps:

1. Launch the Pick Outbound Delivery app.

2. In the **Delivery** field, enter your delivery number from the previous section, as shown in Figure 12.36, and confirm it with the Enter key.

Figure 12.36 Entering the Delivery Number

How to Find the Delivery Number

Click on ⊡ (**Value Help**) in the **Delivery** field to open the value help. Then enter the customer number in the **Ship-To Party** field. After you click the **Go** button, you'll see the numbers of the corresponding deliveries.

3. In the lower part of the **Pick Outbound Delivery** window, in the **Picking Quantity** field (see Figure 12.37), enter the same quantity that is displayed in the **Delivery Quantity** field, and press Enter.

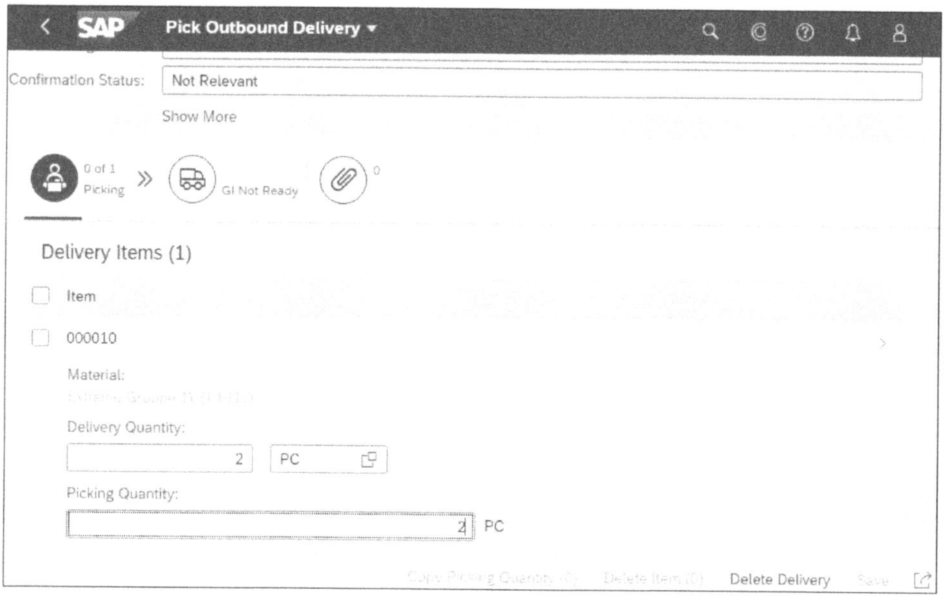

Figure 12.37 Entering the Pick Quantity

4. Click the **Save** button in the footer.

> **Is There a Separate Picking Document?**
>
> By saving the pick quantity, you're only adding to the outbound delivery document; you're not creating a new document.

5. This will leave you in the same window. The message **GI Ready** appears, as shown at the top in Figure 12.38.

   Click the **Post GI** button in the footer (GI stands for goods issue).

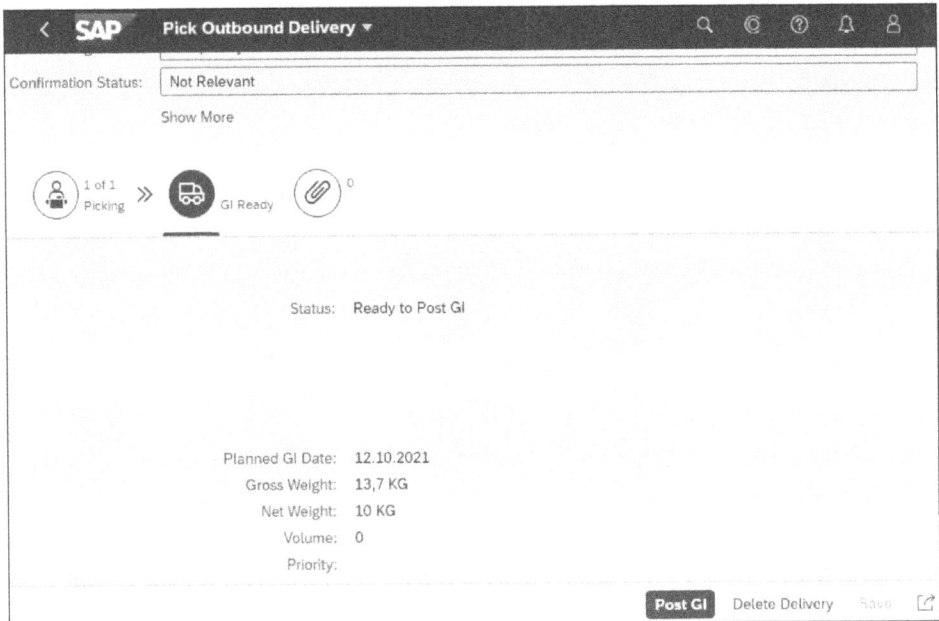

Figure 12.38 Goods Issue: Ready

6. The success message **GI Posted** appears.

Click **SAP** to switch back to the SAP Fiori launchpad start page.

By clicking on the **Post GI** button, you've created two new documents at once:

- **Material document**
  Documents the goods issue and stock reduction in logistics in the materials management module.
- **Accounting document**
  Posts an inventory reduction to the appropriate general ledger account in financial accounting so that the correct inventory value is later shown in the balance sheet.

This general ledger account is also called an *inventory account* and is located in material accounting.

When the goods issue is posted, the stock level is reduced, and, at the same time, the requirements are reduced as a result.

Now you're almost there: the last step in logistics is still invoicing, and then we meet again in finance.

12.10   Create the Outgoing Invoice

In the sales and distribution module, a billing document, that is, an outgoing invoice, is created with reference to the outbound delivery, as shown in Figure 12.39. You don't have to do any typing because the data from the outbound delivery is transferred to the invoice. Great! If there were several outbound deliveries, a billing document can of course also be created with reference to several outbound delivery documents.

Figure 12.39 Process Diagram before Billing with Order and Delivery

To create the accounts receivable invoice, we use the Create Billing Documents legacy app, which corresponds to Transaction VF04 (Maintain Billing Due List) in SAP GUI.

Follow these steps:

1. Start the Create Billing Documents app.
2. Enter the customer number in the **Sold-To Party** field on the screen shown in Figure 12.40. Further down, the **Delivery-Related** field should be selected in the **Documents to Be Selected** area.
3. Click on the **Display Billing List** button in the footer.

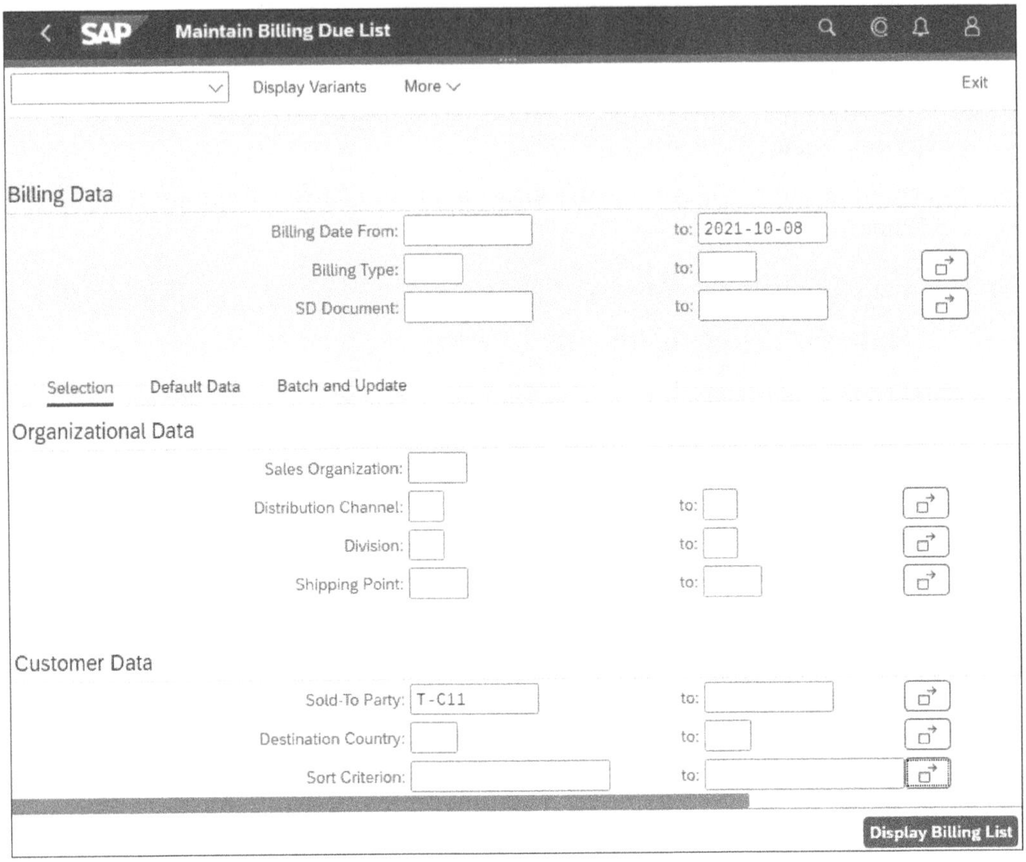

Figure 12.40 Create Invoices App Entries

4. Now you're in the *billing due list*, as shown in Figure 12.41. If necessary, select the line with your outbound delivery from the preceding section. You'll find the outbound delivery number in the **Document** column.

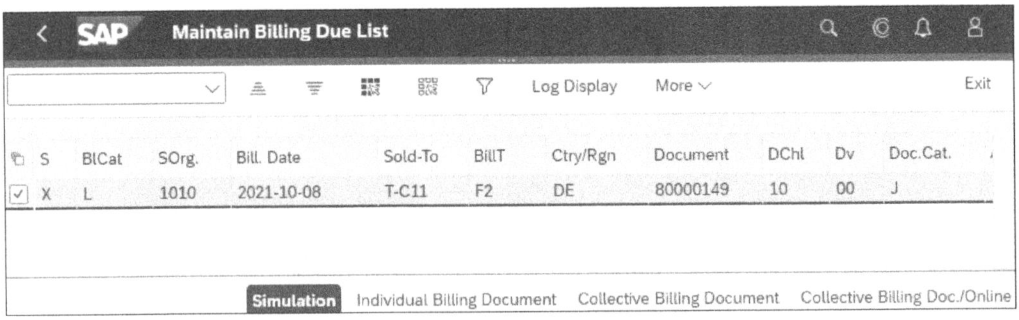

Figure 12.41 Billing Due List

What Options Does the Billing Due List Offer?

In the billing due list, you'll find several buttons in the footer:

- **Individual Billing Document**
  This button creates a single invoice for each delivery.
- **Collective Billing Document**
  This button combines several deliveries into one invoice if they include the same customer as the payer.
- **Simulation**
  This button gives you an overview of the possible billing documents. SAP S/4HANA tries to output the deliveries with the same customer on a collective invoice, just as with a collective billing document.

5. Click the **Individual Billing Document** button in the footer. The **Overview of Billing Items** window appears with information about material, quantity, and amount, as shown in Figure 12.42.

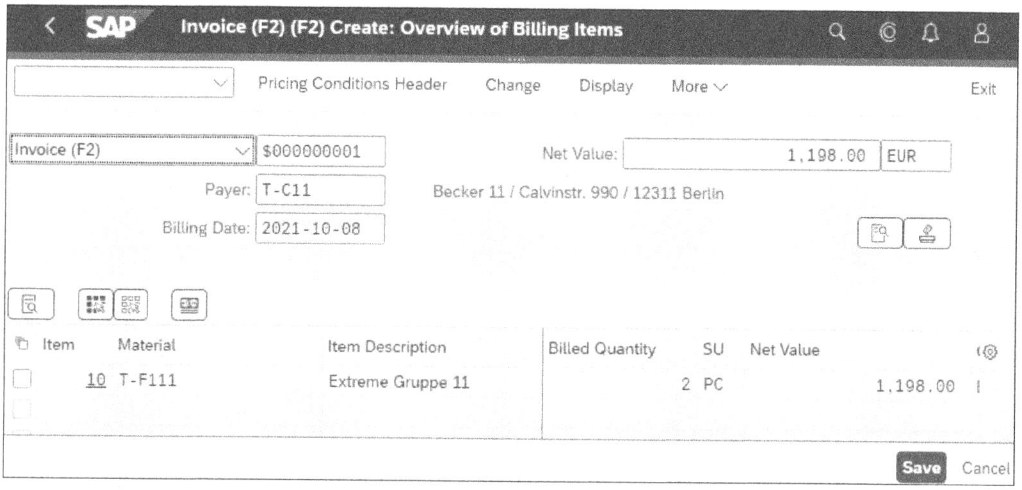

Figure 12.42 Overview of Billing Items

6. We promised you that you don't have to type anything. That's why you click on the **Save** button in the footer right now. After that, you'll find the confirmation that saving has been done in the bottom-left corner with the document number.

7. Click **SAP** to switch back to the SAP Fiori launchpad start page.

By saving the billing document, you've created two documents:

- An invoice document in the sales module.
- An accounting document with an open item, that is, an unpaid invoice item, which financial accounting can use to create a reminder and post the payment. This

accounting document simultaneously documents the sales revenue for the profit and loss (P&L) statement.

You can also see this in the process diagram shown in Figure 12.43.

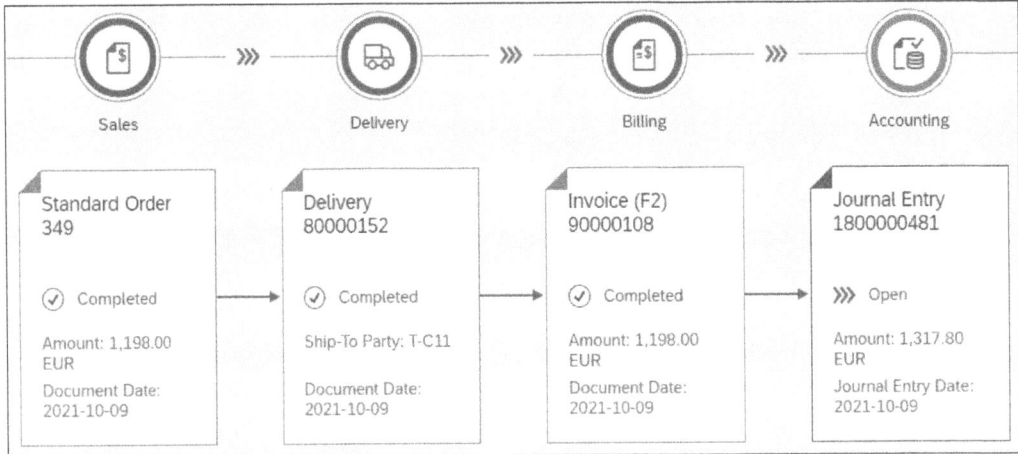

Sales	Delivery	Billing	Accounting
Standard Order 349	Delivery 80000152	Invoice (F2) 90000108	Journal Entry 1800000481
✓ Completed	✓ Completed	✓ Completed	≫ Open
Amount: 1,198.00 EUR	Ship-To Party: T-C11	Amount: 1,198.00 EUR	Amount: 1,317.80 EUR
Document Date: 2021-10-09	Document Date: 2021-10-09	Document Date: 2021-10-09	Journal Entry Date: 2021-10-09

Figure 12.43 Billing Document with the Invoice Document from Sales and the Accounting Document for Financial Accounting

> **How do I Invoice a Service?**
>
> This is a legitimate question! In the case of services, there are no deliveries. Therefore, here invoicing is done with reference to the corresponding sales order.

12.11   Display the Invoice and Output It as a PDF

You want to send the invoice as a PDF via email to the customer? No problem. This PDF is created automatically in the SAP Live Access training system based on a corresponding preset.

We'll show you how to display and download this invoice in the following instructions. With the real Manage Billing Documents app, you can see captured invoices first in the list and also in detail after you click the corresponding invoice number. And that's exactly where the PDF is waiting to be displayed and downloaded by you. In SAP GUI, to display a single invoice, use Transaction VF03 (Display Invoice).

Follow these steps:

1. Launch the Manage Billing Documents app.
2. Enter the customer number in the **Sold-To Party** field, and click on the **Go** button to arrive at the screen shown in Figure 12.44.

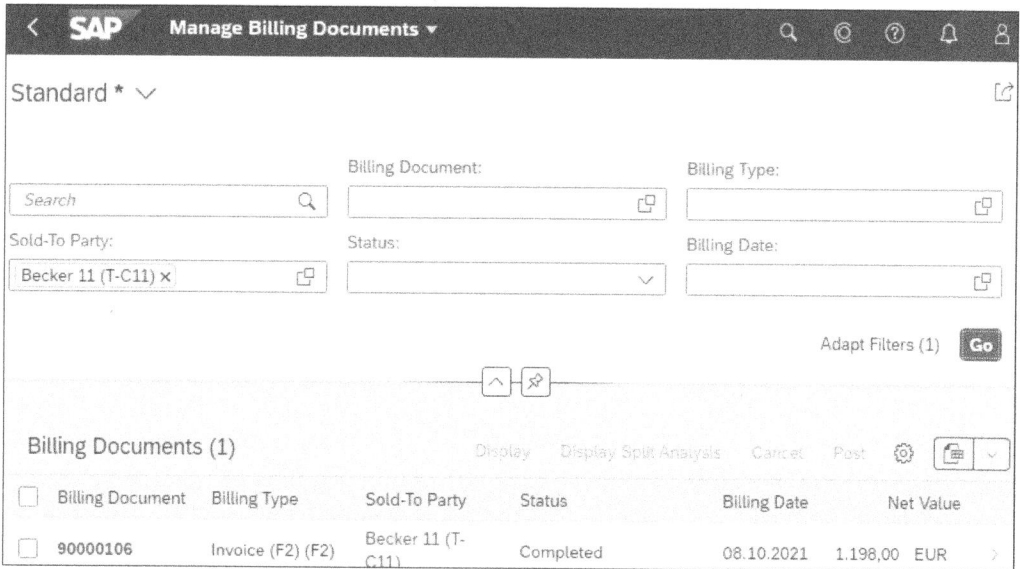

Figure 12.44 Display Billing Documents

3. By clicking on the number of the invoice on the far left, which, in our case, starts with "90", SAP S/4HANA displays some links for further actions.

4. Click the **Display Billing Documents** link. This is equivalent to switching to the legacy app of Transaction VF03 (Display Billing Document), as shown in Figure 12.45.

5. Click ⬇ (**Print**) on the far right.

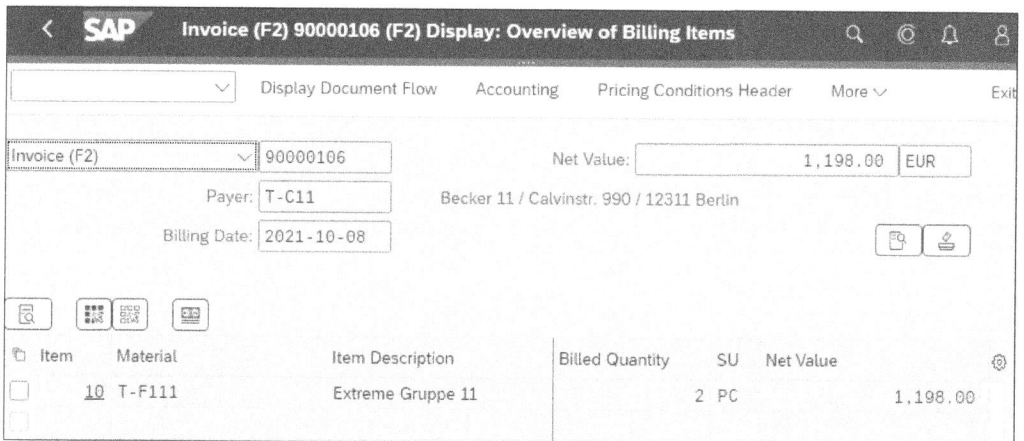

Figure 12.45 Display Billing Document

6. In the output window (see Figure 12.46), click the **Display Document** button in the header line to open the output invoice display.

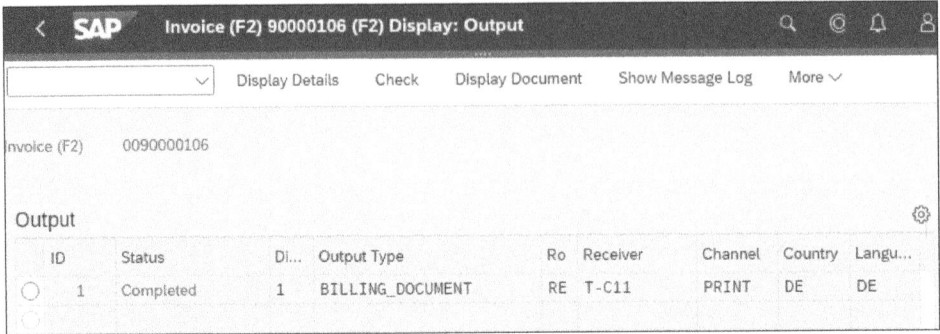

Figure 12.46 Output Window

7. You can see the PDF open, as shown in Figure 12.47. Downloading can also be done from the PDF display.

8. Close the PDF window, and click **SAP** to return to the SAP Fiori launchpad home-page.

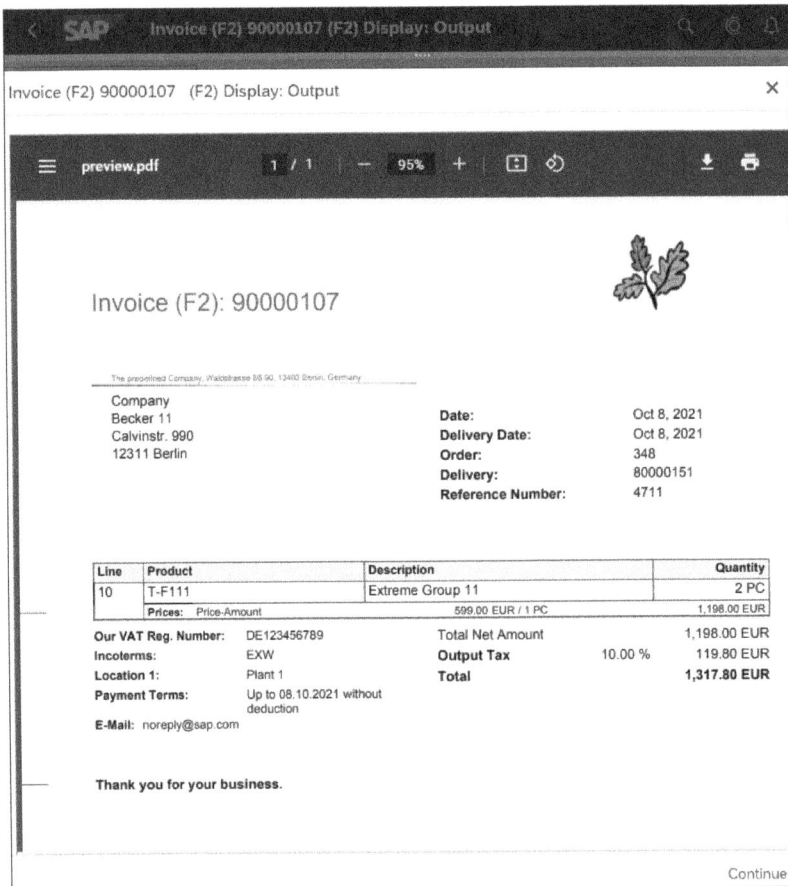

Figure 12.47 Viewing the Invoice as a PDF

12.12   Summary

Done! This was the last step in logistics. We've discussed the order-to-cash process and walked through key daily activities, including working with customer data, sales orders, outbound deliveries, goods issues, and invoices.

In the next chapter, we'll continue with financial accounting.

12

Chapter 13

Financial Accounting: Posting and Evaluating Business Transactions

In 1972—the year SAP was born—former employees of IBM founded the company, not knowing that this foundation would one day become the world's leading provider of standard software for companies. And back then, it all started with accounting software.

Accounting with T-accounts and posting records with debits and credits? You rarely need that in the SAP system and not at all in this chapter because, in most cases, the system automatically knows which amount belongs on the debit side and which on the credit side. In other words, you'll benefit from this chapter even without in-depth accounting knowledge. You probably won't understand every detail, but you can click along with the instructions because we've documented every click and entry for you, as we did in the previous chapters.

After getting to know the basic organizational units and the general ledger, you'll post an outgoing invoice and the subsequent incoming payment in accounts receivable. Afterwards, you'll take a look at accounts payable.

What You'll Learn

- Which organizational units are important in finance
- How to display a balance sheet with profit and loss (P&L)
- How to display general ledger accounts in the chart of accounts
- How to display master data of customers and vendors
- How to enter an outgoing invoice
- How to display receivables
- How to post an incoming payment
- How to enter an incoming invoice

13.1 Organizational Units

You already know the most important organizational unit of financial accounting from the introduction: the *company code*. In addition, in this section we'll introduce you to organizational units from controlling (see Figure 13.1): *controlling area, operating concern*, and *profit center*.

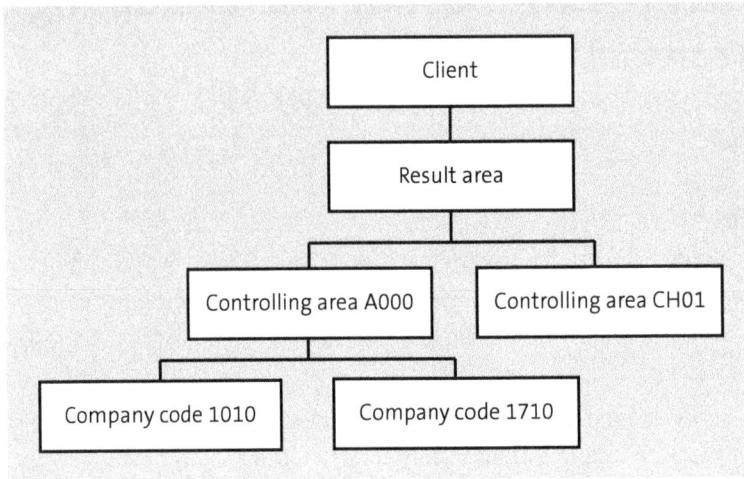

Figure 13.1 Organizational Units of Accounting in the SAP Live Access Training System

13.1.1 Company Code

The company code plays an important role in financial accounting: A company code corresponds to closed accounting in which the *balance sheet* and *income statement* (P&L) are created. In other words, each company that submits a balance sheet to the tax office must be managed as a separate company code in SAP S/4HANA. Therefore, subsidiaries within a group usually have their own company codes.

13.1.2 Controlling Area

Often there are companies with several company codes. In controlling, evaluations are carried out that include several company codes. For this purpose, there are *controlling areas* that can contain several company codes. However, these company codes must have the same chart of accounts and the same fiscal year.

Within a controlling area, costs are collected on *cost centers*. However, cost centers aren't organizational units, but master data.

13.1.3 Operating Concern

If there are several controlling areas in a company, they can be combined in one *operating concern*. This is the highest level for evaluations in controlling.

13.1.4   Profit Center

Profit centers are organizational units that can be used to calculate partial results below a company code, for example, for individual stores. For these financial statements and analyses below the company code, both the cost-of-sales accounting and the period accounting methods can be used.

13.2   Components of the Financial System

In this section, we'll introduce you to the main components of the financial accounting module.

In the *general ledger* component, the financial statements for the company codes are created. These financial statements are the balance sheet and the P&L statement (income statement), as shown in Figure 13.2.

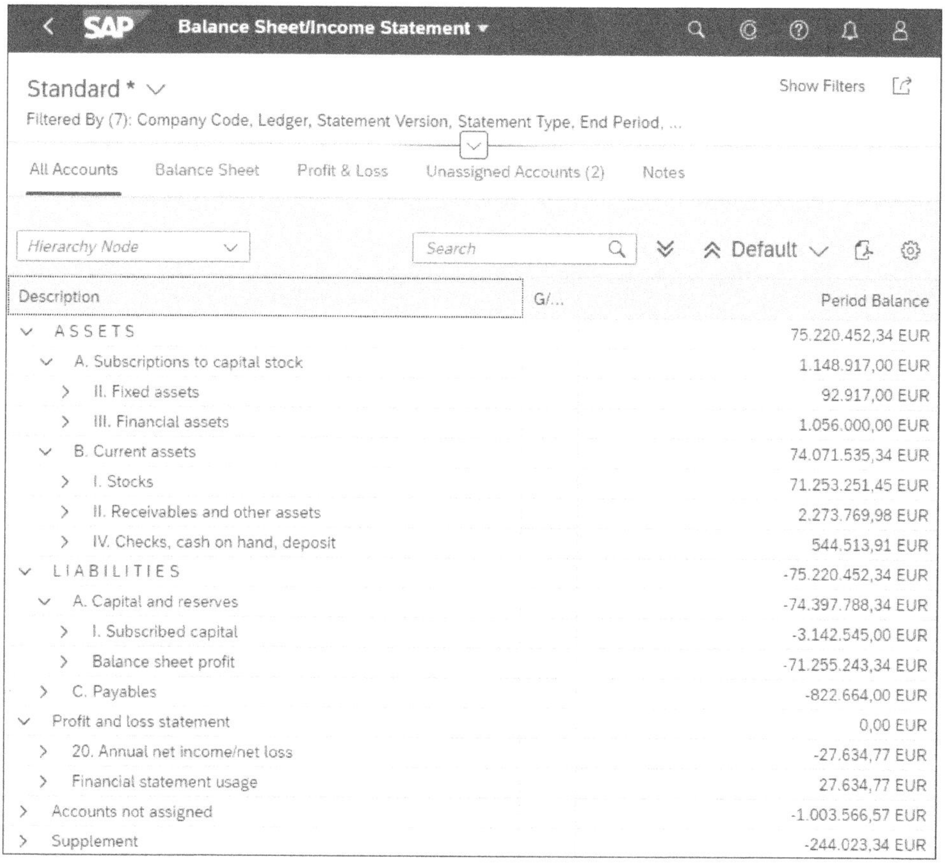

Figure 13.2 Extracts of a Balance Sheet and P&L Statement in the Balance Sheet/Income Statement App

In the balance sheet and income statement, you can see the balances of the *general ledger accounts*. For example, one general ledger account is the account for office supplies; another general ledger account is an account for sales revenue.

Financial accounting has the following *subledgers*, and each subledger is a separate component:

- **Accounts receivable**
  The accounts receivable component receives the outgoing invoices (billing documents, customer invoices) from the sales and distribution module. These are cleared in accounts receivable when payment is received or reminders are sent in case of delay. Outgoing invoices can also be posted directly in financial accounting without the sales and distribution module.

- **Accounts payable**
  According to the same principle, the materials management module delivers the incoming invoices (vendor invoices) to accounts payable. During invoice verification, the incoming invoices are created, which are cleared in accounts payable as open items by an outgoing payment. Incoming invoices can be entered directly in financial accounting even without the materials management module.

- **Asset accounting**
  The asset accounting component records assets and depreciates their value. Assets are, for example, machines, company vehicles, or buildings. Here, SAP S/4HANA calculates depreciation and transfers it as an expense to general ledger accounting.

Table 13.1 lists the main components of the financial accounting module.

Component	Abbreviation	Function Examples
General ledger accounting	FI-GL	General ledger accounts, chart of accounts, balance sheet and income statement
Accounts receivable	FI-AR	Outgoing invoices, reminders, incoming payments
Accounts payable	FI-AP	Incoming invoices, outgoing payments
Asset accounting	FI-AA	Asset management, depreciation

Table 13.1 Components of the Financial Accounting Module

13.3   Default Settings for Financial Accounting and Controlling

For the financial accounting and controlling modules, too, you first make some default settings that reduce your typing effort. (We've already presented the method for setting default values for "real" SAP Fiori apps in Chapter 10, Section 10.3.)

Follow these steps:

1. Click ⌂ (**User Menu**) to open the user menu in SAP Fiori, and then click the ⚙ (**Settings**) menu item. The **Settings** window opens.

2. Click **Default Values** in the lower-left corner, as shown in Figure 13.3.

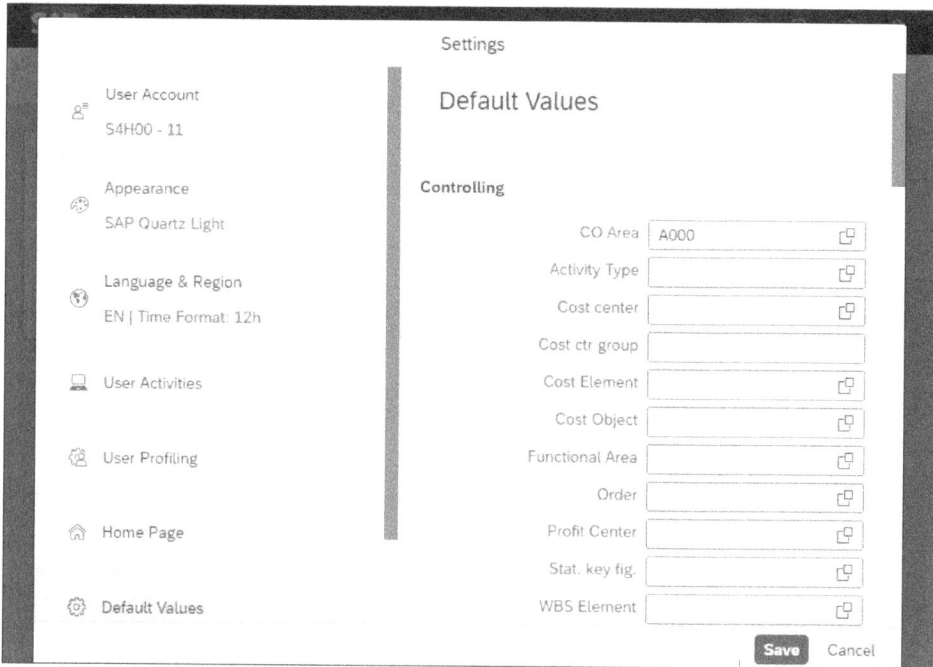

Figure 13.3 Opening Default Values in Settings

3. You'll now see the fields for which you can set default values. If you use the SAP Live Access training system, use the following values:

   – **CO Area**: "A000" (controlling area)
   – **Company Code**: "1010"
   – **Calculate Tax**: "Yes"
   – **Chart of Accts**: "YCOA" (chart of accounts)
   – **Display Currency**: "EUR"

   With ⌗ (**Value Help**), you can search for values using the value help.

4. Save your presets by clicking the **Save** button.

As in logistics, the same is true in finance: legacy apps ignore the default values defined in SAP Fiori. Instead, you can define your default values using SAP GUI Transaction SU3 (Maintain User Own Data).

For this chapter, also refer to our notes on training data in Chapter 11, Section 11.3.

[+]

There Is Still Much to Discover

For practical purposes, note that, in this book, we show you only the tip of the iceberg of possible presets. Many more settings can be made either within transactions or apps, or using module-specific transactions.

As a finance employee, you should definitely take a look at the powerful Transaction FB00 (Accounting Editing Options) if you use SAP GUI or legacy apps.

13.4   General Ledger

In the *general ledger* component, in this section, you'll first display a balance sheet and then the general ledger accounts in the chart of accounts. In addition, you'll learn how the subledgers for accounts receivable and accounts payable are linked to the general ledger.

13.4.1   Show Balance Sheet and Profit and Loss

In the balance sheet, use the following instructions to display the totals of receivables and payables. In each case, these are the totals from the outgoing or incoming invoices that haven't yet been paid. In addition, you'll find information about the revenue accounts in the income statement.

Follow these steps:

1. Launch the real Balance Sheet/Income Statement app.
2. If you're using the SAP Live Access training system, enter the values from Figure 13.4, as far as they aren't already preassigned. In other systems, use 🗗 (**Value Help**). Use the current year and month in the **End Period** field.

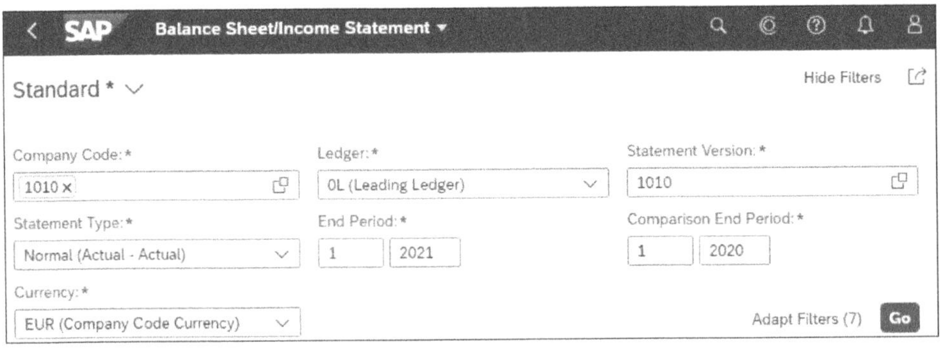

Figure 13.4 Balance Sheet/Income Statement App Entries

3. Click the **Go** button, and you'll see the main folders **ASSETS** and **LIABILITIES**, which belong to the balance sheet.
4. Click on [^] (**Collapse Header**) to close the header with the filters.
5. Open the following folders in the tree structure (see Figure 13.5): **Assets • B. Current Assets • II. Receivables and Other Assets • 1. Accounts Receivable • Residual Maturity Less 1 Year**.

   The row with the trade receivables domestic reconciliation account shows the general ledger balance for the receivables, which is the sum of the open items from accounts receivable.

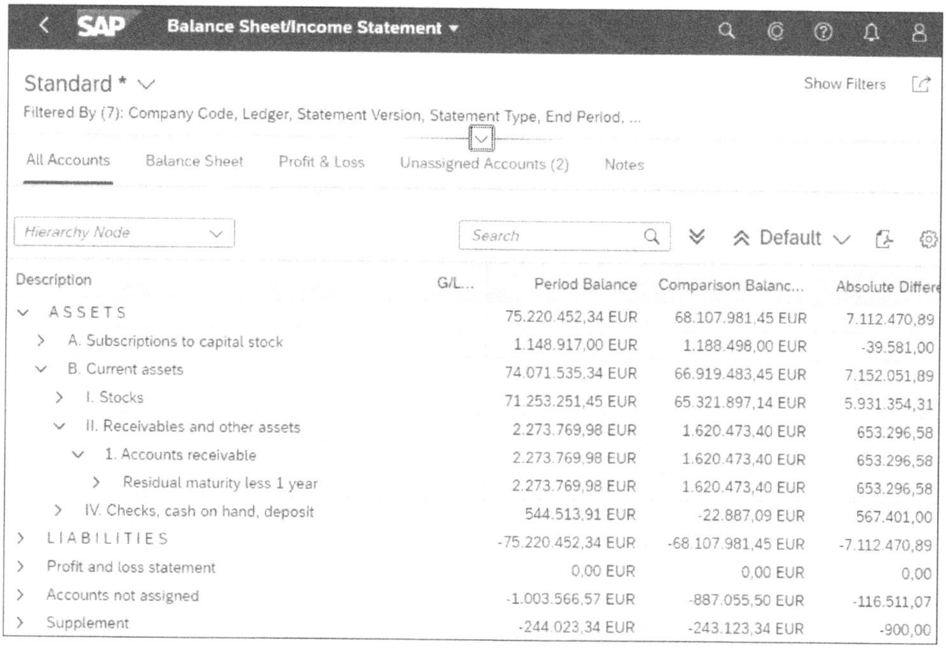

Balance Sheet/Income Statement
Standard *
Show Filters
Filtered By (7): Company Code, Ledger, Statement Version, Statement Type, End Period, ...
All Accounts
Balance Sheet
Profit & Loss
Unassigned Accounts (2)
Notes
Hierarchy Node
Search
Default
Description | G/L... | Period Balance | Comparison Balanc... | Absolute Differe
ASSETS | 75.220.452,34 EUR | 68.107.981,45 EUR | 7.112.470,89
A. Subscriptions to capital stock | 1.148.917,00 EUR | 1.188.498,00 EUR | -39.581,00
B. Current assets | 74.071.535,34 EUR | 66.919.483,45 EUR | 7.152.051,89
I. Stocks | 71.253.251,45 EUR | 65.321.897,14 EUR | 5.931.354,31
II. Receivables and other assets | 2.273.769,98 EUR | 1.620.473,40 EUR | 653.296,58
1. Accounts receivable | 2.273.769,98 EUR | 1.620.473,40 EUR | 653.296,58
Residual maturity less 1 year | 2.273.769,98 EUR | 1.620.473,40 EUR | 653.296,58
IV. Checks, cash on hand, deposit | 544.513,91 EUR | -22.887,09 EUR | 567.401,00
LIABILITIES | -75.220.452,34 EUR | -68.107.981,45 EUR | -7.112.470,89
Profit and loss statement | 0,00 EUR | 0,00 EUR | 0,00
Accounts not assigned | -1.003.566,57 EUR | -887.055,50 EUR | -116.511,07
Supplement | -244.023,34 EUR | -243.123,34 EUR | -900,00

Figure 13.5 Navigate to Receivables

6. Now you want to see the **Payables**. To do this, you could click through the tree structure again via the **Liabilities** folder. But it's faster to enter "Paya" in the **Hierarchy Node** field above the table, select **C. Payables** in the list, and confirm with [Enter].
7. Immediately, SAP S/4HANA displays the two reconciliation accounts—**Trade Payables Domestic** and **Trade Payables Foreign**—with their balances, as shown in Figure 13.6.

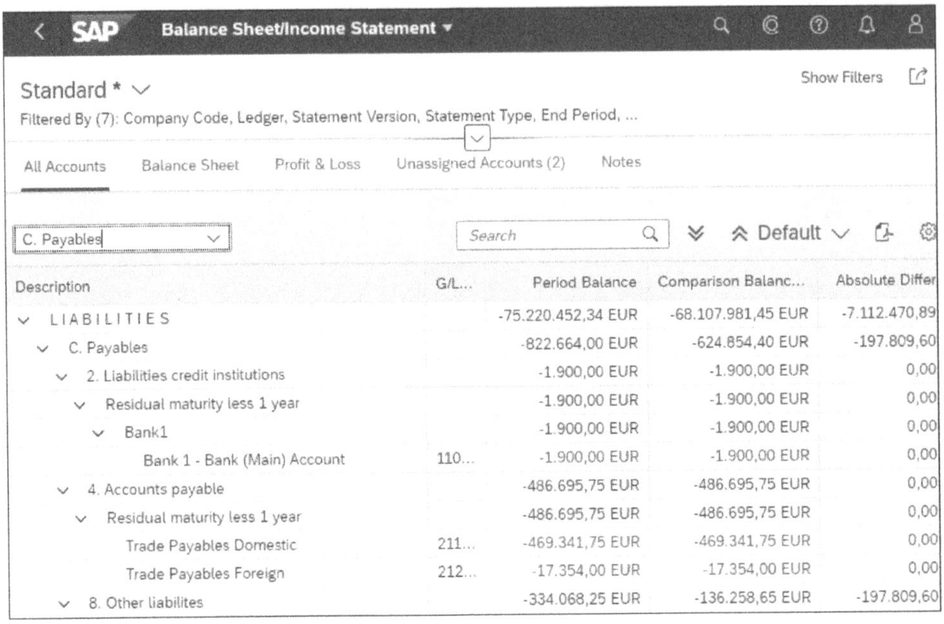

Description	G/L...	Period Balance	Comparison Balanc...	Absolute Differ
∨ LIABILITIES		-75.220.452,34 EUR	-68.107.981,45 EUR	-7.112.470,89
∨ C. Payables		-822.664,00 EUR	-624.854,40 EUR	-197.809,60
∨ 2. Liabilities credit institutions		-1.900,00 EUR	-1.900,00 EUR	0,00
∨ Residual maturity less 1 year		-1.900,00 EUR	-1.900,00 EUR	0,00
∨ Bank1		-1.900,00 EUR	-1.900,00 EUR	0,00
Bank 1 - Bank (Main) Account	110...	-1.900,00 EUR	-1.900,00 EUR	0,00
∨ 4. Accounts payable		-486.695,75 EUR	-486.695,75 EUR	0,00
∨ Residual maturity less 1 year		-486.695,75 EUR	-486.695,75 EUR	0,00
Trade Payables Domestic	211...	-469.341,75 EUR	-469.341,75 EUR	0,00
Trade Payables Foreign	212...	-17.354,00 EUR	-17.354,00 EUR	0,00
∨ 8. Other liabilites		-334.068,25 EUR	-136.258,65 EUR	-197.809,60

Figure 13.6 Navigating to Payables

8. The *revenue account* is hidden in the P&L, and you'll post to it later when you enter an outgoing invoice. To see the revenue accounts, type "Reve" in the **Hierarchy Node** field above the table, select **1. Revenues** in the list, and confirm with [Enter].

9. Now you can see the general ledger balances of the revenue accounts.

10. Click SAP from the shell bar to go back to the start page.

[+]

More Options in the Balance Sheet/Income Statement App

Clicking on the balance of an account opens a window called **Related Apps**, from which you can branch to the general ledger line items. The 🗎 (**Create PDF Preview**) icon allows you to create a PDF file.

13.4.2 Reconciliation Accounts

In the subledgers of accounts receivable and accounts payable, you post incoming and outgoing invoices, credit memos, or incoming and outgoing payments. The amounts from these postings must also be taken into account in the general ledger in the balance sheet and income statement. But how do these amounts get from the subledgers to the general ledger and thus to the balance sheet and P&L?

The amounts are also posted automatically to reconciliation accounts in general ledger accounting. This applies to basically all postings in accounts receivable, accounts payable, and asset accounting. How this works is explained in the box using the example of an outgoing invoice from accounts receivable.

[Ex]

Posting an Outgoing Invoice

An outgoing invoice with a value of EUR 119, including EUR 19 value-added tax (VAT), is entered in accounts receivable.

The amount of EUR 119 remains in the accounts receivable accounting as a receivable in the corresponding customer account. This is the open item that will be cleared later by paying the invoice.

In addition, the net amount of 100 EUR is automatically transferred to the reconciliation account in the general ledger as a receivable, so that this amount is included in the balance sheet as a receivable. In addition, 100 EUR appears on a revenue account in the general ledger to represent this amount in the P&L statement, and 19 EUR ends up in the account for output tax.

A *reconciliation account* is a general ledger account that collects either receivables from customers or payables to vendors.

In the standard SAP S/4HANA system, for example, there is a reconciliation account called domestic receivables in which all posting amounts for domestic customers are collected. For example, if there are 3,000 domestic customers in our company, every posting for these customers is also posted to the one reconciliation account called domestic accounts receivable in the general ledger. There you can see at any time the current status of the sum of receivables for all domestic customers. At the same time, you can see this amount in the balance sheet.

For example, if the reconciliation account had a balance of EUR 20,000 before posting an outgoing invoice of EUR 1,000, it will have a balance of EUR 21,000 after posting this outgoing invoice.

[«]

Reconciliation Accounts for Subledgers

Just like accounts receivable, accounts payable and asset accounting are also linked to the general ledger via reconciliation accounts. In the SAP Live Access training system, for example, the following reconciliation accounts are in the chart of accounts called YCOA:

- **12100000 – Rcvbls Domestic**: For domestic receivables from accounts receivable accounting.
- **21100000 – Paybls Domestic**: For domestic payables from accounts payable.
- **21200000 – Paybls Foreign**: For foreign payables from accounts payable.

When you enter an invoice, you don't need to specify a reconciliation account. The amount is automatically posted to the correct reconciliation account in the general ledger.

Now there is only one mystery left to solve: How does the system "know" which recon-ciliation account to post to? The solution to this riddle can be found later in Section 13.5, because an entry in the business partner master record is responsible for this.

13.4.3   Display General Ledger Accounts in Chart of Accounts

In the balance sheet and in the income statement, you've seen general ledger accounts such as the trade receivables domestic account. General ledger accounts are structured using charts of accounts. A *chart of accounts* contains all general ledger accounts used by one or more company codes. It's therefore also called a general ledger chart of accounts.

[»]

Does Each Company Code Have Its Own Chart of Accounts?

This doesn't have to be the case. A chart of accounts can also be assigned to several company codes. If there are several subsidiaries in a client, they can basically use the same chart of accounts.

However, each company code is always assigned exactly one chart of accounts. Con-versely, a chart of accounts can be used in several company codes.

The chart of accounts assigned to a company code is also called an *operational chart of accounts*. In the SAP Live Access training system, the operational chart of accounts YCOA is used for company code 1010.

Follow these steps:

1. To display general ledger accounts in a chart of accounts, launch the real Manage G/L Account Master Data app.
2. In the **Chart of Accounts** field at the top, enter "YCOA" if this field isn't already pre-populated by this default value (see Figure 13.7). In the **Short Text** field enter "Rcvbls", and for the **View** field, select the **Chart of Accounts View** entry from the dropdown menu.

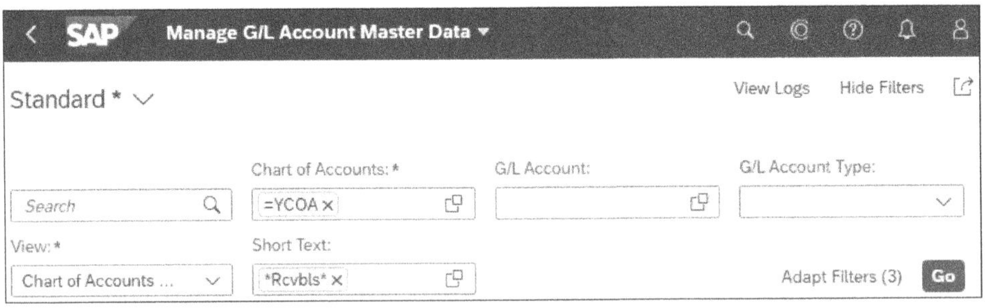

Figure 13.7 Chart of Accounts Entries

[«]

> **What Other Views Are Available for General Ledger Accounts?**
>
> With the *company code view*, you display only the accounts that are used in a given company code. A chart of accounts is a list of all general ledger accounts that can be shared by several company codes, and not every company code has to use all accounts.
>
> In the *controlling area view*, you only see the accounts for a given controlling area that are relevant for controlling. This means that general ledger accounts for neutral expenses or revenues aren't displayed, as these are only posted to in financial accounting.

3. Click the **Go** button to see all the selected accounts in the form of a list on the screen, as shown in Figure 13.8.

   Here you'll also find the domestic receivable account with the number **12100000**, which you've seen in Section 13.4.1.

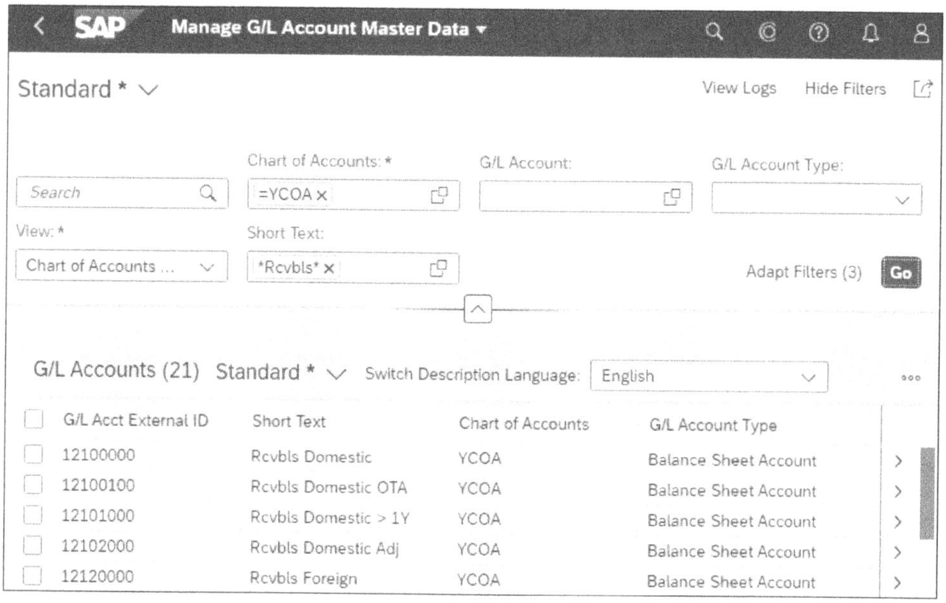

< SAP	Manage G/L Account Master Data ▾		Q © ⑦ ♫ ៱

Standard * ⌄   View Logs   Hide Filters   [↗]

	Chart of Accounts: *	G/L Account:	G/L Account Type:
Search 🔍	=YCOA ✕ 🗗	🗗	⌄
View: *	Short Text:		
Chart of Accounts ... ⌄	*Rcvbls* ✕ 🗗		Adapt Filters (3) **Go**

G/L Accounts (21)   Standard * ⌄   Switch Description Language: | English ⌄ | ₀₀₀

	G/L Acct External ID	Short Text	Chart of Accounts	G/L Account Type	
☐	12100000	Rcvbls Domestic	YCOA	Balance Sheet Account	>
☐	12100100	Rcvbls Domestic OTA	YCOA	Balance Sheet Account	>
☐	12101000	Rcvbls Domestic > 1Y	YCOA	Balance Sheet Account	>
☐	12102000	Rcvbls Domestic Adj	YCOA	Balance Sheet Account	>
☐	12120000	Rcvbls Foreign	YCOA	Balance Sheet Account	>

Figure 13.8 List of Accounts

4. Do you want to see more account data? Click $>$ on the far right in the row with the account in question. In our example, we use the account **Rcvbls Domestic** with the number **12100000**.

   Now you've reached the "sacred data" of the general ledger accounts, which may only be changed by financial specialists (see Figure 13.9). In **G/L Account Type**, you can see that this account is a balance sheet account and therefore appears in the balance sheet. By clicking on the **Where Used** tab, experts can see in which balance sheet or P&L structures the account is used.

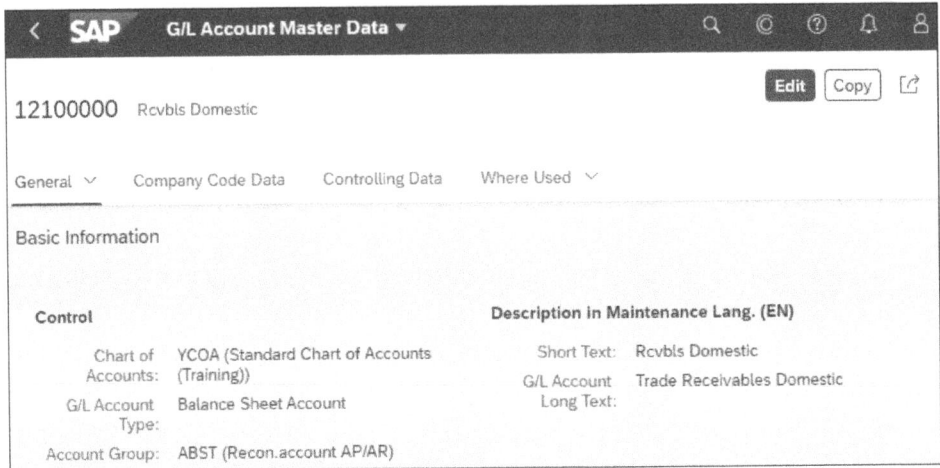

Figure 13.9 General Ledger Account Data

5. Click **SAP** to go back to the SAP Fiori launchpad homepage.

That was an excursion into the deepest jungle of accounting. We promise readers with no previous accounting experience that the following sections can be mastered even without in-depth commercial knowledge.

13.5   Business Partner in Finance

In this chapter, we assume that you're basically familiar with the structure of the business partner master record from Chapter 12, Section 12.4, and branch directly to the data that is important for accounts receivable and accounts payable. Customer and vendor accounts are thus part of accounting and are used when posting invoices or payments.

> **Further Use of Customer Accounts**
>
> If the financial accounting components cash management or credit management are used, they also use the customer accounts.
>
> Cash management enables a liquidity forecast that takes into account expected cash inflows and outflows. Credit management is designed to prevent bad debts. For example, the credit limit is checked when a transaction document is created.

In the following steps, we'll now display the accounts receivable master data of financial accounting. In doing so, we'll also clear up the mystery of how SAP S/4HANA automatically posts invoices from subledgers to the correct reconciliation account. We'll use the Maintain Business Partner legacy app, which corresponds to Transaction BP (Maintain Business Partner) in SAP GUI.

Follow these steps:

1. Start the Maintain Business Partner app.

2. When you see the **Worklist** and **Find** tabs on the left, as shown in Figure 13.10, click the **Locator On/Off** button at the top left. From then on, you'll see the same display as in the book.

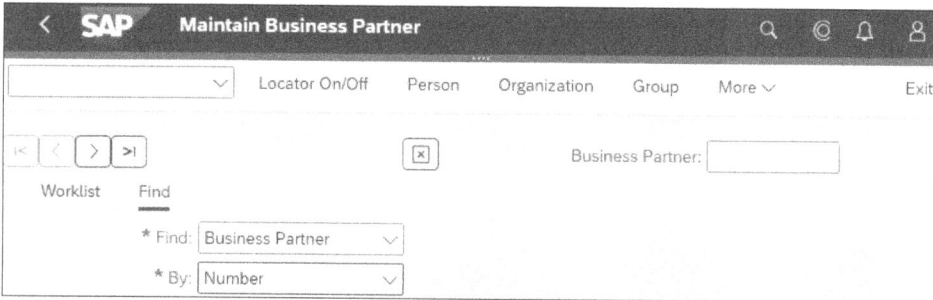

Figure 13.10 Worklist and Search Tabs

3. Enter a customer number in the **Business Partner** field, as shown in Figure 13.11, or use ⌗ (**Value Help**) to find a number.

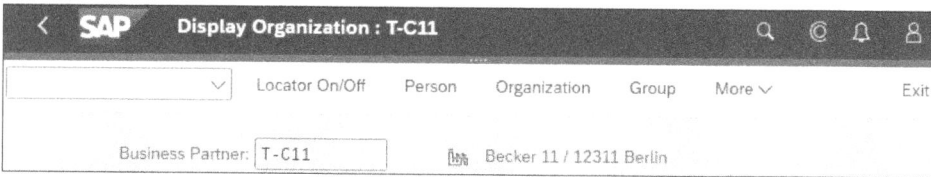

Figure 13.11 Entering a Business Partner Number

4. Confirm the number with the ⎡Enter⎤ key. You're now in the screen with the detailed data, as shown in Figure 13.12.

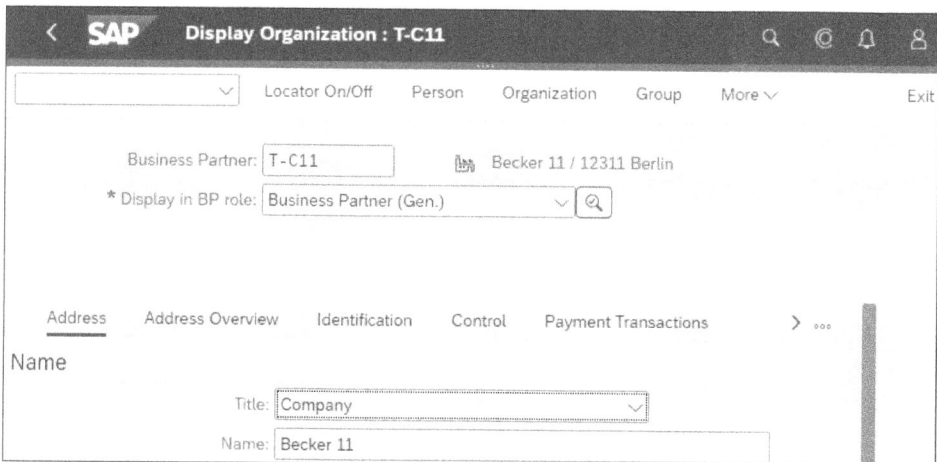

Figure 13.12 General Data

5. First, you'll see the general data here, such as the address. To see the financial data instead, click ☑ to the right of the **Display in BP Role** field, and select the **FI Customer** entry from the list.

6. In the header, click the **Company Code** button. (If this button isn't displayed, as in our illustration, the window is too narrow. In this case, choose **More • Company Code**.)

   In both cases, you'll now see several company code–specific tabs such as **Customer: Account Management** or **Customer: Payment Transactions**.

   This brings us to our goal: On the **Customer: Account Management** tab, as shown in Figure 13.13, the account **12100000** with the designation **Trade Receivables Domestic** has been entered for this customer in the **Reconciliation Acct** field. This account from the general ledger will be posted to automatically as soon as you save an invoice or an incoming payment in accounts receivable, so that this amount flows into the balance sheet. Your advantage is that you never need to enter this account number manually when entering invoices or incoming payments. This is a rule without exceptions.

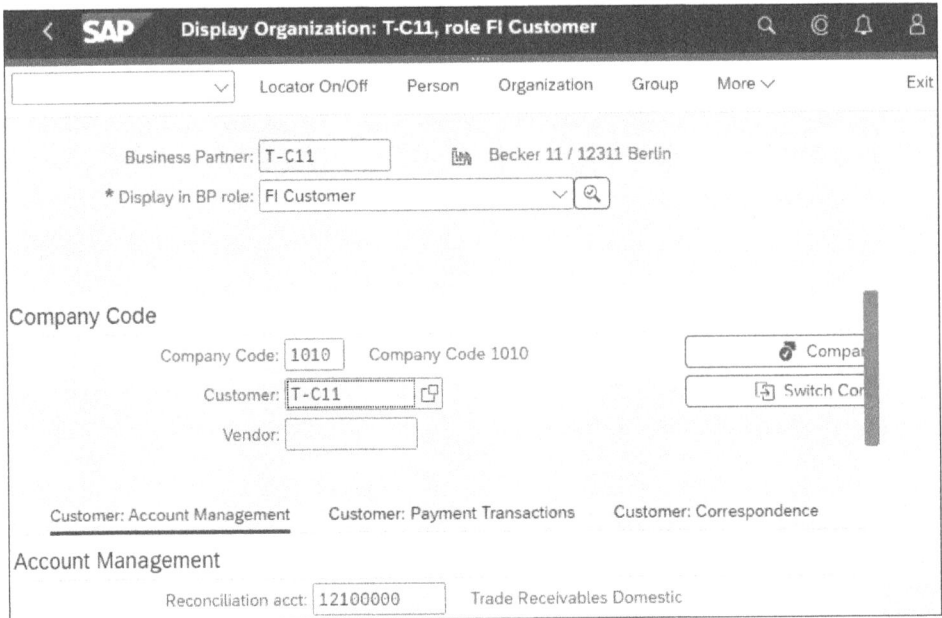

Figure 13.13 Customer: Account Management Tab

[»]

Can a Customer Have Different Data in Different Company Codes?

Yes, that is possible. For example, different payment terms apply to the customers of your group's subsidiaries from other countries. The data of the **FI Customer** role can be maintained separately for each subsidiary. With a click on the **Company Codes** button, you can see the company codes that have already been created for this customer.

7. Click on the **Customer: Payment Transactions** tab, as shown in Figure 13.14. This tab shows the **Payment Terms** field. The payment term **0001** means **Pay Immediately w/o Deduction**, and this specification will be included in the outgoing invoices that are posted in accounts receivable.

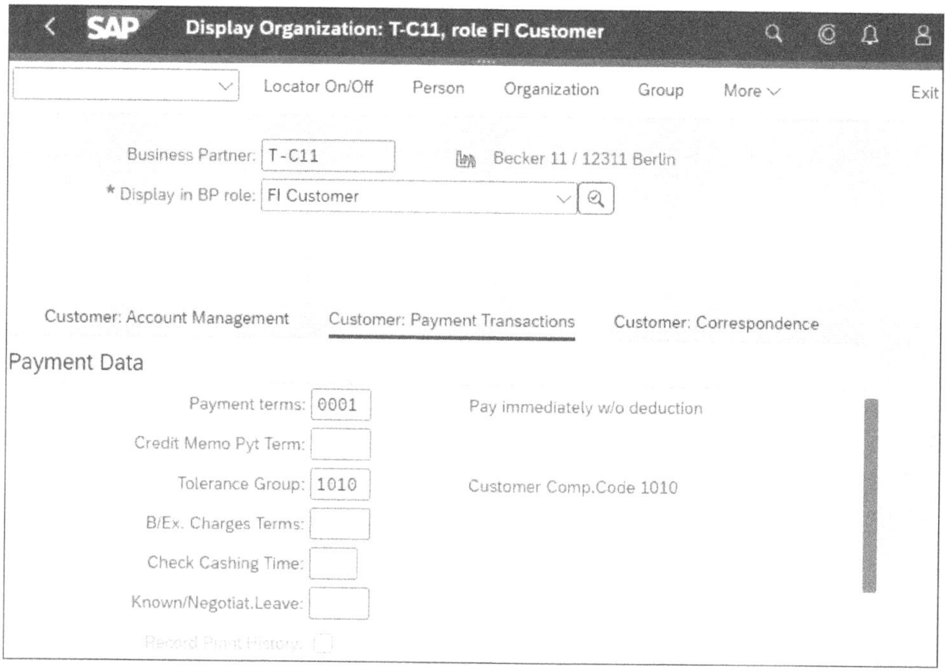

Figure 13.14 Customer: Payment Transactions Tab

8. Click SAP to go back to the SAP Fiori launchpad start page.

Display Vendor Master Data of Financial Accounting

The same procedure applies for displaying vendor master data as for displaying customer master data. The only difference is that you select the **FI Vendor** role in the **Display in BP role** field.

13.6   Accounts Receivable

In Chapter 12, you already posted to financial accounting. When posting the outgoing invoice in sales and distribution, SAP S/4HANA created a document for financial accounting in addition to the logistics document (see Chapter 12, Section 12.10).

Have you clicked through these instructions in the previous chapter? If so, you'll encounter the invoice posted there in this chapter as well. If not, it's also okay. An invoice can also be posted directly in financial accounting, which is exactly what you'll do in this section.

13.6.1   Display an Item List

An *item list* shows already posted invoices or payments. The real Manage Customer Line Items app roughly corresponds to Transaction FBL5N (Customers Line Items) in SAP GUI.

Follow these steps:

1. Launch the Manage Customer Line Items app.

2. Enter the company code and select **Open Items** in the **Status** field, as shown in Figure 13.15.

3. Click the **Go** button. This will show the unpaid invoice items (see Figure 13.16).

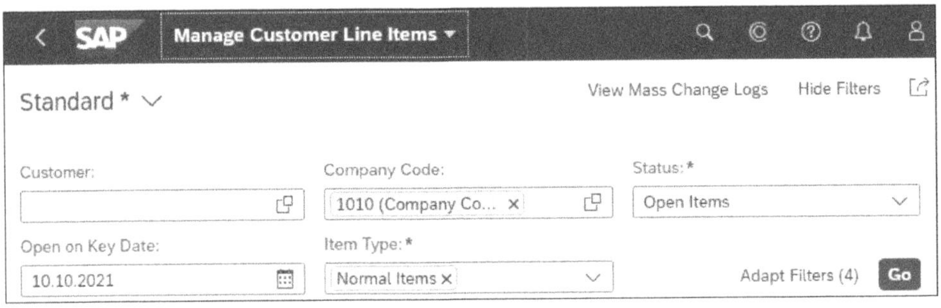

Figure 13.15 Accounts Receivable Items Entries

How Do I Select Items?

In practice, of course, you'll specify a customer number in the **Customer** field when a customer reports a late payment or a reminder.

For example, if you have someone on the phone who claims to have paid the invoice, select one of the **Cleared Items** or **All Items** entries in the **Status** field to display the paid invoices as well.

If you want to process only and exclusively the open items, we recommend the real Process Receivables app, which we'll introduce to you in more detail later in this section.

If you've posted an accounts receivable invoice in the order-to-cash process, you'll also see it in the list further to the right. You can recognize invoices posted in the sales and distribution module in the **Journal Entry Type** column by the abbreviation **RV** (sales invoice). Invoices entered in financial accounting have the abbreviation **DR** (customer invoice).

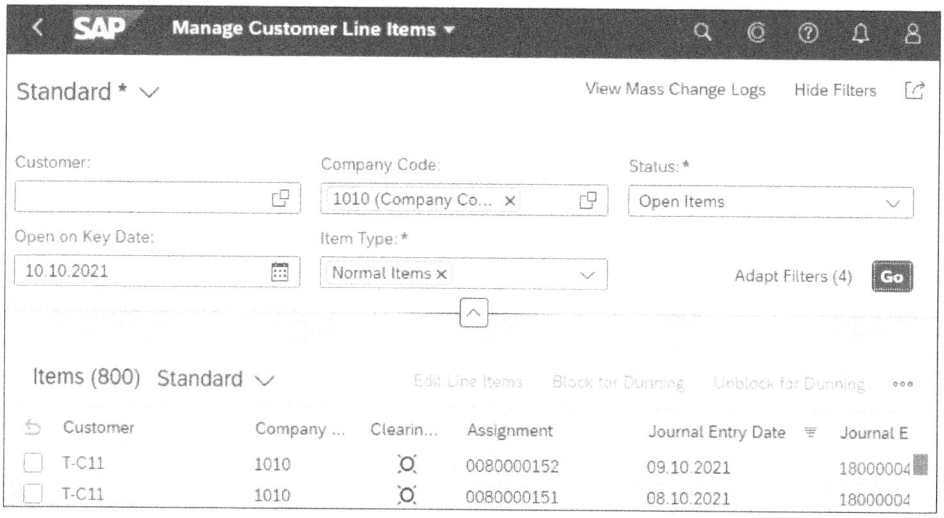

Figure 13.16 Unpaid Invoice Items

4. Click on a document number in the **Journal Entry** column and click on the document number again in a small additional window that appears to see details about this posting.

   You've now switched to the Manage Journal Entries app, as shown in Figure 13.17. Here you don't see the whole document, but only one item, which shows the general ledger account **12100000 (Rcvbls Domestic)**.

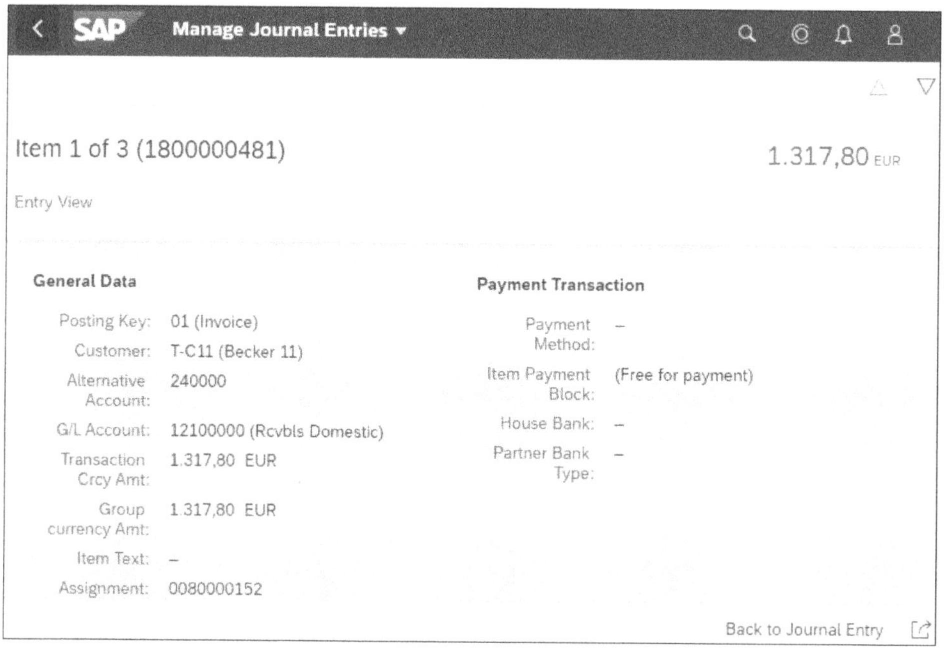

Figure 13.17 Item with the General Ledger Account 12100000 Rcvbls Domestic

5. At least readers from the accounting department would now like to see the complete document. Gladly! Click the **Back to Journal Entry** button at the bottom right. Now scroll down to see all the line items in the accounting document, as shown in Figure 13.18.

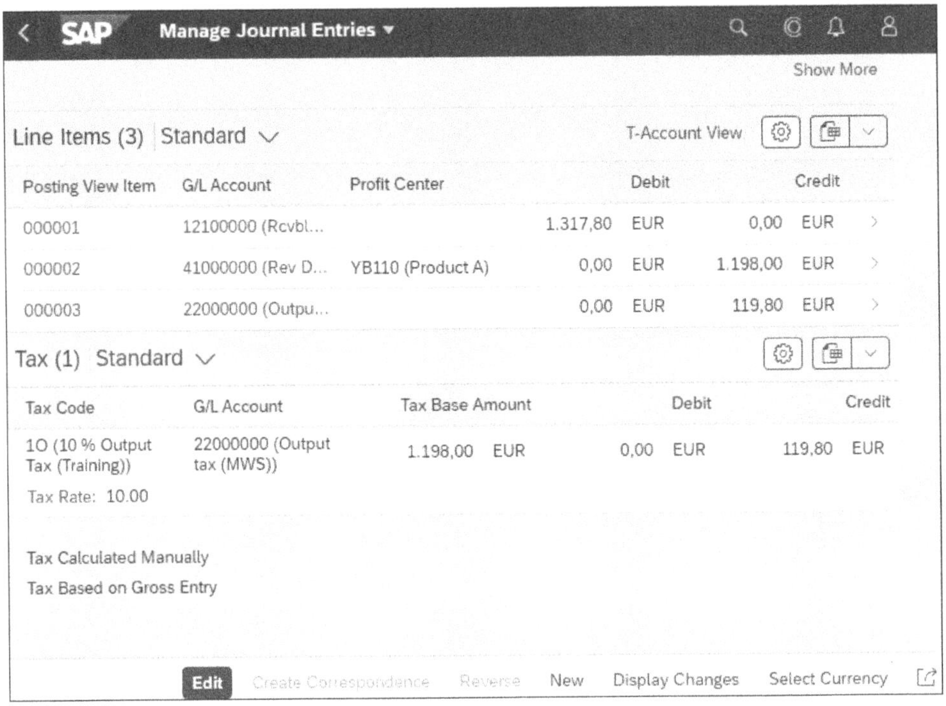

Posting View Item	G/L Account	Profit Center	Debit	Credit	
000001	12100000 (Rcvbl...		1.317,80 EUR	0,00 EUR	>
000002	41000000 (Rev D...	YB110 (Product A)	0,00 EUR	1.198,00 EUR	>
000003	22000000 (Outpu...		0,00 EUR	119,80 EUR	>

Tax (1) Standard ∨

Tax Code	G/L Account	Tax Base Amount	Debit	Credit
1O (10 % Output Tax (Training))	22000000 (Output tax (MWS))	1.198,00 EUR	0,00 EUR	119,80 EUR

Tax Rate: 10.00

Tax Calculated Manually

Tax Based on Gross Entry

Edit   Create Correspondence   Reverse   New   Display Changes   Select Currency

Figure 13.18 Display for the Accounting Document

6. Click **SAP** to go back to the homepage.

[+]

If You Don't Have SAP Live Access

Note here the corresponding revenue account from the second posting item for Section 13.6.3.

13.6.2   Documents and Journal Entry Types

In contrast to the predecessor system SAP ERP, there is only one document per business transaction in accounting in SAP S/4HANA because no additional document is created here for the controlling module.

To find and evaluate documents later, a *journal entry type* is assigned to each posted document. The journal entry type is used to better distinguish and sort the multitude of documents that arise in accounting, for example, according to invoices and credit memos.

In Table 13.2, you can see the most important standard journal entry types of accounts receivable, each with its abbreviation, which always consists of two letters.

Journal Entry Type	Meaning
DG	Customer credit memo
DR	Customer invoice posted in the financial accounting module
DZ	Customer payment
RV	Billing document transfer, sales invoice created in the sales and distribution module for a billing document.

Table 13.2 Journal Entry Types in Financial Accounting

13.6.3   Post an Accounts Receivable Invoice

Soon you'll post an incoming payment for an invoice. However, for this incoming payment, you need an accounts receivable invoice . You could use the invoice from Chapter 12 from the order-to-cash process for this. But we're nice writers, so we won't force you to click through the entire sales process to create the invoice. That's why we'll now show you how to create an invoice directly in financials.

But wait! Aren't accounts receivable invoices created in the sales and distribution module as part of billing, and accounting automatically gets a copy of the sales document? Usually yes, if a company uses the sales and distribution module as well. However, sometimes there are accounts receivable invoices that are only created in financials and not in sales and distribution for an order. You can also see this in our process flow shown in Figure 13.19.

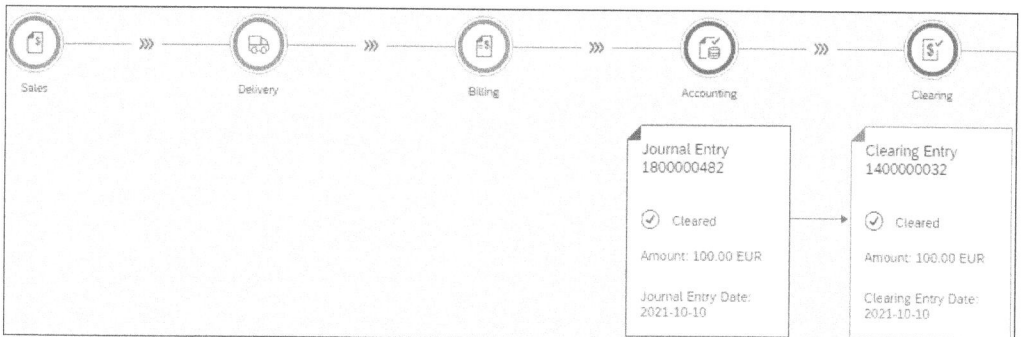

Figure 13.19 Process Flow for an Accounts Receivable Invoice Entered in Accounts Receivable

So, to create an invoice directly in financials, use the Create Outgoing Invoices legacy app, which corresponds to Transaction FB70 (Enter Outgoing Invoices). As promised, you don't need to know anything about debits and credits or T-accounts for this entry.

Follow these steps:

1. Launch the Create Outgoing Invoices app.

2. Make the following entries in the upper part of the window in the **Basic Data** tab, as shown in Figure 13.20:

   - **Transactn**: Select the **Invoice** entry here if it's not already preset.
   - **Customer**: Enter the customer number; if you don't have a customer number, grab one from the item list in Section 13.6.1.
   - **Invoice Date**: Enter the current date using the F4 and F2 key sequence.
   - **Reference**: Specify a number.
   - **Amount**: Enter the invoice amount in this field.
   - **Calculate Tax**: Place a checkmark here, and specify "1O (10% Output Tax (Training))" for the output tax.

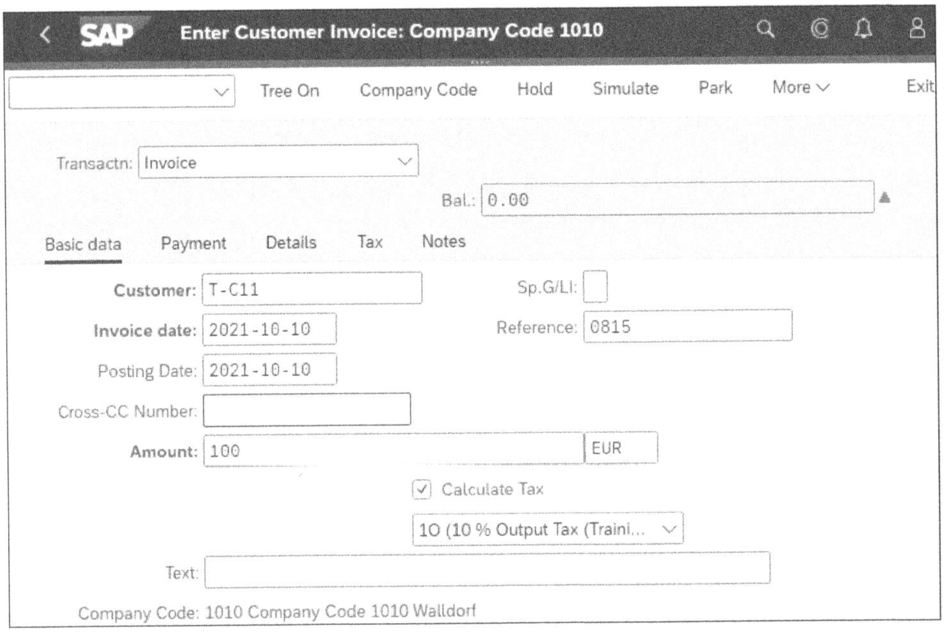

Figure 13.20 Basic Data Tab

[»]

For Accountants: Where Is Debit and Where Is Credit?

You've just entered the debit side of the entry. The credit side comes now, as you now enter the general ledger account in the lower item table. The sales tax will be posted automatically. The official posting record is "Customer to Revenues and Tax." The reconciliation account from the general ledger isn't displayed because it's posted in the background according to the specifications in the customer master record.

3. Confirm your entries with the [Enter] key. Now SAP S/4HANA checks the data you've entered so far and completes it. In the right part of the screen, you'll then see information about the supplier.

4. Continue in the lower part of the window in the **Items** table, as shown in Figure 13.21. Here the following information is important:

   – In the **G/L Acct** field, you need an account number for a revenue account from the general ledger. If you're using the SAP Live Access training system, enter the account number "41000500" here for the account "Sales Revenue w/o CO-PA". If not, follow the procedure in Section 13.6.1 to find a revenue account in a posting document.

   – If, as in our case, there is only one item in the **Items** table, SAP S/4HANA can independently determine the amount from the amount entered. To do this, simply enter "*" in the **Amount in Doc.Curr.** field.

   – SAP S/4HANA takes the tax code from the upper part of the window.

   – If you scroll further to the right, you'll see fields for account assignment objects such as **Business Area**, **Cost Center**, **Order**, or **Profit Center**, which are optional in our example in the SAP Live Access training system.

St...	G/L acct	Short Text	D/C	Amount in doc.curr.	Loc.curr.amount	T...	Tax jurisdictn code
	41000500		Credit ∨	*	0.00	10	
			Credit ∨		0.00	10	
			Credit ∨		0.00	10	
			Credit ∨		0.00	10	
			Credit ∨		0.00	10	

0 Items (No entry variant selected)

Post   Cancel

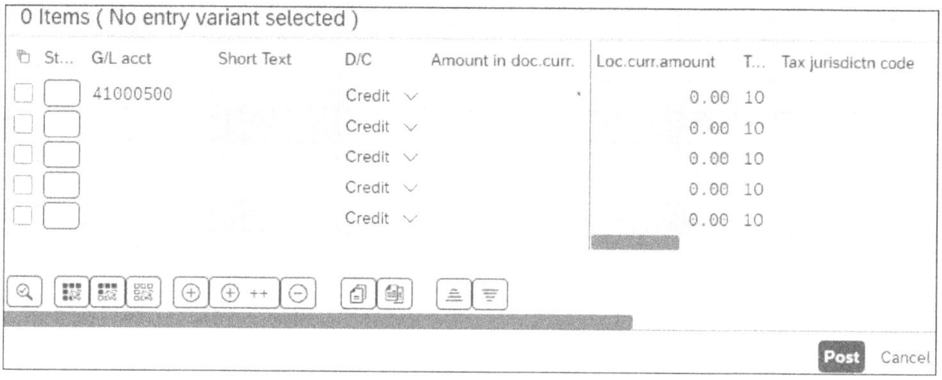

Figure 13.21 Item Table Entries

5. Confirm your entries with the [Enter] key. What happens? At the top right of the **Balance** display field, the traffic light symbol changes to green. Note the changes in the item table:

   – A green checkmark appears on the far left.

   – The account name is displayed in the **Short Text** column.

   – In the **Amount in doc.curr.** column, the asterisk (*) is replaced with the real amount.

6. You could click the **Post** button right now. However, if you want to see the posting before, click the **Simulate** button in the header. The **Document Overview** screen appears, as shown in Figure 13.22:

- In the upper-left corner, you'll find the journal entry type **DR (Customer Invoice)**. The document type corresponds to the journal entry type.
- In the item data, SAP S/4HANA shows the additional item line for output tax.

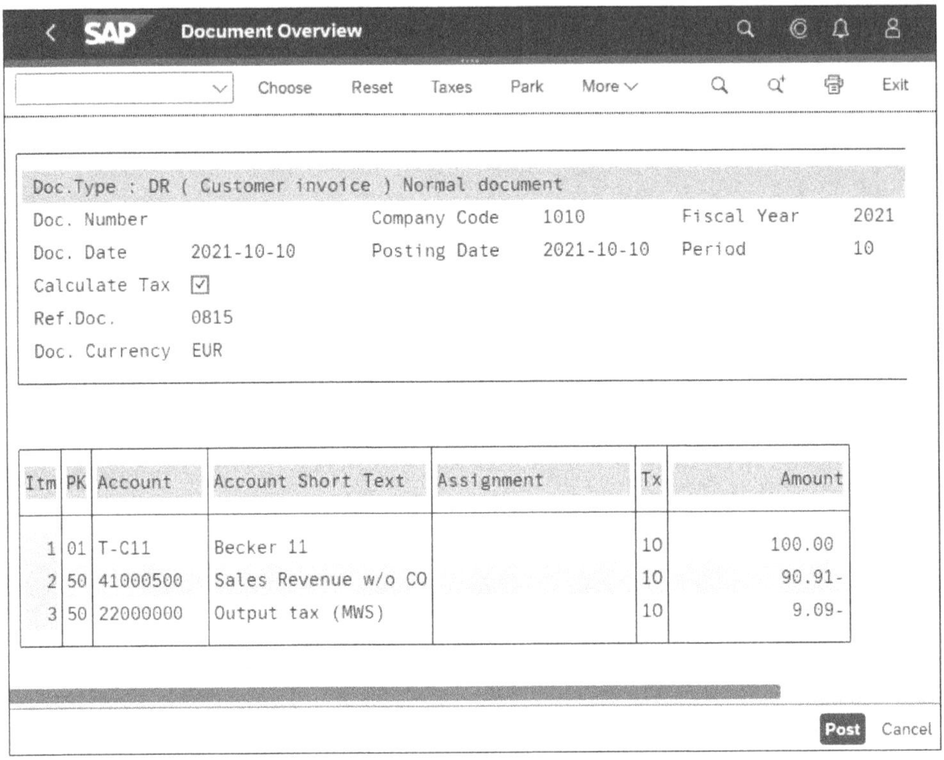

```
<   SAP     Document Overview                         Q   ©  Δ  8

              ∨   Choose   Reset   Taxes   Park   More ∨      Q   Q'   🖶   Exit


Doc.Type : DR ( Customer invoice ) Normal document
Doc. Number                 Company Code   1010      Fiscal Year   2021
Doc. Date      2021-10-10    Posting Date   2021-10-10  Period        10
Calculate Tax  ☑
Ref.Doc.       0815
Doc. Currency  EUR


Itm PK Account    Account Short Text  Assignment    Tx           Amount

 1 01 T-C11       Becker 11                         10           100.00
 2 50 41000500    Sales Revenue w/o CO              10            90.91-
 3 50 22000000    Output tax (MWS)                  10             9.09-


                                                          Post   Cancel
```

Figure 13.22 Document Overview Screen

7. Everything in order? Then post the invoice by clicking the **Post** button in the footer line.
8. Click SAP to return to the SAP Fiori launchpad homepage.

[+] **Check Booking**

Use the following SAP Fiori apps, which you learned about earlier in this chapter, to verify the posting of the accounts receivable invoice:

- **Manage Customer Line Items**
  In this app, you'll find the additional document for the customer invoice.
- **Balance Sheet/Income Statement**
  In this app, you can see that the receivables balance on the reconciliation account and the balance on the revenue account have increased.

13.6.4   Process Receivables

Although you're actually sure, you want to check whether the invoice has been posted. To do this, you could display the item list already discussed in the Manage Customer Line Items app. But there is also a more elegant method: the Process Receivables app. In this app, you not only see an overview of all open invoices but also some master data about the customer and a due date diagram.

Follow these steps:

1. Launch the Process Receivables app.

2. Enter the customer number in the **Customer** field at the top left, as shown in Figure 13.23.

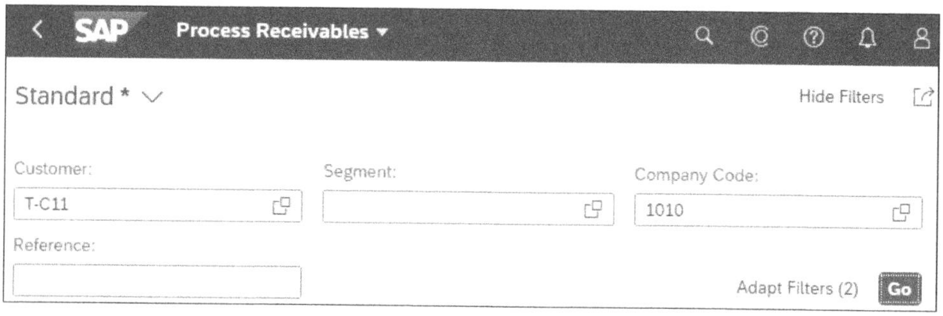

Figure 13.23 Entering the Customer Number

3. Click the **Go** button to see all selected open items on the screen, as shown in Figure 13.24.

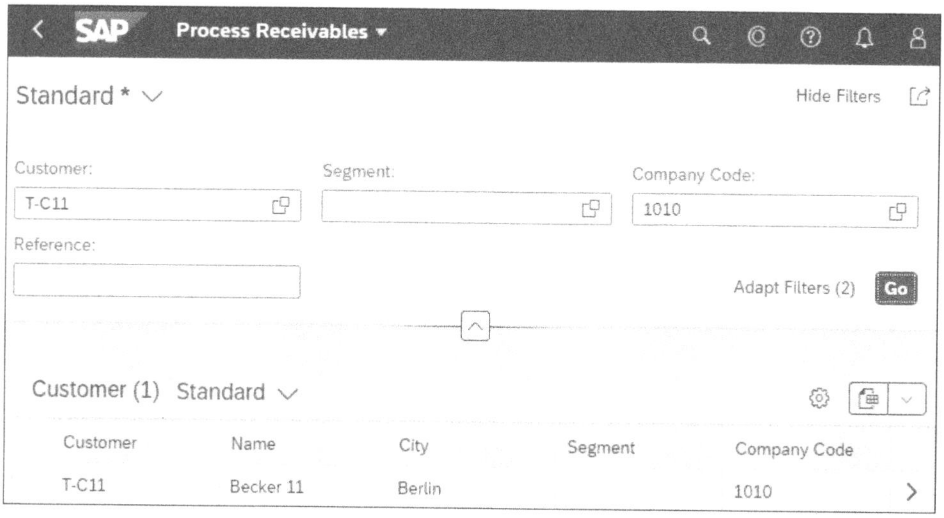

Figure 13.24 Viewing the Selected Open Items

4. You want detailed information? Click on the $\boxed{>}$ icon on the far right in the respective table row.

5. You'll see the **Due Date Grid** in the upper-right corner, as shown in Figure 13.25, which you can enlarge by clicking on it.

   Above the table, you'll find the **Create Dispute** and **Create Promise** buttons, with which you can edit the invoices.

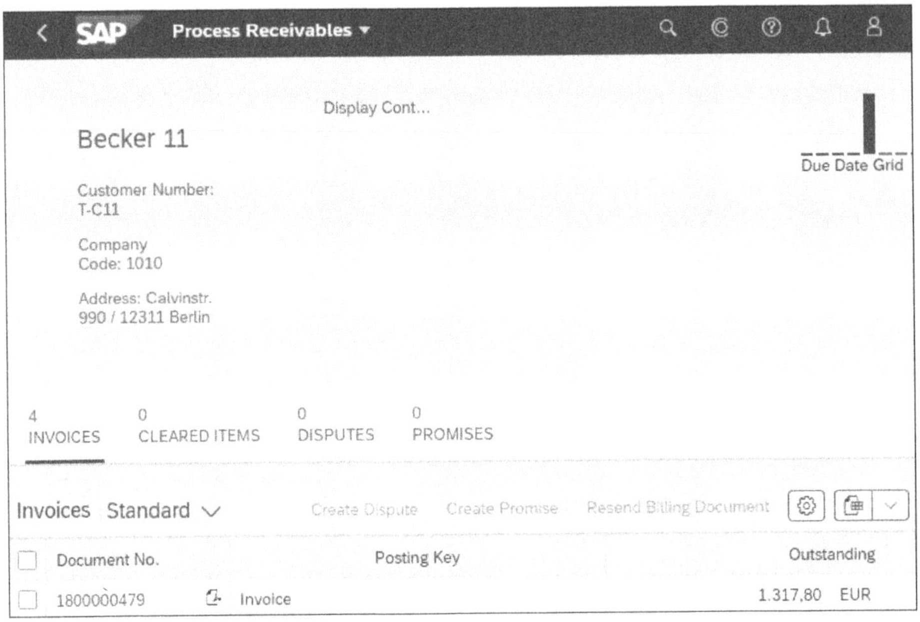

Figure 13.25 Invoices and Due Date Grid

6. Click $\boxed{\text{SAP}}$ to go back to the SAP Fiori launchpad homepage.

13.6.5   Enter an Incoming Payment

Did you post the accounts receivable invoice in Section 13.6.3? Then you now have an open item and thus an incomplete transaction: an unpaid invoice. For a transaction to be complete, it must be cleared. To do this, you now make a *clearing entry*, which turns the open item into a cleared item.

In our example, you enter an incoming payment in the Post Incoming Payments app, which corresponds to SAP GUI Transaction F-28 (Post Incoming Payments).

Follow these steps:

1. Launch the Post Incoming Payments app.

2. In the upper part of the window, the **Company Code** field should already be filled in thanks to your default settings. Make the following additional entries, as shown in Figure 13.26:

- Fill both the **Journal Entry Date** and **Value Date** fields with the current date.
- For the **Journal Entry Type** field, select **Customer Payment**.
- In the **G/L Account** field, specify a bank account; in the SAP Live Access training system, specify the account with the number "11100125".
- In the **Amount** field, enter "100.00", the gross amount of the accounts receivable invoice from Section 13.6.3.

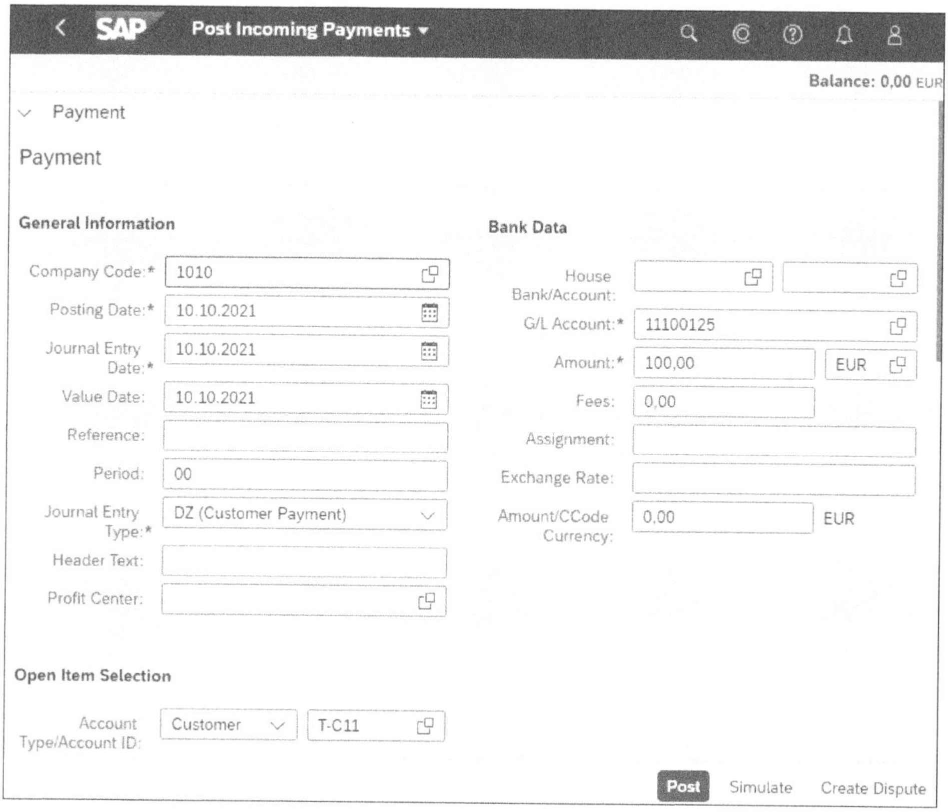

Figure 13.26 Post Incoming Payments App Entries

3. Further to the right, or further down if the screen resolution is larger, select **Customer** for the **Account Type/Account ID** field, and enter the number of the customer in the **Account ID** field.

4. Click the **Propose Items** button below it.

5. In the lower part of the window, you'll now see the invoices that haven't been paid yet on the left under **Open Items**, as shown in Figure 13.27. In the row of the item that should be cleared, click the **Clear** button.

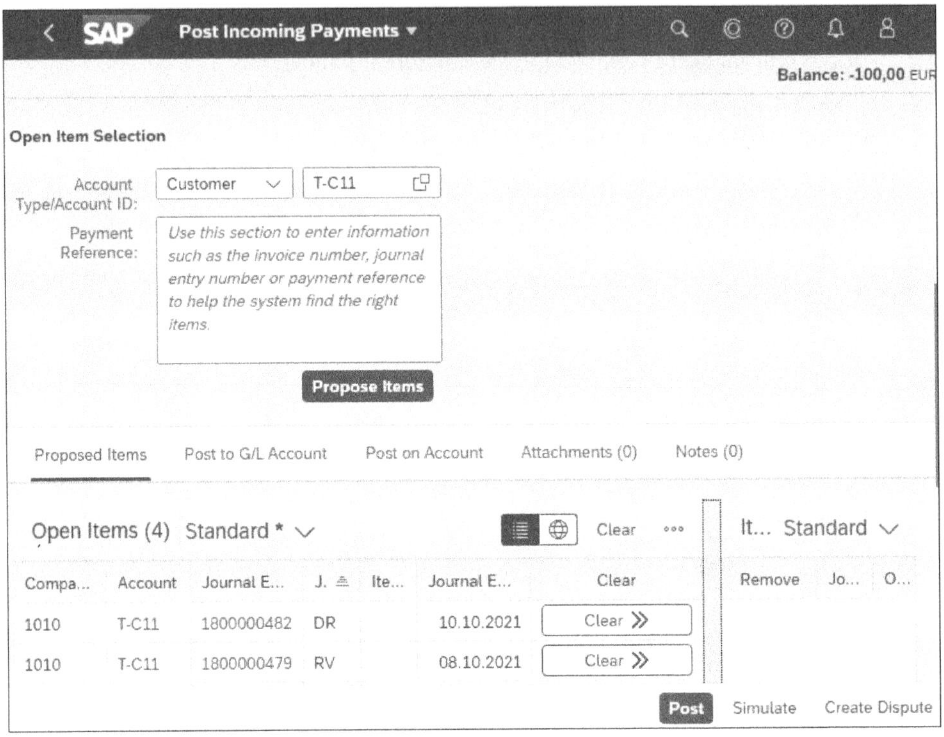

Figure 13.27 Unpaid Invoices under Open Items

6. You can now see the selected item in the right part of the window under **Items to Be Cleared**, as shown in Figure 13.28. In the left part in the **Amount** field (not visible in the illustration), the invoice amount is displayed. In the upper-right corner of the **Balance** field, you should now see **0,00 EUR**.

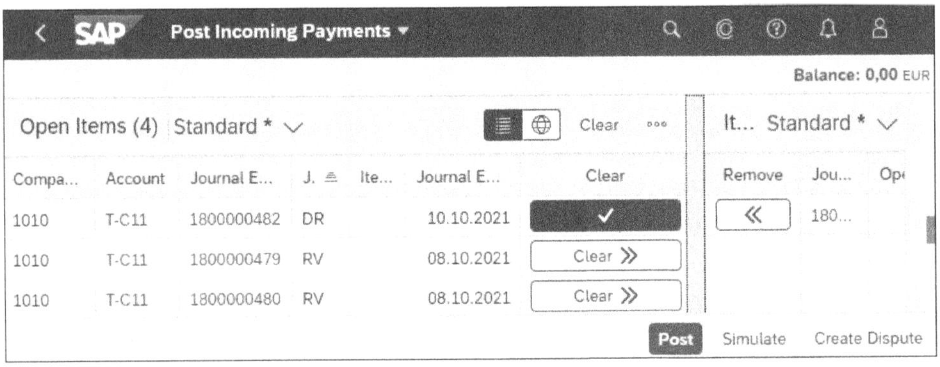

Figure 13.28 Items to Be Cleared

7. Click the **Post** button in the footer line, and a message box appears, as shown in Figure 13.29. Congratulations, the clearing is done! SAP S/4HANA informs you that it has created another new document and shows you the corresponding document number.

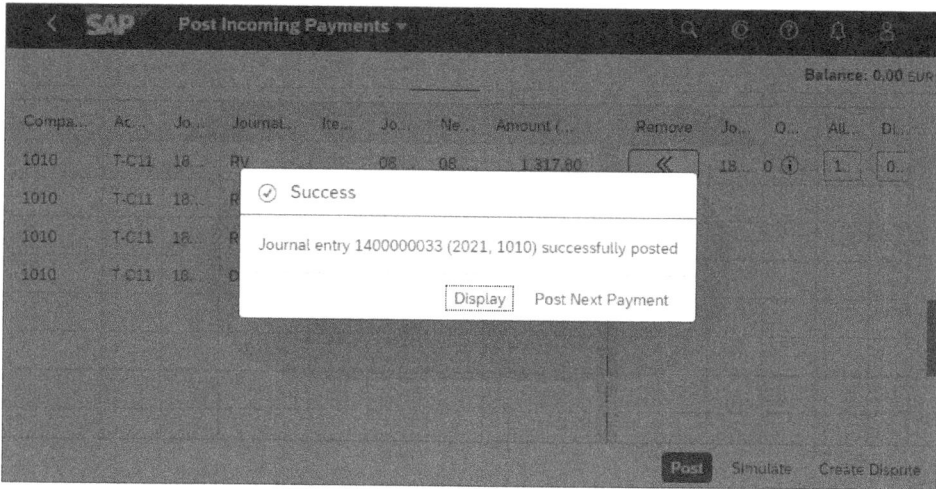

Figure 13.29 Success Message

8. Click the **Display** button in this message to see the posting document, as shown in Figure 13.30.

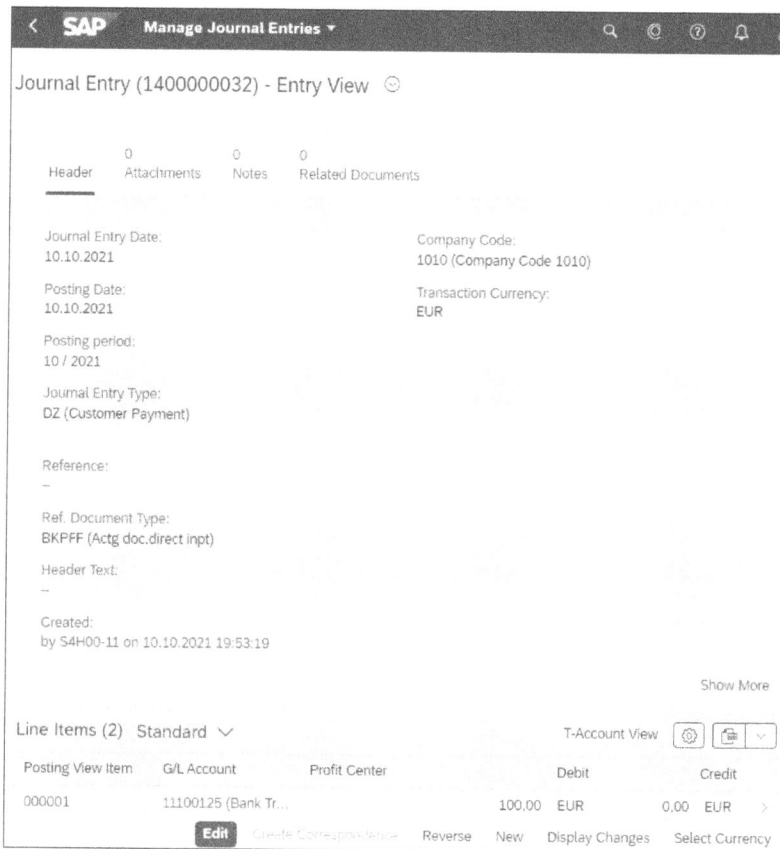

Figure 13.30 Viewing the Posting Document

9. In the posting document, you can see the items in a table further down. If you love the T-accounts of accounting as much as we do, you may now click once on the **T-Account View** button above the table, as shown in Figure 13.31.

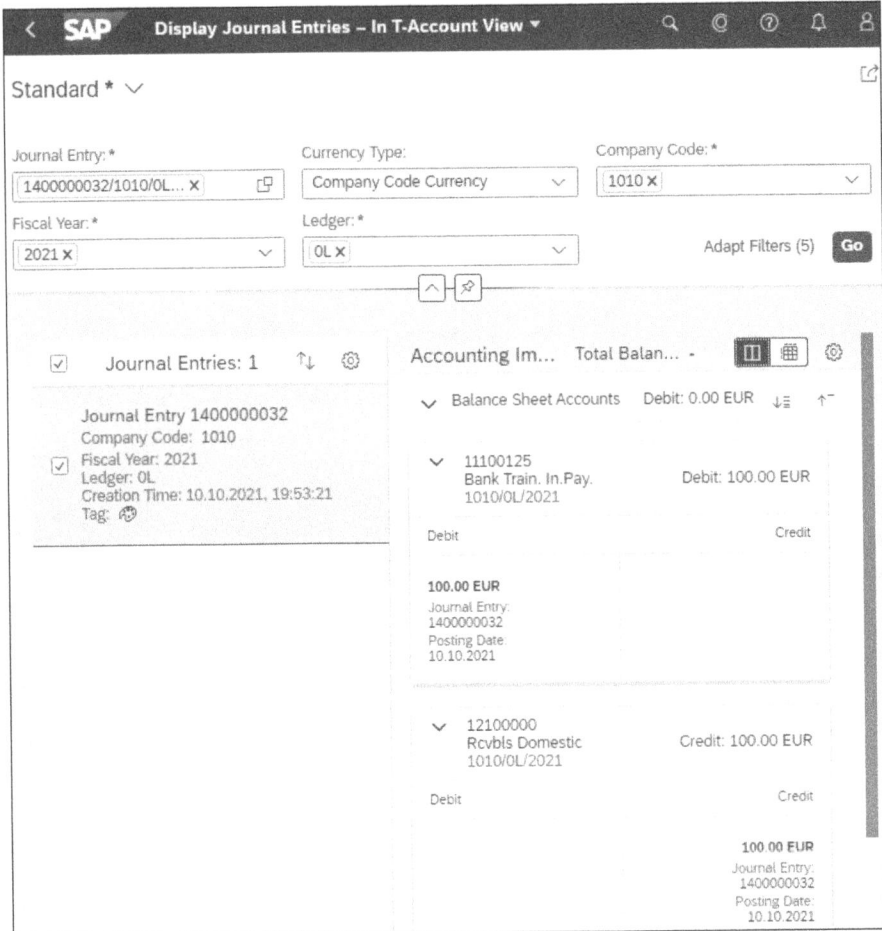

Figure 13.31 T-Account View

10. You see in Figure 13.31 the two T-accounts in each case with the correct **Debit** and **Credit**. We wonder if this function was made especially for the commercial vocational school lessons?

11. Anyway, last but not least, click ◄**SAP** to go back to the SAP Fiori launchpad homepage.

SAP S/4HANA has created a separate clearing document for the clearing entry. In Figure 13.32, you can see both documents (invoice and clearing document) in the process overview.

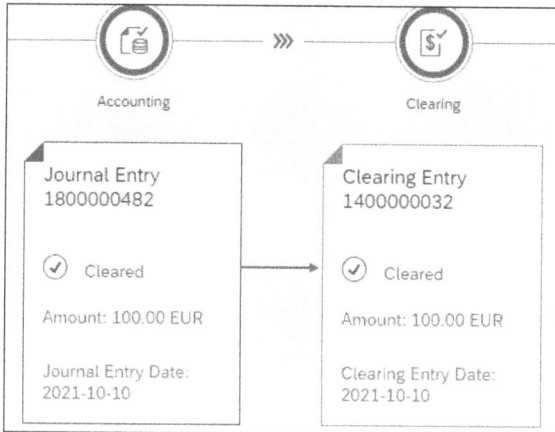

Figure 13.32 Invoice and Clearing Document in the Process Overview

---

Check Booking [+]

13

Use the following SAP Fiori apps, which you've already learned about in this chapter, to verify the posting of the incoming payment:

- **Manage Customer Line Items**
  In this app, you'll find the additional document for the customer payment.
- **Balance Sheet/Income Statement**
  In this app, you can see that the receivables balance in the reconciliation account has decreased, and the cash balance in the posted bank account has increased.
- **Process Receivables**
  In this app, the cleared invoice is now missing after this process.

13.7 Accounts Payable

Accounts payable works almost exactly like accounts receivable—apps and transactions follow the same pattern. Instead of an outgoing invoice, you post an incoming invoice, and instead of an incoming payment, you post an outgoing payment. That's why we won't show you the complete program here, but only the creation of an incoming invoice.

Incoming invoices are usually created in the materials management module as part of invoice verification, and accounting automatically receives a copy of the document there. Sometimes, however, there are also incoming invoices that are only created in financial accounting and not in materials management for a purchase order. You can also see this in our process flow shown in Figure 13.33.

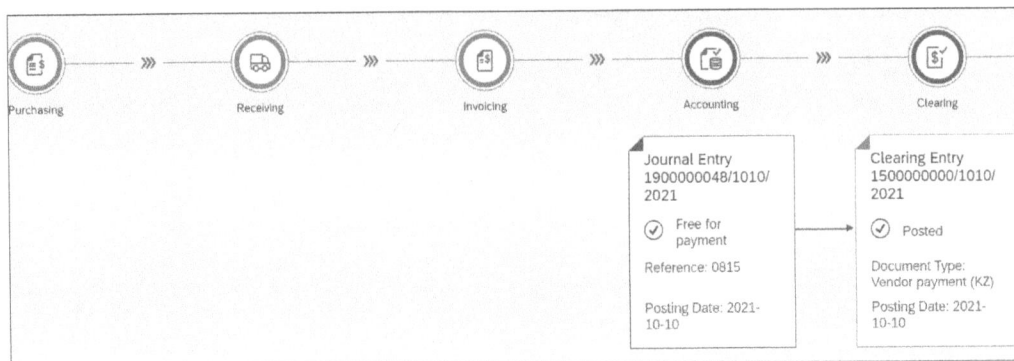

Figure 13.33 Process Flow with an Incoming Invoice Entered in Accounts Payable and with the Subsequent Outgoing Payment as Settlement

To tell a little story about our example again: a trainee has caused a window breakage in your company, so a glazier is urgently needed and ordered by phone. In contrast to the purchase-to-pay process, in this case, there is no purchase order and no invoice verification in SAP S/4HANA.

13.7.1 Post an Incoming Invoice

In the following instructions, you'll notice a small but subtle difference to the entry of the outgoing invoice: you make an account assignment of the amount to a cost center. To do this, you use the Create Incoming Invoices legacy app, which corresponds to Transaction FB60 (Enter Vendor Invoice).

Follow these steps:

1. Start the Create Incoming Invoices app.
2. Make the following entries in the upper part of the window on the **Basic Data** tab, as shown in Figure 13.34:
   - In the **Transactn** field, select the **Invoice** entry if it's not already preset.
   - In the **Vendor** field, enter the vendor number.
   - In the **Invoice Date** field, use the F4 and F2 key sequence to insert the current date.
   - In the **Reference** field, enter any number as the document number of the supplier invoice.
   - In the **Amount** field, add the invoice amount.
   - Check the **Calculate Tax** box if it's not already preset, and enter "1I (10% Input tax (Training))" for the input tax if it's not already preset.

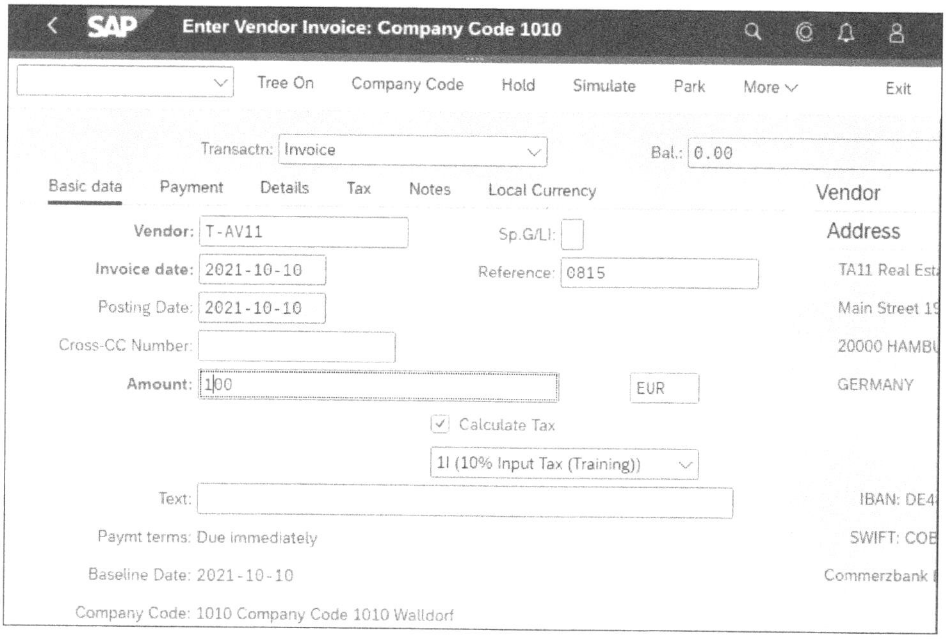

Figure 13.34 Basic Data Tab

3. Confirm your entries with the Enter key. Now SAP S/4HANA checks the data you've entered so far and completes it. In the right part of the screen, you'll then see details of the supplier.

4. Let's move on to the **Items** table, as shown in Figure 13.35. Now it's getting exciting:

   – In the **G/L Acct** field, you need an account number for an expense account from the general ledger. If you're using the SAP Live Access training system, enter account number "63003000" for the "Other maintenance" account. Otherwise, find a suitable expense account using ☐ (**Value Help**).

   – Enter "*" in the **Amount in Doc.Curr.** field to accept the amount already entered.

5. Finally, confirm your entries with the Enter key. But wait, you get an error message, as shown in Figure 13.35.

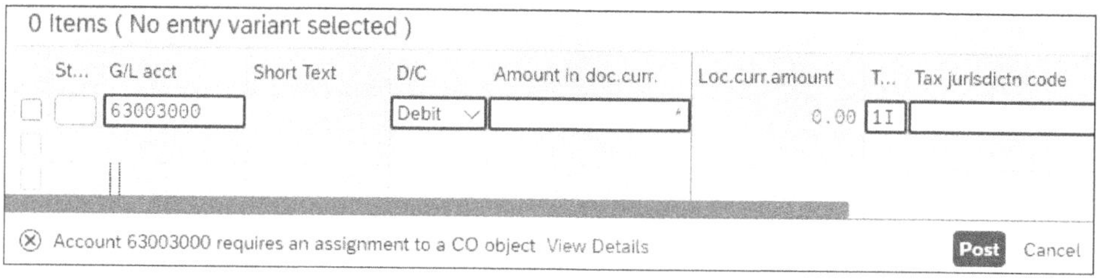

Figure 13.35 Item Table Entries with Error Message

403

This time, you're posting an expense, and an expense, unlike a revenue, must always be assigned to an account assignment object relevant to cost accounting. As account assignment object, you simply use a cost center (for more information, see the next section).

6. Scroll to the right until the **Cost Center** field appears. Search for a cost center here using ☐ (**Value Help**), and confirm the entry with the ⌨Enter key. Now it works, as shown in Figure 13.36.

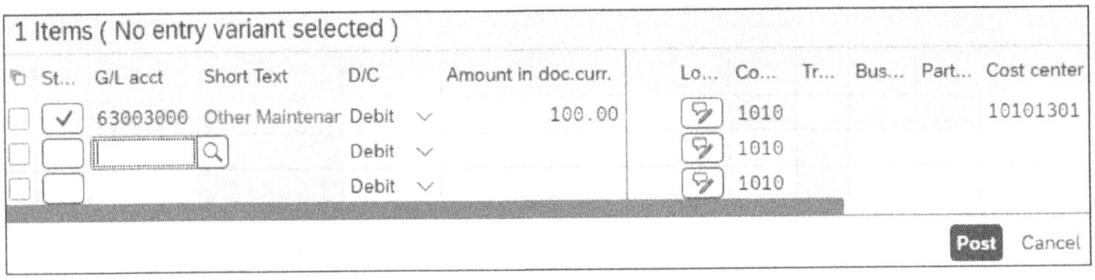

	St...	G/L acct	Short Text	D/C	Amount in doc.curr.		Lo...	Co...	Tr...	Bus...	Part...	Cost center
☐	✓	63003000	Other Maintenar	Debit ∨	100.00		🗂	1010				10101301
☐			🔍	Debit ∨			🗂	1010				
☐				Debit ∨			🗂	1010				

1 Items (No entry variant selected)

Post Cancel

Figure 13.36 Item Table Entries with Cost Center Field

7. Post the invoice by clicking the **Post** button in the footer row. After posting, you'll see the document number in the footer row.

8. Click ⬛SAP⬛ to return to the home screen.

[+]

> **Check Booking**
>
> You can use the following SAP Fiori apps to verify the posting of the accounts payable invoice:
>
> ■ **Manage Supplier Line Items**
> This app shows the document for the vendor invoice.
> ■ **Balance Sheet/Income Statement**
> This app shows the balance of liabilities in the reconciliation account and the balance in the expense account have both increased.

13.7.2 Account Assignment Objects

In the **Items** table, you've assigned the invoice to a cost center. This allows the controlling and the cost managers to keep an eye on the budgets. In addition to the cost centers, there are other possible account assignment objects, for example, profit centers, business areas, or internal orders.

Profit centers are organizational units that can be used to calculate partial results below a company code, for example, for individual stores.

A *business area* is used to prepare internal balance sheets and partial balance sheets or P&Ls. Thus, the rules for externally published balance sheets or P&Ls don't have to be taken into account for these evaluations.

An *internal order* is created in controlling and can be used as a cost collector. It can be used for special actions, for example, for a trade fair. In this case, when entering the incoming invoices for the trade fair costs, you specify the internal order as the account assignment object in each case. In controlling, this allows the corresponding cost reports to be created at the push of a button.

The corresponding costs can be charged to one or more cost centers according to a previously defined *accounting rule*. This is also called *real posting. Statistical postings*, where no settlement takes place, are the opposite of real postings. Statistical postings can be made not only for internal orders but also for cost centers. Ultimately, the account assignment object determines whether a real or statistical posting is to be made.

13.7.3   Journal Entry Types and Document Numbers

When you use the Manage Supplier Line Items app to check your entries, you'll see the journal entry types here, as shown in Figure 13.37. As already mentioned in accounts receivable, the journal entry type is used to better distinguish and sort the multitude of documents that arise in accounting, for example, according to invoices or credit memos.

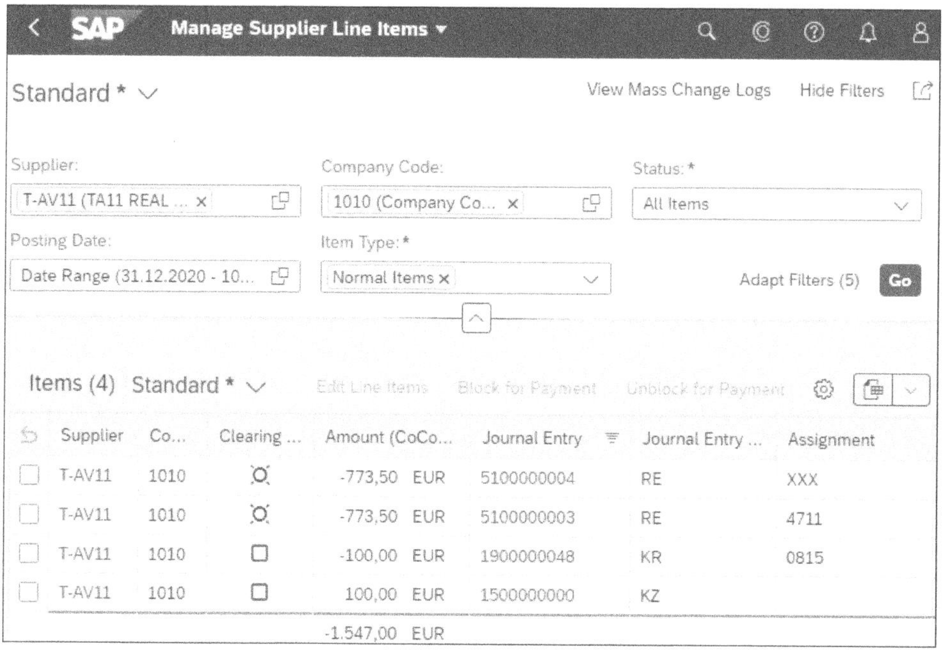

Figure 13.37 Journal Entry Types in the Manage Supplier Line Items App

In Table 13.3, you'll find the most important standard journal entry types for accounts payable, each with its abbreviation, which always consists of two letters.

Journal Entry Type	Meaning
KG	Creditor credit memo
KR	Vendor invoice posted in financial accounting
KZ	Vendor payment
RE	Vendor invoice created in materials management during invoice verification

Table 13.3 Journal Entry Types for Accounts Payable

A number range is assigned to each journal entry type. Journal entry types thus control document number assignment. In the SAP Live Access training system, vendor invoices (journal entry type KR) start with "19", and outgoing payments (journal entry type KZ) start with "15".

In SAP S/4HANA, each document is assigned a number when it's posted. Each number is assigned only once per company code and fiscal year. You can therefore use this document number as a search criterion if you want to find a particular document again at a later date.

And how exactly are document numbers assigned? The usual method is *internal number assignment*. SAP S/4HANA automatically assigns a consecutive number in ascending order.

Very rarely, the system is set to require you to type in the number yourself. This is referred to as *external number assignment*, which takes place outside SAP S/4HANA, so to speak. However, to prevent a number from being assigned twice by mistake, SAP S/4HANA at least checks whether the number you entered already exists and rejects duplicate assignment if necessary. External number assignment is rarely encountered in documents.

13.8   Summary

This wraps up our discussion of financial accounting activities for SAP S/4HANA users. We covered organizational units, default settings, the general ledger, business partners, accounts receivable, and accounts payable.

We've now reached the end of this book. We would have liked to continue, but because we've already extended the deadline to the maximum possible, we'll make a stop here.

We had a lot of fun writing this book, just as SAP S/4HANA is a lot of fun for us. And if we were able to contribute to you also looking forward to the start of your SAP system every day, we'll have achieved our goal. Stay curious; there is still a lot to discover!

If you have a comment on this book or have found an exciting trick in SAP Fiori that you would like to tell us about, we would also be happy to hear from you directly via *fitznar@gmx.net*. We wish you much success and even more fun using SAP S/4HANA!

The Authors

Wolfgang Fitznar is an SAP expert and holds a degree in business administration. Since 1992, he has been working on SAP software, first as a product manager for SAP qualifications and later as a project manager for SAP training projects. He is currently a freelance trainer and application consultant in national and international SAP training projects in the areas of logistics and accounting. He is certified both as an SAP solution consultant and as an SAP trainer. Wolfgang's training is characterized by professional competence and practical relevance. He systematically collects tips and tricks that help users with their daily work and passes on his extensive know-how to advanced users and key users so that specialist departments can achieve faster and better results with SAP software. He is the author of a German-language user book, *SAP for Users – Tips & Tricks*.

Dennis Fitznar has a bachelor of arts in business administration and works as a freelance SAP consultant and author. He gained practical experience in implementing and operating SAP Fiori and SAP Fiori apps at a major automotive manufacturer. He has published multimedia SAP training on the topics of reporting with SAP Query, customizing, Legacy System Migration Workbench (LSMW), scripting, and administration.

Wolfgang and Dennis are also happy to help on-site to identify and implement the best practices for the different roles and workplaces. If you would like to benefit from this concept, please use the e-mail address *fitznar@gmx.net* to contact them. They prefer jobs in Bavaria, the Caribbean, Hawaii, and the Canary Islands.

Index

- Learn what SAP S/4HANA offers your company
- Explore key business processes and system architecture
- Consider your deployment options and implementation paths

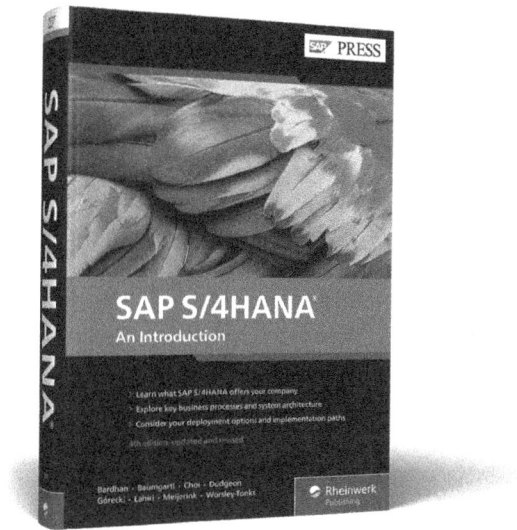

Bardhan, Baumgartl, Choi, Dudgeon, Górecki, Lahiri, Meijerink, Worsley-Tonks

SAP S/4HANA

An Introduction

Interested in what SAP S/4HANA has to offer? Find out with this big-picture guide! Take a tour of SAP S/4HANA functionality for your key lines of business: finance, manufacturing, supply chain, sales, and more. Preview SAP S/4HANA's architecture, and discover your options for reporting, extensions, and adoption. With insights into the latest intelligent technologies, this is your all-in-one SAP S/4HANA starting point!

648 pages, 4th edition, pub. 03/2021
E-Book: $69.99 | **Print:** $79.95 | **Bundle:** $89.99

www.sap-press.com/5232

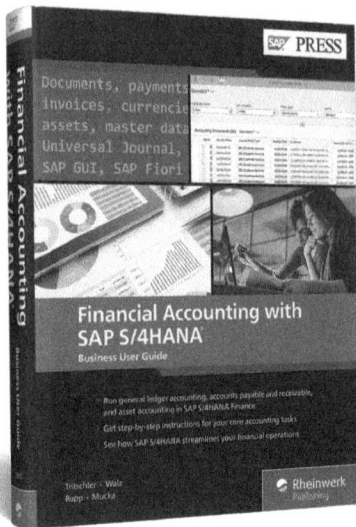

- Run general ledger accounting, accounts payable and receivable, and asset accounting in SAP S/4HANA Finance

- Get step-by-step instructions for your core accounting tasks

- See how SAP S/4HANA streamlines your operations

Tritschler, Walz, Rupp, Mucka

Financial Accounting with SAP S/4HANA: Business User Guide

Finance professionals, it's time to simplify your day-to-day. This book walks through your financial accounting tasks, whether you're using SAP GUI transactions or SAP Fiori apps in your SAP S/4HANA system. For each of your core FI business processes—general ledger accounting, accounts payable, accounts receivable, and fixed asset accounting—learn how to complete key tasks, click by click. Complete your FI operations smoothly and efficiently!

604 pages, pub. 12/2019
E-Book: $69.99 | **Print:** $79.95 | **Bundle:** $89.99

www.sap-press.com/4938

- Master your core controlling tasks in SAP S/4HANA

- Assess overhead, manufacturing, sales, project, and investment costs

- Streamline your operations with both the new and classic user interfaces

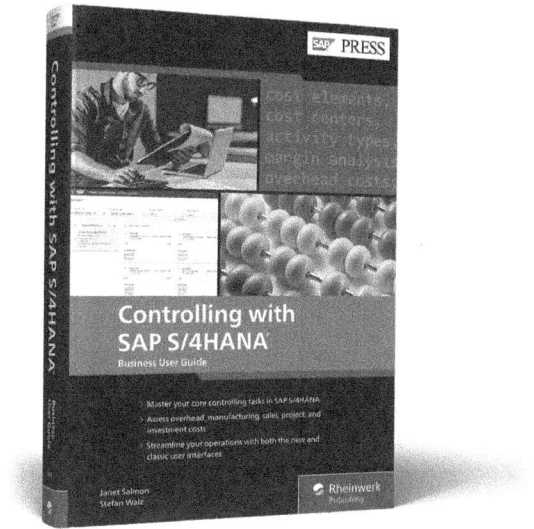

Janet Salmon, Stefan Walz

Controlling with SAP S/4HANA: Business User Guide

SAP S/4HANA brings change to your routine controlling activities. Perform your key tasks in the new environment with this user guide! Get click-by-click instructions for your daily and monthly overhead controlling tasks, and then dive deeper into processes such as make-to-stock/make-to-order scenarios, margin analysis, and investment management. Finally, instructions for inter-company transactions and reporting make this your all-in-one resource!

593 pages, pub. 05/2021
E-Book: $69.99 | **Print:** $79.95 | **Bundle:** $89.99

www.sap-press.com/5282

Rheinwerk Publishing

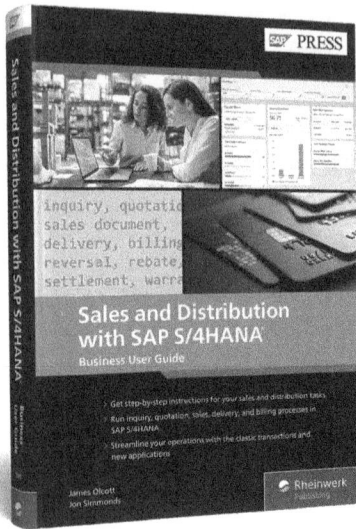

- Get step-by-step instructions for your sales and distribution tasks

- Run inquiry, quotation, sales, delivery, and billing processes in SAP S/4HANA

- Streamline your operations with the classic transactions and new applications

James Olcott, Jon Simmonds

Sales and Distribution with SAP S/4HANA: Business User Guide

Master the ins and outs of running sales and distribution in your SAP S/4HANA system. Follow step-by-step instructions, workflow diagrams, and system screenshots to complete your critical tasks and keep the sales pipeline moving. Learn how to create a quotation, change a sales document, cancel a delivery, and more. Your SAP S/4HANA sales manual is here!

434 pages, pub. 05/2021

E-Book: $69.99 | **Print:** $79.95 | **Bundle:** $89.99

www.sap-press.com/5263